Lecture Notes in Bioinformatics 12074

Subseries of Lecture Notes in Computer Science

T0172260

More information about this series at http://www.springer.com/series/5381

Russell Schwartz (Ed.)

Research in Computational Molecular Biology

24th Annual International Conference, RECOMB 2020
Padua, Italy, May 10–13, 2020
Proceedings

 Springer

Editor
Russell Schwartz
Carnegie Mellon University
Pittsburgh, PA, USA

ISSN 0302-9743   ISSN 1611-3349   (electronic)
Lecture Notes in Bioinformatics
ISBN 978-3-030-45256-8   ISBN 978-3-030-45257-5   (eBook)
https://doi.org/10.1007/978-3-030-45257-5

LNCS Sublibrary: SL8 – Bioinformatics

This Springer imprint is published by the registered company Springer Nature Switzerland AG
The registered company address is: Gewerbestrasse 11, 6330 Cham, Switzerland

Preface

This volume contains 13 extended abstracts and 24 short abstracts representing a total of 37 proceedings papers presented at the 24th International Conference on Research in Computational Molecular Biology (RECOMB 2020), hosted by the University of Padova, Italy, May 10–13, 2020. These 37 contributions were selected during a rigorous peer-review process from 206 submissions to the conference. Each of these 206 submissions received reviews from at least three members of the Program Committee or their designated sub-reviewers. Following an initial process of independent reviewing, all submissions were opened to discussion by their reviewers and the conference program chair through the EasyChair Conference Management System. Final decisions were made based on reviewer assessments with some adjustment to ensure the technical diversity of the conference program.

RECOMB 2020 allowed authors an option to publish their full extended papers in the conference proceedings or to provide short abstracts for the proceedings and pursue alternative arrangements for publishing the full paper. In addition, a select set of accepted papers were invited to submit revised manuscripts, considered for publication in *Cell Systems*. Authors who chose to publish only short abstracts in the proceedings were required to deposit their full papers in the preprint servers arxiv.org or biorxiv.org to be available before the meeting. All other papers that appear as extended abstracts in the proceedings were invited for submission to the RECOMB 2020 special issue of the *Journal of Computational Biology*.

RECOMB 2020 also featured highlight talks of computational biology papers that were published in journals during the previous 18 months. Of the 25 submissions to the Highlights Track, 10 were selected for oral presentation at RECOMB 2020.

In addition to presentations of these contributed papers, RECOMB 2020 featured six invited keynote talks given by leading scientists. The keynote speakers were Manuela Helmer-Citterich (University of Rome Tor Vergata), Eran Segal (Weizmann Institute), Marie-France Sagot (Inria at University of Lyon), Satoru Miyano (University of Tokyo), Pavel Pevzner (University of California at San Diego, Howard Hughes Medical Institute, and U.S. NIH), and Maurizio Corbetta (University of Padova).

RECOMB 2020 also featured two special invited panel discussions. The first panel, was on the Future of Algorithms in Biology, organized by Dan DeBlasio (University of Texas at El Paso), Guillaume Marcais (Carnegie Mellon University), and Carl Kingsford (Carnegie Mellon University). The second panel was on the DREAM challenges, organized by Julio Saez-Rodriguez (Heidelberg University).

In addition, three RECOMB Satellite meetings took place in parallel directly preceding the main RECOMB meeting. The RECOMB Genetics Satellite was co-chaired by Itsik Pe'er (Columbia University), Anna-Sapfo Malaspinas (Swiss Institute of Bioinformatics), Sriram Sankararaman (University of California at Los Angeles), and Gillian Belbin (Mt. Sinai Institute for Genomics Health). The RECOMB Satellite Workshop on Massively Parallel Sequencing (RECOMB-Seq) was co-chaired by

Robert Patro (University of Maryland) and Leena Salmela (University of Helsinki). The RECOMB-Computational Cancer Biology Satellite meeting (RECOMB-CCB) was co-chaired by Ewa Szczurek (University of Warsaw) and Iman Hajirasouliha (Weill Cornell Medicine). We thank them for organizing these great companion meetings, as well as their respective Program Committees for the hard work in making these meetings possible.

The organization of this conference was the work of many colleagues contributing their time, effort, and expertise. I am especially grateful to the Local Organizing Committee, particularly conference chair Fabio Vandin (University of Padova) and co-organizers Matteo Comin (University of Padova), Barbara Di Camillo (University of Padova), and Cinzia Pizzi (University of Padova). I also want to thank the members of Sistema Congressi: Marisa Sartori, Sabrina De Poli, Linda Frasson, and Elisa Quaggio. I am grateful to the many others who volunteered their time and work, including those whose names were not yet known to us at the time of this writing. I also want to thank the conference poster chair, Dario Ghersi (University of Nebraska Omaha) and the conference highlights chair, Itsik Pe'er (Columbia University), for their efforts in ensuring a high-quality technical program. I am also grateful to all of those Program Committee members and sub-reviewers who took time out of their busy schedules to review and discuss submissions on a very tight schedule. I also thank the authors of the proceedings papers, the highlights, and the posters for contributing their work to the meeting and for their attendance at the conference.

Finally, I would like to thank all our conference sponsors for their support, who at press time for this volume included Akamai Technologies, Illumina, the University of Padova, and the Department of Information Engineering (University of Padova), and especially to the sponsors of our student travel awards: the US National Science Foundation (NSF) and the International Society for Computational Biology (ISCB)[1].

May 2020   Russell Schwartz

---

[1] Please note that due to travel restrictions from the COVID-19 pandemic, RECOMB 2020 was postponed and reorganized as an online-only meeting after this proceedings volume was finalized. Details of this reorganization were still in progress as of the production time of the proceedings volume. The text of the preface thus reflects the original conference plans rather than the rescheduled conference as it actually occurred. We regret the inaccuracies in this description.

Organization

General Chair

Fabio Vandin   University of Padova, Italy

Program Chair

Russell Schwartz   Carnegie Mellon University, USA

Steering Committee

Vineet Bafna   University of California, San Diego, USA
Bonnie Berger (Chair)   Massachusetts Institute of Technology, USA
Eleazar Eskin   University of California, Los Angeles, USA
Teresa Przytycka   National Institutes of Health, USA
Cenk Sahinalp   National Institutes of Health, USA
Roded Sharan   Tel Aviv University, Israel

Program Committee

Derek Aguiar   University of Connecticut, USA
Tatsuya Akutsu   Kyoto University, Japan
Can Alkan   Bilkent University, Turkey
Mukul S. Bansal   University of Connecticut, USA
Niko Beerenwinkel   ETH Zurich, Switzerland
Bonnie Berger   Massachusetts Institute of Technology, USA
Sebastian Böcker   Friedrich Schiller University Jena, Germany
Christina Boucher   University of Florida, USA
Tony Capra   Vanderbilt University, USA
Cedric Chauve   Simon Fraser University, Canada
Brian Chen   Lehigh University, USA
Rayan Chikhi   CNRS, France
Leonid Chindelevitch   Simon Fraser University, Canada
Lenore Cowen   Tufts University, USA
Barbara Di Camillo   University of Padova, Italy
Mohammed El-Kebir   University of Illinois at Urbana-Champaign, USA
Nadia El-Mabrouk   University of Montreal, Canada
Xin Gao   King Abdullah University of Science and Technology,
                                    Saudi Arabia
Iman Hajirasouliha   Cornell University, USA
Bjarni Halldorsson   deCODE Genetics and Reykjavik University, Iceland
Fereydoun Hormozdiari   University of Washington, USA

Louxin Zhang   National University of Singapore, Singapore
Marinka Zitnik   Harvard University, USA

Additional Reviewers

Ahmed Abbas
Sergey Aganezov
Vanessa Aguiar-Pulido
Berk Alpay
Aybuge Altay
Federico Altieri
Bayarbaatar Amgalan
Aryan Arbabi
Abbas Roayaei Ardakany
Lars Arvestad
Marzieh Ayati
Xin Bai
Dipankar Ranjan Baisya
Burcu Bakir-Gungor
Marleen Balvert
Zhipeng Bao
Giacomo Baruzzo
Anais Baudot
Brittany Baur
Adi Bejan
Philipp Benner
Tristan Bepler
Jeremy Berg
Marilia Braga
Maria Brbic
Matthew Brendel
Davide Buffelli
Lydia Buntrock
Tomasz Burzykowski
Kevin Butler
Kieran Campbell
Mathias Cardner
Mark Chaisson
Annie Chateau
Chengqian Che
Xing Chen
Ben Chidester
Hyunghoon Cho
Julie Chow

Sarah Christensen
Neo Christopher Chung
A. Ercument Cicek
Ann Cirincione
Thérèse Commes
Tyler Cowman
Miklós Csürös
Sebastian Daberdaku
David Danko
Charlotte Darby
Viraj Deshpande
Alexander Dilthey
Meleshko Dmitrii
Daniel Doerr
Natnatee Dokmai
Sergio Doria
Yoann Dufresne
Kai Dührkop
Mohammad Haghir Ebrahimabadi
Hannes P. Eggertsson
Marcel Ehrhardt
Rebecca Elyanow
Massimo Equi
Oliver Eulenstein
Harry Feng
Markus Fleischauer
Eric Gamazon
Riqiang Gao
Thomas Gaudelet
Dario Ghersi
Omer Gokcumen
Abel Gonzalez-Perez
Ilan Gronau
Barak Gross
Bjarni Gunnarsson
Gisli Halldorsson
Mira Han
Marteinn Harõarson
Md Abid Hasan

Jotun Hein
David Heller
Ralf Herwig
Minh Hoang
Ermin Hodzic
Michael M. Hoffman
Martin Hoffmann
Jan Hoinka
Farhad Hormozdiari
Marjan Hosseini
Yidi Huang
Nitay Itzhacki
Erna Ivarsdottir
Katharina Jahn
Sarath Janga
Xiangying Jiang
Eric Jorgenson
Seong-Hwan Jun
Emre Karakoc
Mael Kerbiriou
Pouya Kheradpour
Parsoa Khorsand
Pegah Khosravi
Jongkyu Kim
Yoo-Ah Kim
Can Kockan
Hazal Koptagel
Krzysztof Koras
Snædís Kristmundsdottir
Jack Kuipers
David Laehnemann
Manuel Lafond
Yinglei Lai
Lisa Lamberti
Vincent Laville
Charles Lecellier
Da-Inn Lee
Anton Levitan
Cindy Li
Pengyuan Li
Qunhua Li
Qihua Liang
Max Libbrecht
Antoine Limasset
Johannes Linder
Markus List

Jie Liu
Kaiyuan Liu
Enrico Longato
Yang Lu
Marcus Ludwig
Lam-Ha Ly
Cong Ma
Ilari Maarala
Salem Malikic
Noel Malod-Dognin
Noël Malod-Dognin
Ion Mandoiu
Igor Mandric
Aniket Mane
Francesco Marass
Camille Marchet
Guillaume Marçais
Farid Rashidi Mehrabadi
Svenja Mehringer
Siavash Mirarab
Kristján Moore
Kingshuk Mukherjee
Dan Munro
Matthew Myers
Sheida Nabavi
Giri Narasimhan
Mohammadreza
    Mohaghegh Neyshabouri
Emmanuel Noutahi
Ibrahim Numanagić
Carlos Oliver
Baraa Orabi
Rachid Ounit
Soumitra Pal
Aaron Palmer
Weihua Pan
Prashant Pandey
Fabio Pardi
Tyler Park
Leonardo Pellegrina
Hedi Peterson
Pavlin Poličar
Susana Posada-Céspedes
Kaustubh Prabhu
Mattia Prosperi
Simon Puglisi

Gunnar Pálsson
Chengxiang Qiu
Yutong Qiu
Jianghan Qu
Fernando Racimo
Mathieu Raffinot
Elior Rahmani
René Rahn
Vladimir Reinharz
Jie Ren
Mina Rho
Camir Ricketts
Yonnie Rosenski
Mikhail Rotkevich
Kristoffer Sahlin
Diego Santoro
Roman Sarrazin-Gendron
Ilie Sarrpe
Palash Sashittal
Itay Sason
Gryte Satas
Natalie Sauerwald
Andrea Sboner
Jacob Schreiber
Roman Schulte-Sasse
Tizian Schulz
Celine Scornavacca
Alexander Sczyrba
Enrico Seiler
Saleh Sereshki
Yihang Shen
Junha Shin
Alireza Fotuhi Siahpirani
Pijus Simonaitis
Ritambhara Singh
Pavel Skums
Yan Song
Camila de Souza
Daniel Standage
Virginie Stanislas
Elena Sugis
Polina Suter
Garõar Sveinbjörnsson
Gergely Szöllősi
Zhengzheng Tang
Eric Tannier

Erica Tavazzi
Daniel Tello
Genki Terashi
Anna Thibert
Alexandru I. Tomescu
Andrea Tonon
Hosein Toosi
Tuan Trieu
Chittaranjan Tripathy
Viachaslau Tsyvina
Laura Tung
Lin Wan
Beibei Wang
Lei Wang
Liangliang Wang
Xin Wang
Yijie Wang
Yuchuan Wang
Zuoheng Wang
Catherine Welsh
Catie Welsh
Joshua Wetzel
Sam Windels
Jörg Winkler
Roland Wittler
Baolin Wu
Hao Wu
Yi-Chieh Wu
Damian Wójtowicz
Alexandros Xenos
Kyle Xiong
Chencheng Xu
Chen Yanover
Changchuan Yin
Shibu Yooseph
Kaan Yorgancioglu
Markus Kirolos Youssef
Alice Yue
Serhan Yılmaz
Simone Zaccaria
Carme Zambrana
Ron Zeira
Xiangrui Zeng
Gongbo Zhang
Kui Zhang
Ruochi Zhang

Contents

Short Papers

Extended Abstracts

Computing the Rearrangement Distance of Natural Genomes

Leonard Bohnenkämper[ID], Marília D. V. Braga[ID], Daniel Doerr[ID],
and Jens Stoye[(✉)][ID]

Faculty of Technology and Center for Biotechnology (CeBiTec),
Bielefeld University, Bielefeld, Germany
`jens.stoye@uni-bielefeld.de`

Abstract. The computation of genomic distances has been a very active
field of computational comparative genomics over the last 25 years. Substantial results include the polynomial-time computability of the inversion distance by Hannenhalli and Pevzner in 1995 and the introduction
of the double-cut and join (DCJ) distance by Yancopoulos, Attie and
Friedberg in 2005. Both results, however, rely on the assumption that
the genomes under comparison contain the same set of unique *markers* (syntenic genomic regions, sometimes also referred to as *genes*). In
2015, Shao, Lin and Moret relax this condition by allowing for duplicate
markers in the analysis. This generalized version of the genomic distance problem is NP-hard, and they give an ILP solution that is efficient
enough to be applied to real-world datasets. A restriction of their approach is that it can be applied only to *balanced* genomes, that have equal
numbers of duplicates of any marker. Therefore it still needs a delicate
preprocessing of the input data in which excessive copies of unbalanced
markers have to be removed.

In this paper we present an algorithm solving the genomic distance
problem for *natural* genomes, in which any marker may occur an arbitrary number of times. Our method is based on a new graph data structure, the *multi-relational diagram*, that allows an elegant extension of the
ILP by Shao, Lin and Moret to count *runs* of markers that are under- or
over-represented in one genome with respect to the other and need to be
inserted or deleted, respectively. With this extension, previous restrictions on the genome configurations are lifted, for the first time enabling
an uncompromising rearrangement analysis. Any marker sequence can
directly be used for the distance calculation.

The evaluation of our approach shows that it can be used to analyze
genomes with up to a few ten thousand markers, which we demonstrate
on simulated and real data.

Keywords: Comparative genomics · Genome rearrangement ·
DCJ-indel distance

R. Schwartz (Ed.): RECOMB 2020, LNBI 12074, pp. 3–18, 2020.
https://doi.org/10.1007/978-3-030-45257-5_1

1   Introduction

The study of genome rearrangements has a long tradition in comparative genomics. A central question is how many (and what kind of) mutations have occurred between the genomic sequences of two individual genomes. In order to avoid disturbances due to minor local effects, often the basic units in such comparisons are syntenic regions identified between the genomes under study, much larger than the individual DNA bases. We refer to such regions as *genomic markers*, or simply *markers*, although often one also finds the term *genes*.

Following the initial statement as an edit distance problem [15], a comprehensive trail of literature has addressed the problem of computing the number of rearrangements between two genomes in the past 25 years. In a seminal paper in 1995, Hannenhalli and Pevzner [12] introduced the first polynomial time algorithm for the computation of the inversion distance of transforming one chromosome into another one by means of segmental inversions. Later, the same authors generalized their results to the HP model [11] which is capable of handling multi-chromosomal genomes and accounts for additional genome rearrangements. Another breakthrough was the introduction of the double cut and join (DCJ) model [2,18], that is able to capture many genome rearrangements and whose genomic distance is computable in linear time. The model is based on a simple operation in which the genome sequence is cut twice between two consecutive markers and re-assembled by joining the resulting four loose cut-ends in a different combination.

A prerequisite for applying the DCJ model in practice to study rearrangements in genomes of two related species is that their genomic marker sets must be identical and that any marker occurs exactly once in each genome. This severely limits its applicability in practice. Linear time extensions of the DCJ model allow markers to occur in only one of the two genomes, computing a genomic distance that minimizes the sum of DCJ and insertion/deletion (indel) events [5,9]. Still, markers are required to be *singleton*, i.e., no duplicates can occur. When duplicates are allowed, the problem is more intricate and all approaches proposed so far are NP-hard, see for instance [1,6,7,14,16,17]. From the practical side, more recently, Shao *et al.* [17] presented an integer linear programming (ILP) formulation for computing the DCJ distance in presence of duplicates, but restricted to *balanced genomes*, where both genomes have equal numbers of duplicates. A generalization to unbalanced genomes was presented by Lyubetsky *et al.* [13], but their approach does not seem to be applicable to real data sets, see Sect. 4.1 for details.

In this paper we present the first feasible exact algorithm for solving the NP-hard problem of computing the distance under a general genome model where any marker may occur an arbitrary number of times in any of the two genomes, called *natural genomes*. Specifically, we adopt the *maximal matches* model where only markers appearing more often in one genome than in the other can be deleted or inserted. Our ILP formulation is based on the one from Shao *et al.* [17], but with an efficient extension that allows to count *runs* of markers that are under- or over-represented in one genome with respect to the other, so that the pre-existing model of minimizing the distance allowing DCJ and indel operations [5] can be adapted to our problem. With this extension, once we have the genome markers, no other restriction on the genome configurations is imposed.

The evaluation of our approach shows that it can be used to analyze genomes with up to a few ten thousand markers, provided the number of duplicates is not too large.

An extended version of this paper containing omitted proofs and additional results appeared as an arxiv preprint [3]. The complete source code of our ILP implementation and the simulation software used for generating the benchmarking data in Sect. 4.2 are available from https://gitlab.ub.uni-bielefeld.de/gi/ding.

2 Preliminaries

A *genome* is a set of *chromosomes* and each chromosome can be linear or circular. Each *marker* in a chromosome is an oriented DNA fragment. The representation of a marker m in a chromosome can be the symbol m itself, if it is read in direct orientation, or the symbol $\overline{m}$, if it is read in reverse orientation. We represent a chromosome S of a genome A by a string s, obtained by the concatenation of all symbols in S, read in any of the two directions. If S is circular, we can start to read it at any marker and the string s is flanked by parentheses.

Given two genomes A and B, let $\mathcal{U}$ be the set of all markers that occur in both genomes. For each marker $m \in \mathcal{U}$, let $\Phi_A(m)$ be the number of occurrences of m in genome A and $\Phi_B(m)$ be the number of occurrences of m in genome B. We can then define $\Delta\Phi(m) = \Phi_A(m) - \Phi_B(m)$. If both $\Phi_A(m) > 0$ and $\Phi_B(m) > 0$, m is called a *common marker*. We denote by $\mathcal{G} \subseteq \mathcal{U}$ the set of common markers of A and B. The markers in $\mathcal{U}\backslash\mathcal{G}$ are called *exclusive markers*. For example, if we have two unichromosomal linear genomes $A = \{1\,3\,2\,\overline{5}\,\overline{4}\,3\,5\,4\}$ and $B = \{1\,6\,2\,3\,1\,7\,3\,4\,1\,3\}$, then $\mathcal{U} = \{1, 2, 3, 4, 5, 6, 7\}$ and $\mathcal{G} = \{1, 2, 3, 4\}$. Furthermore, $\Delta\Phi(1) = 1 - 3 = -2$, $\Delta\Phi(2) = 1 - 1 = 0$, $\Delta\Phi(3) = 2 - 3 = -1$, $\Delta\Phi(4) = 2 - 1 = 1$, $\Delta\Phi(5) = 2$, and $\Delta\Phi(6) = \Delta\Phi(7) = -1$.

2.1 The DCJ-Indel Model

A genome can be transformed or *sorted* into another genome with the following types of mutations:

- A *double-cut-and-join* (DCJ) is the operation that cuts a genome at two different positions (possibly in two different chromosomes), creating four open ends, and joins these open ends in a different way. This can represent many different rearrangements, such as inversions, translocations, fusions and fissions. For example, a DCJ can cut linear chromosome $1\,2\,\overline{4}\,3\,5\,6$ before and after $\overline{4}\,3$, creating the segments $1\,2\bullet$, $\bullet\overline{4}\,3\bullet$ and $\bullet5\,6$, where the symbol $\bullet$ represents the open ends. By joining the first with the third and the second with the fourth open end, we invert $\overline{4}\,3$ and obtain $1\,2\,3\,4\,5\,6$.

- Since the genomes can have distinct multiplicity of markers, we also need to consider *insertions* and *deletions* of segments of contiguous markers [5,9, 19]. We refer to insertions and deletions collectively as *indels*. For example, the deletion of segment 5262 from linear chromosome 12352624 results in 1234. Indels have two restrictions: (i) only markers that have positive $\Delta\Phi$ can be deleted; and (ii) only markers that have negative $\Delta\Phi$ can be inserted.

In this paper, we are interested in computing the *DCJ-indel distance* between two genomes A and B, that is denoted by $d_{DCJ}^{id}(A, B)$ and corresponds to the minimum number of DCJs and indels required to sort A into B. We separate the instances of the problem in three types:

1. *Singular genomes*: the genomes contain no duplicate markers, that is, each common marker[1] is singular in each genome. Formally, we have that, for each $m \in \mathcal{G}$, $\Phi_A(m) = \Phi_B(m) = 1$. The distance between singular genomes can be easily computed in linear time [2,5,9].
2. *Balanced genomes*: the genomes contain no exclusive markers, but can have duplicates, and the number of duplicates in each genome is the same. Formally, we have $\mathcal{U} = \mathcal{G}$ and, for each $m \in \mathcal{U}$, $\Phi_A(m) = \Phi_B(m)$. Computing the distance for this set of instances is NP-hard, and an ILP formulation was given in [17].
3. *Natural genomes*: these genomes can have exclusive markers and duplicates, with no restrictions on the number of copies. Since these are generalizations of balanced genomes, computing the distance for this set of instances is also NP-hard. In the present work we present an efficient ILP formulation for computing the distance in this case.

2.2 DCJ-Indel Distance of Singular Genomes

First we recall the problem when common duplicates do not occur, that is, when we have singular genomes. We will summarize the linear time approach to compute the DCJ-indel distance in this case that was presented in [5], already adapted to the notation required for presenting the new results of this paper.

Relational Diagram. For computing a genomic distance it is useful to represent the relation between two genomes in some graph structure [2,4,5,10,11]. Here we adopt a variation of this structure, defined as follows. For each marker m, denote its two extremities by m^t (tail) and m^h (head). Given two singular genomes A and B, the *relational diagram* $R(A, B)$ has a set of vertices $V = V(A) \cup V(B)$, where $V(A)$ has a vertex for each extremity of each marker of genome A and $V(B)$ has a vertex for each extremity of each marker of genome B. Due to the 1-to-1 correspondence between the vertices of $R(A, B)$ and the occurrences of marker extremities in A and B, we can identify each extremity with its corresponding vertex. It is convenient to represent vertices in $V(A)$ in an upper line,

---

[1] The exclusive markers are not restricted to be singular, because it is mathematically trivial to transform them into singular markers when they occur in multiple copies.

respecting the order in which they appear in each chromosome of A, and the vertices in $V(B)$ in a lower line, respecting the order in which they appear in each chromosome of B.

If the marker extremities γ_1 and γ_2 are adjacent in a chromosome of A, we have an *adjacency edge* connecting them. Similarly, if the marker extremities γ_1' and γ_2' are adjacent in a chromosome of B, we have an adjacency edge connecting them. Marker extremities located at chromosome ends are called *telomeres* and are not connected to any adjacency edge. In contrast, each extremity that is not a telomere is connected to exactly one adjacency edge. Denote by E_{adj}^A and by E_{adj}^B the adjacency edges in A and in B, respectively. In addition, for each common marker $m \in \mathcal{G}$, we have two *extremity edges*, one connecting the vertex m^h from $V(A)$ to the vertex m^h from $V(B)$ and the other connecting the vertex m^t from $V(A)$ to the vertex m^t from $V(B)$. Denote by E_γ the set of extremity edges. Finally, for each occurrence of an exclusive marker in $\mathcal{U}\backslash\mathcal{G}$, we have an *indel edge* connecting the vertices representing its two extremities. Denote by E_{id}^A and by E_{id}^B the indel edges in A and in B. Each vertex is then connected either to an extremity edge or to an indel edge.

All vertices have degree one or two, therefore $R(A, B)$ is a simple collection of cycles and paths. A path that has one endpoint in genome A and the other in genome B is called an *AB-path*. In the same way, both endpoints of an *AA-path* are in A and both endpoints of a *BB-path* are in B. A cycle contains either zero or an even number of extremity edges. When a cycle has at least two extremity edges, it is called an *AB-cycle*. Moreover, a path (respectively cycle) of $R(A, B)$ composed exclusively of indel and adjacency edges in one of the two genomes corresponds to a whole linear (respectively circular) chromosome and is called a *linear* (respectively *circular*) *singleton* in that genome. Actually, linear singletons are particular cases of *AA*-paths or *BB*-paths. An example of a relational diagram is given in Fig. 1.

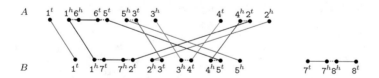

Fig. 1. For genomes $A = \{1\bar{6}53, 42\}$ and $B = \{172345, 7\bar{8}\}$, the relational diagram contains one cycle, two *AB*-paths (represented in blue), one *AA*-path and one *BB*-path (both represented in red). Short dotted horizontal edges are adjacency edges, long horizontal edges are indel edges, top-down edges are extremity edges.

The numbers of telomeres and of *AB*-paths in $R(A, B)$ are even. The *DCJ-cost* [5] of a DCJ operation ρ, denoted by $\|\rho\|$, is defined as follows. If it either increases the number of *AB*-cycles by one, or the number of *AB*-paths by two, ρ is *optimal* and has $\|\rho\| = 0$. If it does not affect the number of *AB*-cycles and

AB-paths in the diagram, ρ is *neutral* and has $\|\rho\| = 1$. If it either decreases the number of AB-cycles by one, or the number of AB-paths by two, ρ is *counter-optimal* and has $\|\rho\| = 2$.

Runs and Indel-Potential. The approach that uses DCJ operations to group exclusive markers for minimizing indels depends on the following concepts.

Given two genomes A and B and a component C of $R(A, B)$, a *run* [5] is a maximal subpath of C, in which the first and the last edges are indel edges, and all indel edges belong to the same genome. It can be an A-run when its indel edges are in genome A, or a B-run when its indel edges are in genome B. We denote by $\Lambda(C)$ the number of runs in component C. If $\Lambda(C) \geq 1$ the component C is said to be *indel-enclosing*, otherwise $\Lambda(C) = 0$ and C is said to be *indel-free*.

While sorting components separately with optimal DCJs only, runs can be *merged* (when two runs become a single one), and also *accumulated* together (when all its indel edges alternate with adjacency edges only and the run can be inserted or deleted at once) [5]. The *indel-potential* of a component C, denoted by $\lambda(C)$, is the minimum number of indels derived from C after this process and can be directly computed from $\Lambda(C)$:

$$\lambda(C) = \begin{cases} 0\,, & \text{if } \Lambda(C) = 0 \quad (C \text{ is indel-free}); \\ \left\lceil \frac{\Lambda(C)+1}{2} \right\rceil, & \text{if } \Lambda(C) \geq 1 \quad (C \text{ is indel-enclosing}). \end{cases}$$

Let λ_0 and λ_1 be, respectively, the sum of the indel-potentials for the components of the relational diagram before and after a DCJ ρ. The *indel-cost* of ρ is then $\Delta\lambda(\rho) = \lambda_1 - \lambda_0$, and the *DCJ-indel cost* of ρ is defined as $\Delta d(\rho) = \|\rho\| + \Delta\lambda(\rho)$. While sorting components separately, it has been shown that by using neutral or counter-optimal DCJs one can never achieve $\Delta d < 0$ [5]. This gives the following result:

Lemma 1 (from [2,5]). *Given two singular genomes A and B, whose relational diagram $R(A, B)$ has c AB-cycles and i AB-paths, we have*

$$d_{DCJ}^{id}(A, B) \leq |\mathcal{G}| - c - \frac{i}{2} + \sum_{C \in R(A,B)} \lambda(C).$$

Distance of Circular Genomes. For singular circular genomes, the graph $R(A, B)$ is composed of cycles only. In this case the upper bound given by Lemma 1 is tight and leads to a simplified formula [5]:

$$d_{DCJ}^{id}(A, B) = |\mathcal{G}| - c + \sum_{C \in R(A,B)} \lambda(C).$$

Recombinations and Linear Genomes. For singular linear genomes, the upper bound given by Lemma 1 is achieved when the components of $R(A, B)$ are sorted separately. However, there are optimal or neutral DCJ operations, called *recombinations*, that act on two paths and have $\Delta d < 0$. Such path recombinations are said to be *deducting*. The total number of types of deducting recombinations is relatively small. By exhaustively exploring the space of recombination types, it is possible to identify groups of chained recombinations (listed in Table 3 of the extended version of this manuscript [3]), so that the sources of each group are the original paths of the graph. In other words, a path that is a resultant of a group is never a source of another group. This results in a greedy approach (detailed in [3,5]) that optimally finds the value $\delta \geq 0$ to be deducted.

Theorem 1 (adapted from [5]). *Given two singular linear genomes A and B, whose relational diagram $R(A, B)$ has c AB-cycles and i AB-paths, and letting δ be the value obtained by maximizing deductions of path recombinations, we have*

$$d_{DCJ}^{id}(A, B) = |\mathcal{G}| - c - \frac{i}{2} + \sum_{C \in R(A,B)} \lambda(C) - \delta.$$

3 DCJ-Indel Distance of Natural Genomes

In this work we are interested in comparing two natural genomes A and B. First we note that it is possible to transform A and B into *matched* singular genomes $A^{\ddagger}$ and $B^{\ddagger}$ as follows. For each common marker $m \in \mathcal{G}$, if $\Phi_A \leq \Phi_B$, we should determine which occurrence of m in B matches each occurrence of m in A, or if $\Phi_B < \Phi_A$, which occurrence of m in A matches each occurrence of m in B. The matched occurrences receive the same identifier (for example, by adding the same *index*) in $A^{\ddagger}$ and in $B^{\ddagger}$. Examples are given in Fig. 2 (top and center). Observe that, after this procedure, the number of common markers between any pair of matched genomes $A^{\ddagger}$ and $B^{\ddagger}$ is

$$n_* = \sum_{m \in \mathcal{G}} \min\{\Phi_A(m), \Phi_B(m)\}.$$

Let $\mathbb{M}$ be the set of all possible pairs of matched singular genomes obtained from natural genomes A and B. The DCJ-indel distance of A and B is then defined as

$$d_{DCJ}^{id}(A, B) = \min_{(A^{\ddagger}, B^{\ddagger}) \in \mathbb{M}} \{d_{DCJ}^{id}(A^{\ddagger}, B^{\ddagger})\}.$$

3.1 Multi-relational Diagram

While the original relational diagram clearly depends on the singularity of common markers, when they appear in multiple copies we can obtain a data structure that integrates the properties of all possible relational diagrams of matched genomes. The *multi-relational diagram* $MR(A, B)$ of two natural genomes A and B also has a set $V(A)$ with a vertex for each of the two extremities of each marker occurrence of genome A and a set $V(B)$ with a vertex for each of the two extremities of each marker occurrence of genome B.

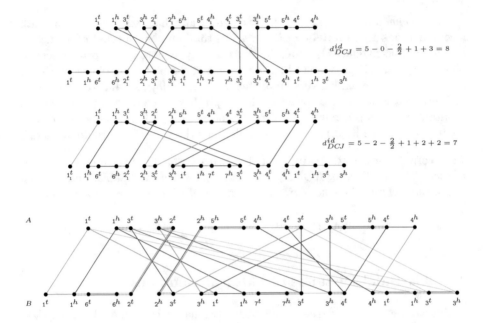

Fig. 2. Natural genomes $A = 132\overline{5}\overline{4}354$ and $B = 1623173413$ can give rise to many distinct pairs of matched singular genomes. The relational diagrams of two of these pairs are represented here, in the top and center. In the bottom we show the multi-relational diagram $MR(A, B)$. The decomposition that gives the diagram in the top is represented in red/orange. Similarly, the decomposition that gives the diagram in the center is represented in blue/cyan. Edges that are in both decompositions have two colors.

Again, sets E_{adj}^A and E_{adj}^B contain adjacency edges connecting adjacent extremities of markers in A and in B. But here the set E_γ contains, for each marker $m \in \mathcal{G}$, an extremity edge connecting each vertex in $V(A)$ that represents an occurrence of m^t to each vertex in $V(B)$ that represents an occurrence of m^t, and an extremity edge connecting each vertex in $V(A)$ that represents an occurrence of m^h to each vertex in $V(B)$ that represents an occurrence of m^h. Furthermore, for each marker $m \in \mathcal{U}$ with $\Phi_A(m) > \Phi_B(m)$, the set E_{id}^A contains one indel edge connecting the vertices representing the two extremities of the same occurrence of m in A. Similarly, for each marker $m' \in \mathcal{U}$ with $\Phi_B(m') > \Phi_A(m')$, the set E_{id}^B contains one indel edge connecting the vertices representing the two extremities of the same occurrence of m' in B. An example of a multi-relational diagram is given in Fig. 2 (bottom).

Consistent Decompositions. Note that if A and B are singular genomes, $MR(A, B)$ reduces to the ordinary $R(A, B)$. On the other hand, in the presence of duplicate common markers, $MR(A, B)$ may contain vertices of degree larger than two. A *decomposition* is a collection of vertex-disjoint *components*, that can be cycles and/or paths, covering all vertices of $MR(A, B)$. There can be multiple ways of selecting a decomposition, and we need to find one that allows to match occurrences of a marker in genome A with occurrences of the same marker in genome B.

Let $m_{(A)}$ and $m_{(B)}$ be, respectively, occurrences of the same marker m in genomes A and B. The extremity edge that connects $m^h_{(A)}$ to $m^h_{(B)}$ and the extremity edge that connects $m^t_{(A)}$ to $m^t_{(B)}$ are called *siblings*. A set $E_D \subseteq E_\gamma$ is a *sibling-set* if it is exclusively composed of pairs of siblings and does not contain any pair of incident edges. Thus, a *maximal* sibling-set E_D corresponds to a maximal matching of occurrences of common markers in both genomes.

The set of edges D *induced* by a maximal sibling-set E_D is said to be a *consistent decomposition* of $MR(A, B)$ and can be obtained as follows. In the beginning, D is the union of E_D with the sets of adjacency edges E^A_{adj} and E^B_{adj}. Then, for each indel edge e, if its two endpoints have degree one or zero in D, then e is added to D. Note that the consistent decomposition D covers all vertices of $MR(A, B)$ and is composed of cycles and paths, allowing us to compute the value

$$d^{id}_{DCJ}(D) = n_* - c_D - \frac{i_D}{2} + \sum_{C \in D} \lambda(C) - \delta_D,$$

where c_D and i_D are the numbers of *AB*-cycles and *AB*-paths in D, respectively, and δ_D is the optimal deduction of recombinations of paths from D. Since n_* is constant for any consistent decomposition, we can separate the part of the formula that depends on D, called *weight* of D:

$$w(D) = c_D + \frac{i_D}{2} - \sum_{C \in D} \lambda(C) + \delta_D.$$

Theorem 2. *Given two natural genomes A and B, the DCJ-indel distance of A and B can be computed by the following equation:*

$$d^{id}_{DCJ}(A, B) = \min_{D \in \mathbb{D}} \{d^{id}_{DCJ}(D)\} = n_* - \max_{D \in \mathbb{D}} \{w(D)\},$$

where $\mathbb{D}$ is the set of all consistent decompositions of $MR(A, B)$.

Proof. In the extended version of this manuscript.

A consistent decomposition D such that $d^{id}_{DCJ}(D) = d^{id}_{DCJ}(A, B)$ is said to be *optimal*. Computing the DCJ-indel distance between two natural genomes A and B, or, equivalently, finding an optimal consistent decomposition of $MR(A, B)$ is an NP-hard problem. In Sect. 4 we will describe an efficient ILP formulation to solve it. Before that, we need to introduce a transformation of $MR(A, B)$ that is necessary for our ILP.

3.2   Capping

The ends of linear chromosomes produce some difficulties for the decomposition. Fortunately there is an elegant technique to overcome this problem, called *capping* [11]. It consists of modifying the genomes by adding *artificial* singular common markers, also called *caps*, that circularize all linear chromosomes, so that their relational diagram is composed of cycles only, but, if the capping is optimal, the genomic distance is preserved.

Singular Genomes. An optimal capping that transforms singular genomes A and B into singular circular genomes $A_\circ$ and $B_\circ$ can be obtained after greedily identifying the recombination groups following a top-down order of Table 3 of the extended version of this manuscript [3]. The optimal Δd for each recombination group is achieved by linking the groups as indicated in Table 5 of the extended version, where we also prove the following theorem.

Theorem 3. *Let κ_A and κ_B be, respectively, the total numbers of linear chromosomes in singular genomes A and B. We can obtain an optimal capping of A and B with exactly*

$$p_* = \max\{\kappa_A, \kappa_B\}$$

caps and $a_* = |\kappa_A - \kappa_B|$ artificial adjacencies between caps.

Capped Multi-relational Diagram. We can transform $MR(A, B)$ into the *capped multi-relational diagram* $MR_\circ(A, B)$ as follows. First we need to create $4p_*$ new vertices, named $\circ_A^1, \circ_A^2, \ldots, \circ_A^{2p_*}$ and $\circ_B^1, \circ_B^2, \ldots, \circ_B^{2p_*}$, each one representing a *cap extremity*. Each of the $2\kappa_A$ telomeres of A is connected by an adjacency edge to a distinct cap extremity among $\circ_A^1, \circ_A^2, \ldots, \circ_A^{2\kappa_A}$. Similarly, each of the $2\kappa_B$ telomeres of B is connected by an adjacency edge to a distinct cap extremity among $\circ_B^1, \circ_B^2, \ldots, \circ_B^{2\kappa_B}$. Moreover, if $\kappa_A < \kappa_B$, for $i = 2\kappa_A+1, 2\kappa_A+3, \ldots, 2\kappa_B-1$, connect $\circ_A^i$ to $\circ_A^{i+1}$ by an *artificial adjacency edge*. Otherwise, if $\kappa_B < \kappa_A$, for $j = 2\kappa_B + 1, 2\kappa_B + 3, \ldots, 2\kappa_A - 1$, connect $\circ_B^j$ to $\circ_B^{j+1}$ by an artificial adjacency edge. All these new adjacency edges and artificial adjacency edges are added to E_{adj}^A and E_{adj}^B, respectively.

We also connect each $\circ_A^i$, $1 \le i \le 2p_*$, by a *cap extremity edge* to each $\circ_B^j$, $1 \le j \le 2p_*$, and denote by $E_\circ$ the set of cap extremity edges. A set $E_D' \subseteq E_\circ$ is a *capping-set* if it does not contain any pair of incident edges. A consistent decomposition D of $MR_\circ(A, B)$ is induced by a maximal sibling-set $E_D \subseteq E_\gamma$ and a maximal capping-set $E_D' \subseteq E_\circ$ and is composed of vertex disjoint cycles covering all vertices of $MR_\circ(A, B)$. We then have $d_{DCJ}^{id}(D) = n_* + p_* - w(D)$, where the weight of D can be computed by the simpler formula

$$w(D) = c_D - \sum_{C \in D} \lambda(C).$$

Finally, let $\mathbb{D}_\circ$ be the set of all consistent decompositions of $MR_\circ(A, B)$. Then

$$d_{DCJ}^{id}(A, B) = n_* + p_* - \max_{D \in \mathbb{D}_\circ} \{w(D)\}.$$

Note that the $2p_*$ cap extremities added to each genome correspond to p_* implicit caps. Furthermore, the number of artificial adjacency edges added to the genome with less linear chromosomes is $a_* = |\kappa_A - \kappa_B|$. Since each pair of matched singular genomes $(A^\ddagger, B^\ddagger) \in \mathbb{M}$ can be optimally capped with this number of caps and artificial adjacencies, it is clear that at least one optimal capping of each $(A^\ddagger, B^\ddagger)$ corresponds to a consistent decomposition $D \in \mathbb{D}_\circ$.

4   An Algorithm to Compute the DCJ-Indel Distance of Natural Genomes

An ILP formulation for computing the distance of two balanced genomes A and B was given by Shao *et al.* in [17]. In this section we describe an extension of that formulation for computing the DCJ-indel distance of natural genomes A and B, based on consistent cycle decompositions of $MR_\circ(A, B)$. The main difference is that here we need to address the challenge of computing the indel-potential $\lambda(C)$ for each cycle C of each decomposition. Note that a cycle C of $R(A, B)$ has either 0, or 1, or an even number of runs, therefore its indel-potential can be computed as follows:

$$\lambda(C) = \begin{cases} \Lambda(C), & \text{if } \Lambda(C) \leq 1; \\ \frac{\Lambda(C)}{2} + 1, & \text{if } \Lambda(C) \geq 2. \end{cases}$$

The formula above can be redesigned to a simpler one, that is easier to implement in the ILP. First, let a *transition* in a cycle C be an indel-free segment of C that is between a run in one genome and a run in the other genome and denote by $\aleph(C)$ the number of transitions in C. Observe that, if C is indel-free, then obviously $\aleph(C) = 0$. If C has a single run, then we also have $\aleph(C) = 0$. On the other hand, if C has at least 2 runs, then $\aleph(C) = \Lambda(C)$. Our new formula is then split into a part that simply tests whether C is indel-enclosing and a part that depends on the number of transitions $\aleph(C)$.

Proposition 1. *Given the function $r(C)$ defined as $r(C) = 1$ if $\Lambda(C) \geq 1$, otherwise $r(C) = 0$, the indel-potential $\lambda(C)$ can be computed from the number of transitions $\aleph(C)$ with the formula*

$$\lambda(C) = \frac{\aleph(C)}{2} + r(C).$$

Note that $\sum_{C \in D} r(C) = c_D^r + s_D$, where c_D^r and s_D are the number of indel-enclosing AB-cycles and the number of circular singletons in D, respectively. Now, we need to find a consistent decomposition D of $MR_\circ(A, B)$ maximizing its weight

$$w(D) = c_D - \sum_{C \in D} \lambda(C) = c_D - \left(c_D^r + s_D + \sum_{C \in D} \frac{\aleph(C)}{2} \right) = c_D^{\tilde{r}} - s_D - \sum_{C \in D} \frac{\aleph(C)}{2},$$

where $c_D^{\tilde{r}} = c_D - c_D^r$ is the number of indel-free AB-cycles in D.

4.1   ILP Formulation

Our formulation (shown in Algorithm 1) searches for an optimal consistent cycle decomposition of $MR_\circ(A, B) = (V, E)$, where the set of edges E is the union of all disjoint sets of the distinct types of edges, $E = E_\gamma \cup E_\circ \cup E_{adj}^A \cup E_{adj}^B \cup E_{id}^A \cup E_{id}^B$.

In the first part we use the same strategy as Shao *et al.* [17]. A binary variable x_e (D.01) is introduced for every edge e, indicating whether e is part of the computed decomposition. Constraint C.01 ensures that adjacency edges are in all decompositions, Constraint C.02 ensures that each vertex of each decomposition has degree 2, and Constraint C.03 ensures that an extremity edge is selected only together with its sibling. Counting the number of cycles in each decomposition is achieved by assigning a unique identifier i to each vertex v_i that is then used to label each cycle with the numerically smallest identifier of any contained vertex (see Constraint C.04, Domain D.02). A vertex v_i is then marked by variable z_i (D.03) as representative of a cycle if its cycle label y_i is equal to i (C.06). However, unlike Shao *et al.*, we permit each variable y_i to take on value 0 which, by Constraint C.05, will be enforced whenever the corresponding cycle is indel-enclosing. Since the smallest label of any vertex is 1 (cf. D.02), any cycle with label 0 will not be counted.

The second part is our extension for counting transitions. We introduce binary variables r_v (D.04) to label runs. To this end, Constraint C.07 ensures that each vertex v is labeled 0 if v is part of an A-run and otherwise it is labeled 1 indicating its participation in a B-run. Transitions between A- and B-runs in a cycle are then recorded by binary variable t_e (D.05). If a transition occurs between any neighboring pair of vertices $u, v \in V$ of a cycle, Constraint C.08 causes transition variable $t_{\{u,v\}}$ to be set to 1. We avoid an excess of co-optimal solutions by canonizing the locations in which such transitions may take place. More specifically, Constraint C.09 prohibits label changes in adjacencies not directly connected to an indel and Constraint C.10 in edges other than adjacencies of genome A, resulting in all A-runs containing as few vertices as possible.

In the third part we add a new constraint and a new domain to our ILP, so that we can count the number of circular singletons. Let K be the circular chromosomes in both genomes and E_{id}^k be the set of indel edges of a circular chromosome $k \in K$. For each circular chromosome we introduce a decision variable s_k (D.06), that is 1 if k is a circular singleton and 0 otherwise. A circular chromosome is then a singleton if all its indel edges are set (see Constraint C.11).

The objective of our ILP is to maximize the weight of a consistent decomposition, that is equivalent to maximizing the number of indel-free cycles, counted by the sum over variables z_i, while simultaneously minimizing the number of transitions in indel-enclosing AB-cycles, calculated by half the sum over variables t_e, and the number of circular singletons, calculated by the sum over variables s_k.

Implementation. We implemented the construction of the ILP as a python application, available at https://gitlab.ub.uni-bielefeld.de/gi/ding.

Comparison to the Approach by Lyubetsky *et al.* As mentioned in the Introduction, another ILP for the comparison of genomes with unequal content

---

Algorithm 1. ILP for the computation of the DCJ-indel distance of natural genomes

---

Objective:

$$\text{Maximize} \sum_{1 \leq i \leq |V|} z_i - \frac{1}{2} \sum_{e \in E} t_e - \sum_{k \in K} s_k$$

Constraints:

(C.01) $x_e = 1$ $\forall\, e \in E^A_{adj} \cup E^B_{adj}$

(C.02) $\sum_{\{u,v\} \in E} x_{\{u,v\}} = 2$ $\forall\, u \in V$

(C.03) $x_e = x_d$ $\forall\, e, d \in E_\gamma$ such that e and d are siblings

(C.04) $y_i \leq y_j + i(1 - x_{\{v_i, v_j\}})$ $\forall\, \{v_i, v_j\} \in E\,,$

(C.05) $y_i \leq i(1 - x_{\{v_i, v_j\}})$ $\forall\, \{v_i, v_j\} \in E^A_{id} \cup E^B_{id}$

(C.06) $i \cdot z_i \leq y_i$ $\forall\, 1 \leq i \leq |V|$

(C.07) $r_v \leq 1 - x_{\{u,v\}}$ $\forall\, \{u,v\} \in E^A_{id}\,,$

        $r_{v'} \geq x_{\{u',v'\}}$ $\forall\, \{u', v'\} \in E^B_{id}$

(C.08) $t_{\{u,v\}} \geq r_v - r_u - (1 - x_{\{u,v\}})$ $\forall\, \{u,v\} \in E$

(C.09) $\sum_{\{v,w\} \in E^A_{id}} x_{\{v,w\}} - t_{\{u,v\}} \geq 0$ $\forall\, \{u,v\} \in E^A_{adj}$

(C.10) $t_e = 0$ $\forall\, e \in E \setminus E^A_{adj}$

(C.11) $\sum_{e \in E^k_{id}} x_e - |k| \leq s_k$ $\forall k \in K$

Domains:

(D.01) $x_e \in \{0,1\}$ $\forall\, e \in E$

(D.02) $0 \leq y_i \leq i$ $\forall\, 1 \leq i \leq |V|$

(D.03) $z_i \in \{0,1\}$ $\forall\, 1 \leq i \leq |V|$

(D.04) $r_v \in \{0,1\}$ $\forall\, v \in V$

(D.05) $t_e \in \{0,1\}$ $\forall\, e \in E$

(D.06) $s_k \in \{0,1\}$ $\forall\, k \in K$

---

and paralogs was presented by Lyubetsky *et al.* [13]. In order to compare our method to theirs, we ran our ILP using CPLEX on a single thread with the two small artificial examples given in that paper on page 8. The results show that both ILPs give the same correct distances and our ILP runs much faster, as shown in Table 1.

Table 1. Comparison of running times and memory usage to the ILP in [13].

Dataset	#Markers	#Marker occurrences	Running time as reported in [13]	Our running time	Our peak memory
Example 1	5/5	9/9	"About 1.5 h"	.16 s	13200 kb
Example 2	10/10	11/11	"About 3 h"	.05 s	13960 kb

4.2 Performance Benchmark

For benchmarking purposes, we used Gurobi 9.0 as solver. In all our experiments, we ran Gurobi on a single thread. Details on how the simulated data is generated are given in the extended version of this manuscript.

In order to evaluate the impact of the number of duplicate occurrences on the running time, we keep the number of simulated DCJ events fixed to 10,000 and vary parameters that affect the number of duplicate occurrences.

Our ILP solves the decomposition problem efficiently for real-sized genomes under small to moderate numbers of duplicate occurrences: the solving times for genome pairs with less than 10,000 duplicate occurrences (∼50% of the genome size) shown in Fig. 3 are with few exceptions below 5 min and exhibit a linear increase, but the solving time is expected to boost dramatically with higher numbers of duplicate occurrences. To further exploit the conditions under which the ILP is no longer solvable with reasonable compute resources we continued the experiment with even higher amounts of duplicate occurrences and instructed Gurobi to terminate within 1 h of computation. We then partitioned the simulated data set into 8 intervals of length 500 according to the observed number of duplicate occurrences. For each interval, we determined the average as well as the maximal multiplicity of any duplicate marker and examined the average *optimality gap*, i.e., the difference in percentage between the best primal and the best dual solution computed within the time limit. The results are shown in Table 2 and emphasize the impact of duplicate occurrences on the solving time: below 14,000 duplicate occurrences, the optimality gap remains small and sometimes even the exact solution is computed, whereas above that threshold the gap widens very quickly.

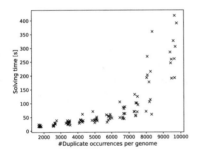

Fig. 3. Solving times for genomes with varying number of duplicate occurrences, totaling 20,000 marker occurrences per genome.

Table 2. Average optimality gap for simulated genome pairs grouped by number of duplicate occurrences after 1 h of running time.

#Dupl. occurrences	Avg. mult. of dupl. markers	Max. multiplicity	Avg. opt. gap (%)
11500..11999	2.206	8	0.000
12000..12499	2.219	8	0.031
12500..12999	2.217	7	0.025
13000..13499	2.233	9	0.108
13500..13999	2.247	8	0.812
14000..14499	2.260	8	1.177
14500..14999	2.274	8	81.865
15000..15499	2.276	9	33.102

Additionally, we ran three experiments, in each varying one of the following parameters while keeping the others fixed: (i) genome size, (ii) number of simulated DCJs and indels, and (iii) number of linear chromosomes. The results, given in the extended version of this manuscript, indicate that the number of linear chromosomes also has a considerable impact in the running time, while the other two have minor effect.

4.3   Real Data Analysis

Recently, the first three high-resolution haplotype-resolved human genomes have been published [8]. The study reports an average number of 156 inversions per genome, of which 121 are characterized as simple and 35 as copy-variable inversions. Here, we demonstrate the applicability of our approach to the study of real data by calculating the DCJ-indel distance between one of these haplotypes (HG00514.h0) and the human reference sequence (GRCh38). After the construction of a genomic marker set, we represented each chromosome of both genomes as marker sequence, with the largest chromosome (chr. 1) comprising close to 18,000 markers. We then ran our ILP for the computation of the DCJ-indel distance on each pair of chromosomes independently. We were able to obtain exact solutions for 17 chromosomes within few minutes and two more within a few days. However, the remaining four comparisons did not complete within a timelimit of 3 days. Still, after that time, their optimality gaps were below 0.1%. The calculated DCJ-indel distances ranged between 1.3% and 7.7% of the length of the marker sequences, with the number of runs accounting for at least 48.7% of the distance. Further details on the data set, the construction of the genomic markers, and the calculated DCJ-indel distances are described in Appendix A of the extended version of this paper.

5   Conclusion

By extending the DCJ-indel model to allow for duplicate markers, we introduced a rearrangement model that is capable of handling *natural genomes*, i.e., genomes that contain shared, individual, and duplicated markers. In other words, under this model genomes require no further processing nor manipulation once genomic markers and their homologies are inferred. The DCJ-indel distance of natural genomes being NP-hard, we presented a fast method for its calculation in form of an integer linear program. Our program is capable of handling real-sized genomes, as evidenced in simulation and real data experiments. It can be applied universally in comparative genomics and enables uncompromising analyses of genome rearrangements. We hope that such analyses will provide further insights into the underlying mutational mechanisms. Conversely, we expect the here presented model to be extended and specialized in future to reflect the insights gained by these analyses.

References

1. Angibaud, S., Fertin, G., Rusu, I., Thévenin, A., Vialette, S.: On the approximability of comparing genomes with duplicates. J. Graph Algorithms Appl. **13**(1), 19–53 (2009). A preliminary version appeared in Proceedings of WALCOM 2008
2. Bergeron, A., Mixtacki, J., Stoye, J.: A unifying view of genome rearrangements. In: Bücher, P., Moret, B.M.E. (eds.) WABI 2006. LNCS, vol. 4175, pp. 163–173. Springer, Heidelberg (2006). https://doi.org/10.1007/11851561_16
3. Bohnenkämper, L., Braga, M.D.V., Doerr, D., Stoye, J.: Computing the rearrangement distance of natural genomes. arXiv:2001.02139 (2020)
4. Braga, M.D.V.: An overview of genomic distances modeled with indels. In: Bonizzoni, P., Brattka, V., Löwe, B. (eds.) CiE 2013. LNCS, vol. 7921, pp. 22–31. Springer, Heidelberg (2013). https://doi.org/10.1007/978-3-642-39053-1_3

5. Braga, M.D.V., Willing, E., Stoye, J.: Double cut and join with insertions and deletions. J. Comput. Biol. **18**(9), 1167–1184 (2011). A preliminary version appeared in Proceedings of WABI 2010
6. Bryant, D.: The complexity of calculating exemplar distances. In: Sankoff, D., Nadeau, J.H. (eds.) Comparative Genomics, pp. 207–211. Kluwer Academic Publishers, Dordrecht (2000)
7. Bulteau, L., Jiang, M.: Inapproximability of (1,2)-exemplar distance. IEEE/ACM Trans. Comput. Biol. Bioinf. **10**(6), 1384–1390 (2013). A preliminary version appeared in Proceedings of ISBRA 2012
8. Chaisson, M.J.P., et al.: Multi-platform discovery of haplotype-resolved structural variation in human genomes. Nat. Commun. **10**(1), 1–16 (2019)
9. Compeau, P.E.C.: DCJ-indel sorting revisited. Algorithms Mol. Biol. **8**, 6 (2013). A preliminary version appeared in Proceedings of WABI 2012
10. Friedberg, R., Darling, A.E., Yancopoulos, S.: Genome rearrangement by the double cut and join operation. In: Keith, J.M. (ed.) Bioinformatics, Volume I: Data, Sequence Analysis, and Evolution, Methods in Molecular Biology, vol. 452, pp. 385–416. Humana Press, Totowa (2008)
11. Hannenhalli, S., Pevzner, P.A.: Transforming men into mice (polynomial algorithm for genomic distance problem). In: Proceedings of the 36th Annual Symposium of the Foundations of Computer Science (FOCS 1995), pp. 581–592. IEEE Press (1995)
12. Hannenhalli, S., Pevzner, P.A.: Transforming cabbage into turnip: polynomial algorithm for sorting signed permutations by reversals. J. ACM **46**(1), 1–27 (1999). A preliminary version appeared in Proceedings of STOC 1995
13. Lyubetsky, V., Gershgorin, R., Gorbunov, K.: Chromosome structures: reduction of certain problems with unequal gene contemnt and gene paralogs to integer linear programming. BMC Bioinform. **18**, 537 (2017)
14. Martinez, F.V., Feijão, P., Braga, M.D.V., Stoye, J.: On the family-free DCJ distance and similarity. Algorithms Mol. Biol. **10**, 13 (2015). A preliminary version appeared in Proceedings of WABI 2014
15. Sankoff, D.: Edit distance for genome comparison based on non-local operations. In: Apostolico, A., Crochemore, M., Galil, Z., Manber, U. (eds.) CPM 1992. LNCS, vol. 644, pp. 121–135. Springer, Heidelberg (1992). https://doi.org/10.1007/3-540-56024-6_10
16. Sankoff, D.: Genome rearrangement with gene families. Bioinformatics **15**(11), 909–917 (1999)
17. Shao, M., Lin, Y., Moret, B.M.E.: An exact algorithm to compute the double-cut-and-join distance for genomes with duplicate genes. J. Comput. Biol. **22**(5), 425–435 (2015). A preliminary version appeared in Proceedings of RECOMB 2014
18. Yancopoulos, S., Attie, O., Friedberg, R.: Efficient sorting of genomic permutations by translocation, inversion and block interchange. Bioinformatics **21**(16), 3340–3346 (2005)
19. Yancopoulos, S., Friedberg, R.: DCJ path formulation for genome transformations which include insertions, deletions, and duplications. J. Comput. Biol. **16**(10), 1311–1338 (2009). A preliminary version appeared in Proceedings of RECOMB-CG 2008

Deep Large-Scale Multi-task Learning Network for Gene Expression Inference

Kamran Ghasedi Dizaji[1], Wei Chen[3], and Heng Huang[1,2(✉)]

[1] Department of Electrical and Computer Engineering,
University of Pittsburgh, Pittsburgh, USA
`henghuanghh@gmail.com`
[2] Department of Biomedical Informatics, School of Medicine,
University of Pittsburgh, Pittsburgh, USA
[3] Department of Pediatrics, UPMC Children's Hospital of Pittsburgh,
University of Pittsburgh, Pittsburgh, USA

Abstract. Gene expressions profiling empowers many biological studies in various fields by comprehensive characterization of cellular status under different experimental conditions. Despite the recent advances in high-throughput technologies, profiling the whole-genome set is still challenging and expensive. Based on the fact that there is high correlation among the expression patterns of different genes, the above issue can be addressed by a cost-effective approach that collects only a small subset of genes, called landmark genes, as the representative of the entire genome set and estimates the remaining ones, called target genes, via the computational model. Several shallow and deep regression models have been presented in the literature for inferring the expressions of target genes. However, the shallow models suffer from underfitting due to their insufficient capacity in capturing the complex nature of gene expression data, and the existing deep models are prone to overfitting due to the lack of using the interrelations of target genes in the learning framework. To address these challenges, we formulate the gene expression inference as a multi-task learning problem and propose a novel deep multi-task learning algorithm with automatically learning the biological interrelations among target genes and utilizing such information to enhance the prediction. In particular, we employ a multi-layer sub-network with low dimensional latent variables for learning the interrelations among target genes (*i.e.* distinct predictive tasks), and impose a seamless and easy to implement regularization on deep models. Unlike the conventional complicated multi-task learning methods, which can only deal with tens or hundreds of tasks, our proposed algorithm can effectively learn the interrelations from the large-scale ($\sim$10,000) tasks on the gene expression inference problem, and does not suffer from cost-prohibitive operations. Experimental results indicate the superiority of our method compared to the existing gene expression inference models and alternative multi-task learning algorithms on two large-scale datasets.

This work was partially supported by NSF IIS 1836938, DBI 1836866, IIS 1845666, IIS 1852606, IIS 1838627, IIS 1837956, and NIH AG049371.

R. Schwartz (Ed.): RECOMB 2020, LNBI 12074, pp. 19–36, 2020.
https://doi.org/10.1007/978-3-030-45257-5_2

1   Introduction

Characterizing the cellular status under various states such as disease conditions, genetic perturbations and drug treatments is a fundamental problem in biological studies. Gene expression profiling provides a powerful tool for comprehensive analysis of the cellular status by capturing the gene expression patterns. The recent advances in high-throughput technologies make it possible to collect extensive gene expression profiles in versatile cellular conditions, providing invaluable large-scale databases of gene expressions for various biomedical studies [5,9]. For instance, Van *et al.* recognized the effective genes on the breast cancer by studying gene expression patterns of different patients [43]. Stephens *et al.* analyzed the relations between and within different cancer types by investigating the correlations of gene expression data among distinct types of tumors [41]. Richiardi *et al.* examined the gene expression data in a post meortem brain tissue, and showed correlation between resting-state functional brain networks and activity of genes [34]. Radical change in expression levels of several immune-related genes is identified in mice susceptible to influenza A virus infection using a microarray analysis [45]. The gene expression patterns in response to drug effects are also investigated on different tasks such as drug-target network construction [48] and drug discovery [33].

Despite recent developments on gene expression profiling, constructing large-scale gene expression archives under different experimental conditions is still challenging and expensive [30]. But previous studies have shown high correlations between gene expressions, indicating that the genes have similar functions in response to various conditions [14,31,37]. The clustering analysis of single cell RNA-Seq also shows similar expression pattern between intra-cluster genes across different cellular states [31]. Based on this fact, a small group of informative genes can be considered as the representative set of whole-genome data. The researchers in the Library of Integrated Network-based Cell-Signature (LINCS) program[1] used this assumption and employed principle component analysis (PCA) to choose ~1000 genes, which contain ~80% of the information in the entire set of genes. Note that profiling these ~1000 genes, called landmark genes, instead of the whole-genomes drastically reduces the collection costs (~$5 per profile) [32]. Hence, a cost-effective strategy in profiling of large-scale gene expressions data is to collect the landmark genes and predict the remaining genes (*i.e.* target genes) using a computational model.

The linear regression models with different regularizations are the first candidate models for predicting target genes. Later there were some attempts to use non-linear model to better capture the complex patterns of the gene expression profiles [13]. Deep models generally have shown remarkable flexibility in capturing the non-linear nature of biomedical data and high scalability in dealing with the large-scale datasets. Following the successful application of deep models on multiple biological problems [1,26,38,40,49], a few deep regression models have been also introduced for the gene expression inference problem [7,11]. However,

---

[1] http://www.lincsproject.org/.

these deep regression models do not utilize the interrelations among the target genes. These models usually consist of multiple shared layers among the genes, followed by a specific layer for each gene at the top. Therefore, these models ignore the biological information related to the gene interactions in their training process which leads to their suboptimal predictions.

To address the above challenges and utilize the interrelations between target genes to enhance the prediction task, we formulate the expression inference of target genes from the landmark ones as a multi-task learning problem. Multi-task learning algorithms generally aim to improve the generalization of multiple task predictors using the knowledge transferred across the related tasks through a joint learning framework [6]. We consider each gene expression prediction as a learning task and employ the multi-task learning model to automatically learn the interrelations of all tasks (*i.e.* all target genes) and utilize such information to enhance the prediction. Although there are multiple studies in literature on designing multi-task learning algorithms for deep models [35], they are designed and applied to tens or hundreds of tasks, and are not effective and scalable to deal with large number of tasks like the gene expression inference problem with about 10,000 tasks.

In this paper, we propose a novel multi-task learning algorithm for training a deep regression network with automatically learning the task interrelations of the gene expression data. Our deep large-scale multi-task learning method, denoted as Deep-LSMTL, can effectively learn the task interrelations from a large number of tasks, and is also efficient without suffering from the cost-prohibitive computational operations. In particular, our Deep-LSMTL model learns tasks interrelations using subspace clustering of task-specific parameters. Considering this clustering as the reconstruction of each task parameters by linear and sparse combination of other task-specific parameters, Deep-LSMTL provides a seamless regularization on deep models by approximating the reconstruction loss in the stochastic learning paradigms (*e.g.* stochastic gradient descent). Deep-LSMTL employs two-layer sub-network with low-dimension bottleneck to learn non-linear low-rank representations of task interrelations. Meanwhile, as a multi-task learning model, Deep-LSMTL can transfer asymmetric knowledge across the tasks to avoid the negative transfer issue, and enforce the task interrelations through the latent variables instead of the model parameters. All these advantages help Deep-LSMTL predict the target genes better than conventional approaches. We evaluate Deep-LSMTL with several deep and shallow regression models on two large-scale gene expression datasets. Experimental results indicate that our proposed algorithm has significantly better results compared to the state-of-the-art MTL methods and deep gene expression inference networks disregarding the neural network size and architecture. Furthermore, we gain insights into genes relations by visualizing the relevance of landmark and target genes in our inference model. The main contributions of this paper can be summarized as follows:

- Proposing a novel multi-task learning algorithm for training deep regression models, which is scalable to the large-scale tasks and efficient for the non-image data in the gene expression inference problem.

- Introducing a seamless regularization for deep multi-task models by employing a multi-layer sub-network with low-rank latent variables for learning the task interrelations.
- Outperforming existing gene expression inference models and alternative MTL algorithms by significant margins on two datasets regardless of networks architecture.

The following sections are organized as follows. In Sect. 2, we briefly review the related works on gene expression inference and recent multi-task learning algorithms. In Sect. 3, we start with the general clustering-based multi-task learning method, and then propose our multi-task learning algorithm for deep regression models. Then, we show the experimental results in Sect. 4, and evaluate the effectiveness of our algorithm in comparison with alternative models on multiple experimental conditions. We also plot some visualization figures to confirm the validity of our model. Finally, we conclude the paper in Sect. 5.

2   Related Work

2.1   Gene Expression Inference

Since archiving whole-genome expression profiles under various perturbations and biological conditions is still difficult and expensive [30], finding a way to reduce the costs while preserving the information is an important problem. The previous studies have shown that gene expressions are highly correlated, and even a small set of genes can contain rich information. For instance, Shah *et al.* indicated that a random set of 20 genes contains ~50% of the information of the whole-genome [37]. Moreover, the recent studies in RNA-seq confirm the assumption that a small set of genes is sufficient to indicate the comprehensive information throughout the transcriptome [14,31]. In order to determine the set of most informative genes, researchers of the LINCS program collected *GEO* dataset[2] based on Affymetrix HGU133A microarrays, and analyzed the correlation of gene expression profiles. Given the total number of 12,063 genes, they calculated the maximum percentage of information that can be recovered by a subset genes based on the comparable rank in the Kolmogorov-Smirnov statistic. According to the results of LINCS analysis, a subset of only 978 genes is able recover 82% of the observed connections in the entire transcriptome [20]. These genes are landmark genes and can be utilized to infer the expression of remaining genes referred as target genes.

Considering the gene expression inference as a multi-task regression problem, the shallow models such as linear regression with ℓ_1-norm and ℓ_2-norm regularizations and K-nearest neighbors (KNN) are used to infer the target genes expression from the landmark ones [7,11]. There are also a few attempt to use deep models on detecting and inferring gene expressions [7,11,22,44]. Using the representation power of deep learning models, Chen *et al.* introduced a fully

---

[2] https://cbcl.ics.uci.edu/public_data/D-GEX/.

connected multi-layer perceptron network as a multi-task regression model for the gene expression inference [7]. They justified the effectiveness of their deep model by achieving better experimental results compared to shallow and linear regression models. Recently, Dizaji *et al.* introduced a semi-supervised model, called SemiGAN, based on generative adversarial networks (GAN) for the gene expression inference problem [11]. Assuming a set of landmark genes as the unlabeled data and a set of landmark and their corresponding target genes as the labeled data, SemiGAN learns the joint and marginal distributions of landmark and target genes, and then enhanced the training of a regression model using the estimated target genes for the unlabeled data as pseudo-labels. Although these deep inference models addressed the issue of insufficient capacity in shallow and linear regression models, they did not explore the task interrelations, which indicate the biological knowledge of genes, in their training process. Thus, we formulate the gene expression inference problem as a multi-task learning and propose a new MTL method to explicitly learn the interrelations among the target genes in the learning framework and utilize these information to enhance the prediction results and also improve the generalization of our multi-task inference network.

2.2   Multi-task Learning Algorithms

The main goal of multi-task learning is to enhance the generalization of multiple task predictors using the knowledge transferred across the related tasks in a joint training process [6]. The main assumption in MTL methods is that the parameters of multiple tasks lie in a low-dimensional subspace due to their correlation. Using this assumption, Argyriou *et al.* aimed to have common features across tasks by imposing $\ell_{(2,1)}$-norm regularization on the feature matrix, and solved the convex equivalent of its objective function with this regularization [2]. Kang *et al.* introduced a method to share the features only within group of related tasks rather than all tasks [19]. Because the strict grouping of tasks is infeasible in real-world problems, some studies suggested the overlapping groups of related tasks for sharing the parameters [23,28]. Asymmetric multi-task learning (AMTL) provides a regularization loss by constructing the parameters of each task using the sparse and linear combination of other tasks' parameters, and penalizes the unreliable task predictors with higher loss to have less chance for knowledge transfer compared to the reliable task predictors with lower loss [24]. Furthermore, some works investigated the general idea of regularizing parameters using the task interrelations obtained via clustering-based approaches [4,10,17,42].

The common form of adopting multi-task learning methods on deep neural network is to share multiple layers among all tasks, and stack a specific layer for each task at the top. There are also some studies on designing the shared structure in deep multi-task models [36,46,47]. Lee *et al.* extended AMTL to deep models (Deep-AMTFL) by allowing asymmetric knowledge transfer across tasks through latent features rather than parameters [25]. Our MTL method for deep models differs from the previous studies, since it employs a multi-layer

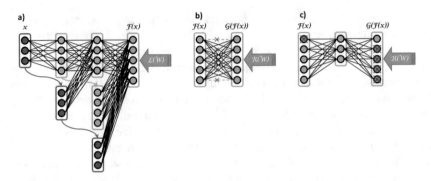

Fig. 1. Deep-LSMTL architecture. (a) This figure illustrates the architecture of our DenseNet ($\mathcal{F}$), where each layer receives the features of all preceding layers as the input. The ℓ_1-norm loss ($\mathcal{L}$) is applied on the output of this network. (b) This network indicates the shallow and linear $\mathcal{G}$ function used on Eq. (4). The crosses on some wights represents the zero diagonal elements constraint. (c) This network shows the two-layer model $\mathcal{G}$ on Eq. (5), where β and $(1 - \beta)$ filters are represented by the cross signs. The regularization loss ($\mathcal{R}$) is applied on the output of this layer.

sub-network with low-dimension latent representations for learning task inter-relations, providing an effective and scalable multi-task learning algorithm for gene expression problem with large number of tasks.

3   Deep Large-Scale Multi-task Learning Network

In the problem of gene expression inference, we consider $\mathcal{D} = \{\mathbf{x}_i, \mathbf{y}_i\}_{i=1}^{N}$ as the training set with N samples, where $\mathbf{x}_i \in \mathbb{R}^D$ and $\mathbf{y}_i \in \mathbb{R}^T$ denoting the landmark and target gene expression profiles for the i-th sample respectively. T shows the number of target genes (*i.e.* output dimension) and D indicates the number of landmark genes (*i.e.* input dimension). Considering that $\mathbf{y}_i \in \mathbb{R}^T$, we have T regression tasks and our goal is to learn a multi-task regression model to estimate the target gene expressions from their corresponding landmark genes. Unless specified otherwise, we use the following notations throughout the paper. The lower and upper case letters denote the scalars (*e.g.* i, T), bold lower case letters indicate vectors (*e.g.* $\mathbf{x}, \mathbf{w}$), the upper case letters represent matrices (*e.g.* $\mathbf{X}, \mathbf{W}$), and calligraphic letters indicate functions, sets and losses.

3.1   Clustered Multi-task Learning

Multi-task learning algorithms generally share the relevant knowledge among tasks by proposing a joint learning framework for the tasks. This joint learning framework usually contains a regularization term to improve generalization of the model as the following objective:

$$\min_{\mathbf{W}} \sum_{i=1}^{N} \sum_{t=1}^{T} \mathcal{L}(\mathbf{w}_t; \mathbf{x}_i, y_{it}) + \mathcal{R}(\mathbf{W}) \tag{1}$$

where the first term $(\mathcal{L})$ is the loss function applied separately on each task, and the second term $(\mathcal{R})$ is the regularization employed to enforce sharing the parameters according to the tasks relations. Note that $\mathbf{w}_t$ shows task-specified parameters as a column of $\mathbf{W} \in \mathbb{R}^{D \times T}$, if we assume a shallow regression network as our model. Although, the mean squared error (MSE) is the first choice for the loss in regression tasks, we empirically find out that the ℓ_1-norm loss function $\mathcal{L}(\mathbf{W}; \mathbf{x}_i, \mathbf{y}_i) = \|\mathbf{y}_i - \mathcal{F}(\mathbf{x}_i)\|_1$ is a better candidate in our objective, where $\mathcal{F}(\mathbf{x}_i) = \mathbf{W}\mathbf{x}_i$ is a regression model. There are also several studies in literature advocating ℓ_1-norm loss rather than MSE in different applications due to its robust performance in dealing with outliers and noisy data.

It has been shown that the shallow MTL models can be extended to deeper models by sharing a set latent features across all tasks as $\mathbf{W} = \mathbf{LS}$, where $\mathbf{L} \in \mathbb{R}^{D \times K}$ shows the shared parameters and $\mathbf{S} \in \mathbb{R}^{K \times T}$ denotes the task-specific weights [2,23]. The same idea can be adopted in deep models to use multiple layers of shared features followed by a task-specific layer. The multi-layer perceptron (i.e. fully connected) network is the simplest form of a deep MTL model as $\mathcal{F}(\mathbf{x}) = \sigma(...\sigma(\sigma(\mathbf{x}\mathbf{W}^{(1)})\mathbf{W}^{(2)})...\mathbf{W}^{(L)})$, where the first $L - 1$ layers are shared across all tasks and the last one is a task-specific layer. However, we employ a more efficient architecture for the shared layers by adopting the densely connected convolutional network [15] in our inference model. Assuming the input for each layer as $\mathbf{x}^{(l)}$ where $l \in \{0, ..., L-1\}$, the output of our DenseNet is computed by $\mathbf{x}^{(l+1)} = \sigma([\mathbf{x}^{(0)}, \mathbf{x}^{(1)}, ..., \mathbf{x}^{(l)}]\mathbf{W}^{(l+1)})$, where $[\mathbf{x}^{(0)}; , \mathbf{x}^{(1)}, ..., \mathbf{x}^{(l)}]$ represents the concatenation of features from all the previous layers. Figure 1(a) shows each layer of DenseNet receiving the features of all preceding layers as the input. The DenseNet has several advantages compared to multi-layer perceptron (MLP) such as reusing the features of previous layers, alleviating the vanishing-gradient issue in deep models, and reducing the number of parameters. The objective function in Eq. (1) can be written for our deep MTL network as follows:

$$\min_{\mathbf{W}^{(1)}, ..., \mathbf{W}^{(L)}} \sum_{i=1}^{N} \|\mathbf{y}_i - \mathcal{F}(\mathbf{x}_i)\|_1 + \mathcal{R}(\mathbf{W}^{(1)}, ..., \mathbf{W}^{(L)}). \tag{2}$$

To regularize the task-specific parameters, we can impose clustering-based constraints according to the task relations [4,10,17,24,42]. While the clustering constraints enforce the related tasks to share information and have similar parameters or features, they do not force all of the tasks to use shared features, and avoid the negative transfer issue where unrelated tasks adversely affect the features of correlated tasks [35]. Grouping the task-specific parameters using subspace clustering is an effective example of the clustering constraints. In the following equation, we replace the regularization term in Eq. (2) by the subspace clustering constraint:

$$\min_{\mathbf{W}^{(1)},...,\mathbf{W}^{(L)},\mathbf{V}} \sum_{i=1}^{N} \|\mathbf{y}_i - \mathcal{F}(\mathbf{x}_i)\|_1 + \lambda\|\mathbf{W}^{(L)} - \mathbf{W}^{(L)}\mathbf{V}\|_F^2 + \gamma\|\mathbf{V}\|_1 \quad (3)$$

where $\mathbf{V} \in \mathbb{R}^{T \times T}$ is the self-representation coefficient matrix with zero diagonal elements (*i.e.* $v_{tt} = 0$), showing the correlation among the T tasks. This regularization encourages the parameters of each task to be reconstructed by the linear and sparse combination of other tasks, and avoids the negative transfer issue by learning asymmetric similarity between the tasks. In order to implement Eq. (3) in deep models seamlessly, we multiply the features of latest hidden layer into the second term loss. Since our last layer has linear activation function, we can reformulate the objective in Eq. (3) as:

$$\min_{\mathbf{W}^{(1)},...,\mathbf{W}^{(L)},\mathbf{V}} \sum_{i=1}^{N} \|\mathbf{y}_i - \mathcal{F}(\mathbf{x}_i)\|_1 + \lambda\|\mathcal{F}(\mathbf{x}_i) - \mathcal{G}(\mathcal{F}(\mathbf{x}_i))\|_F^2 + \gamma\|\mathbf{V}\|_1 \quad (4)$$

where $\mathcal{F}(\mathbf{X}_i) = [\mathbf{x}^{(0)}, \mathbf{x}^{(1)}, ..., \mathbf{x}^{(L-1)}]\mathbf{W}^{(L)}$ is the prediction of our DenseNet model for the i-th sample, and $\mathcal{G}(\mathcal{F}(\mathbf{x}_i)) = \mathcal{F}(\mathbf{X}_i)\mathbf{V}$ can be considered as a layer stacked at the top of our DenseNet. The architecture of this layer is illustrated on Fig. 1(b).

3.2 Deep Large-Scale Multi-task Learning

The introduced model in the previous section has multiple drawbacks. First, it is not scalable to large number of tasks. Specially, this is a critical issue in the gene expression inference problem as the number of target genes (*i.e.* output size) is very large ($\sim$10,000) and consequently the number of parameters in $\mathbf{V}$. Moreover, the shallow and linear layer $\mathcal{G}(.)$ might not capture the complex correlations among the tasks. In addition, while we know that the target genes expressions are highly correlated, there is no explicit constraints to learn a low-dimension manifold for the tasks relations.

In order to address the aforementioned issues, we introduce a new function for $\mathcal{G}(.)$ to better capture the tasks correlations in our MTL algorithm. To increase the capacity of $\mathcal{G}$ function, we replace the linear model with a two-layer network as $\mathcal{G}(\mathcal{F}(\mathbf{x}_i)) = \mathbf{V}^{(2)}\sigma(\mathbf{V}^{(1)}\mathcal{F}(\mathbf{x}_i))$, where $\mathbf{V}^{(1)}$ and $\mathbf{V}^{(2)}$ are the first and second layer parameters respectively. Moreover, we are able to decrease the number of parameters in $\mathcal{G}$ by setting the number of units in its hidden layer smaller than the number of tasks. Specifically, while the shallow linear $\mathcal{G}$ function has T^2 parameters ($\sim 10^4 \times 10^4 = 10^8$), the proposed $\mathcal{G}$ contains $2TK$ free parameters, where $K \ll T$ ($\sim 2\times10^4\times100 = 2\times10^6$). In addition to addressing the scalability issue, a low-dimension bottleneck in $\mathcal{G}$ helps learning a low-rank representation for the tasks relations as shown in the hidden layer of Fig. 1(c). The following equation shows the objective for the proposed method:

$$\min_{\mathbf{W}^{(1)},...,\mathbf{W}^{(L)},\mathbf{V}^{(1)},\mathbf{V}^{(2)}} \|\mathcal{F}(\mathbf{x}) - \mathbf{y}\|_1 + \lambda\|(1 - \beta) \odot \left[\mathcal{F}(\mathbf{x}) - \mathcal{G}(\beta \odot \mathcal{F}(\mathbf{x}))\right]\|_F^2,$$

$$(5)$$

where β is a binary mask, and $\odot$ indicates the element-wise multiplication. The second term of this objective forces each task to be reconstructed by the other tasks, learning the relations among the tasks. Note that reconstructing the output of each task using the other ones in the multi-layer $\mathcal{G}$ function is not as straight-forward as zeroing the diagonal elements of V in the subspace clustering constraint. To solve this problem, we use the random β mask to approximate the reconstruction process in stochastic learning approaches (*e.g.* SGD). In particular, we randomly mask one or a few tasks outputs in each training iteration (*e.g.* $\beta = [1, 0, 0, ..., 0]$), then compute the output of regularization sub-network by $\mathcal{G}(\beta \odot \mathcal{F}(\mathbf{x}))$, and finally apply the reconstruction loss only to the masked tasks via $(1 - \beta)$ filter. Utilizing this approach, we seamlessly adopt the subspace clustering regularization in our deep low-rank MTL network.

4 Experiments

In this section, we evaluate our model compared to the alternative deep and shallow regression methods on multiple datasets. To do so, we first describe the experimental setups, compare Deep-LSMTL with the state-of-the-art models, and investigate the effectiveness of our MTL algorithm on neural network with different architecture. Furthermore, we visualize the relevance of the landmark and target genes in the inference problem, providing insights into the learned knowledge in our model.

4.1 Experimental Setup

Datasets: In our experiments, we include the microarray-based *GEO* dataset, the RNA-Seq-based *GTEx* dataset and the 1000 Genomes (*1000G*) RNA-Seq expression data (see footnote 2). The original *GEO* dataset consists of 129,158 gene expression profiles corresponding to 22,268 probes (978 landmark genes and 21,290 target genes) that are collected from the Affymetrix microarray platform. The original *GTEx* dataset is composed of 2,921 profiles from the Illumina RNA-Seq platform in the format of Reads Per Kilobase per Million (RPKM). The original *1000G* dataset includes 2,921 profiles from the Illumina RNA-Seq platform in the format of RPKM.

Following the data pre-processing in [7], we remove duplicate samples, normalize joint quantile and match cross-platform data. In particular, we first remove duplicated samples. We then map the expression values in the *GTEx* and *1000G* datasets according to the quantile computed in the *GEO* data, after which the expression value has been quantile normalized from 4.11 to 14.97. Finally, we normalize the expression values of each gene to zero mean and unit variance. After pre-processing, 943 landmark genes and 9520 target genes are remained in each profile. Our datasets contain 111,009 profiles in *GEO* dataset, 2,921 profiles in *GTEx* dataset and 462 profiles in the *1000G* dataset.

Following the experimental protocol in [7], we evaluate the methods under two different circumstances. First, we consider 80% of the *GEO* data for training, 10% of the *GEO* data for validation, and the other 10% of the *GEO* data for testing. Second, we use the same 80% of the *GEO* data for training, the *1000G* data for validation, and the *GTEx* data for testing. The second scenario is useful for validating the regression models on cross-platform prediction, since the training, validation and testing belong to the different distributions.

Alternative Methods: The most well-known linear inference model is the least square regression, which has the following objective function: $\min_W \sum_{i=1}^{n} ||\mathbf{x}_i\mathbf{W} - \mathbf{y}_i||_2^2 + \lambda||\mathbf{W}||_p^2$, where $\mathbf{W}$ is the model parameters, and λ represents the regularization hyper-parameter. When $\lambda = 0$, we call the model as the least square regression (LSR). But when $\lambda \neq 0$, we have two other linear models, LSR-L2 with ℓ_2-norm regularization (*i.e.* $p = 2$) and LSR-L1 with ℓ_1-norm regularization (*i.e.* $p = 1$). The regularization terms in LSR-L2 and LSR-L1 help the regression model to alleviate the overfitting issue.

We also include the k-nearest neighbors (KNN) method as a baseline method, where the prediction of a given profile is calculated as the average of its k nearest profiles. In addition, we compare with two deep learning methods, D-GEX [7] and SemiGAN [11], for gene expression inference. Generally, D-GEX model uses a multi-layer perception neural network as the inference model, while SemiGAN is designed based on generative adversarial networks.

We also adopt a few multi-task learning algorithms for training deep inference models in our problem. We review them in the following part very briefly, but refer the readers to the original papers for more details. The CNMTL method aims to cluster the task-specific (*i.e.* last layer) parameters using regularizations based on the wights mean, and between-cluster and within-cluster variances as follows [17]:

$$\min_{\mathbf{W}} \sum_{i=1}^{N}\sum_{t=1}^{T}\mathcal{L}(\mathbf{w}_t, \mathbf{x}_i, \mathbf{y}_i) + \lambda_M||\overline{\mathbf{W}}||_F^2 + \lambda_B \sum_{k=1}^{K}||\overline{\mathbf{W}}_k - \overline{\mathbf{W}}||_F^2 \qquad (6)$$

$$+ \lambda_W \sum_{k=1}^{K}\sum_{j \in \mathcal{J}(k)}||\mathbf{W}_j^{(L)} - \overline{\mathbf{W}}_k||_F^2$$

where the second term is the weights mean regularization with λ_M as the hyper-parameter and $\overline{\mathbf{W}} = 1/T \sum_{t=1}^{T}\mathbf{W}_t^{(L)}$ as the average of last layer weights across tasks, the third term is the between-cluster variance regularization with λ_B as the hyper-parameter and $\overline{\mathbf{W}}_k$ as the average of last layer weights of the k-cluster, and the last term is the within-cluster variance regularization with λ_W as the hyper-parameter and $\mathcal{J}(k)$ representing a set of tasks belonging to the k-th cluster. Note that we set $\mathcal{L}(\mathbf{W}; \mathbf{x}_i, \mathbf{y}_i) = ||\mathbf{y}_i - \mathcal{F}(\mathbf{x}_i)||_1$ for all the alternative models for a fair comparison.

The GO-MTL algorithm imposes ℓ_1-norm regularization on the task-specific parameters and Frobenius-norm regularization on the shared weights [23]:

$$\min_{\mathbf{W}} \sum_{i=1}^{N}\sum_{t=1}^{T} (\mathcal{L})(\mathbf{w}_t, \mathbf{x}_i, \mathbf{y}_i) + \mu\|\mathbf{w}_t^{(L)}\|_1 + \lambda \sum_{l=1}^{L-1} \|\mathbf{W}^{(l)}\|_F^2 \qquad (7)$$

where, μ and λ are the regularization hyper-parameters.

The AMTL method enforces each set of task-specific weights to be reconstructed by the linear combination of other tasks parameters using the following objective [25]:

$$\min_{\mathbf{W},\mathbf{V}} \sum_{i=1}^{N}\sum_{t=1}^{T} \alpha_t \mathcal{L}(\mathbf{w}_t, \mathbf{x}_i, \mathbf{y}_i) + \lambda\|\mathbf{W}^{(L)} - \mathbf{W}^{(L)}\mathbf{V}\|_2^2 \qquad (8)$$

where, λ is the regularization hyper-parameter, and α_t is the coefficient representing the easiness level of the t-th task that makes the outgoing transfer from hard tasks less than the easy tasks.

The AMTFL algorithm extends AMTL to regularize the features rather than the parameters [24]:

$$\min_{\mathbf{W},\mathbf{V}} \sum_{i=1}^{N}\sum_{t=1}^{T} \alpha_t \mathcal{L}(\mathbf{w}_t, \mathbf{x}_i, \mathbf{y}_i) + \mu\|\mathbf{w}_t^{(L)}\|_1 + \gamma\|\mathbf{Z} - \sigma(\mathbf{Z}\mathbf{W}^{(L)}\mathbf{V})\|_F^2 + \lambda \sum_{l=1}^{L-1} \|\mathbf{W}^{(l)}\|_F^2$$
$$(9)$$

where, μ, λ and γ are the regularization hyper-parameters, α_t is the task easiness coefficient, and $\mathbf{Z}$ is the output of the last hidden layer.

Evaluation Metrics: We use mean absolute error (MAE) and concordance correlation (CC) as the evaluation metrics. Given the testing data $\{(\mathbf{x}_i, \mathbf{y}_i)\}_{i=1}^{M}$, for a certain model, we denote the predicted expressions as $\{\hat{\mathbf{y}}_i\}_{i=1}^{M}$. The MAE is then computed using $MAE_t = \frac{1}{M}\sum_{i=1}^{M}|\hat{y}_{it} - y_{it}|$, where MAE_t indicates the mean absolute error for the t-th task, y_{it} shows the ground truth expression value for the t-th target gene in the i-th testing profile, and $\hat{y}_{it}$ represents the corresponding predicted value. The definition of CC is $CC_t = \frac{2\rho\sigma_{\mathbf{y}_t}\sigma_{\hat{\mathbf{y}}_t}}{\sigma_{\mathbf{y}_t}^2 + \sigma_{\hat{\mathbf{y}}_t}^2 + (\mu_{\mathbf{y}_t} - \mu_{\hat{\mathbf{y}}_t})^2}$, where CC_t shows the concordance correlation for the t-th target gene. ρ is the Pearson correlation, and $\mu_{\mathbf{y}_t}$, $\mu_{\hat{\mathbf{y}}_t}$, and $\sigma_{\mathbf{y}_t}$, $\sigma_{\hat{\mathbf{y}}_t}$ are the mean and standard deviation of $\mathbf{y}_t$ and $\hat{\mathbf{y}}_t$ respectively. Note that in addition to the mean values of the absolute error and concordance correlation via $MAE_{mean} = 1/T\sum_{t=1}^{T} MAE_t$ and $CC_{mean} = 1/T\sum_{t=1}^{T} CC_t$, we report the standard deviation across the tasks for each inference model.

Table 1. Comparison of different inference models on *GEO* and *GTEx* datasets based on the MAE and CC evaluation metrics. The results of the shallow regression models in the first part and the previous deep inference networks in the second part are reported from the original papers or running their released codes. The MTL methods in the third part and our proposed models in the forth part use densely connected architecture with different number of hidden units. Better results correspond to lower MAE values or higher CC values.

	Methods	GEO dataset		GTEx dataset	
		MAE	CC	MAE	CC
Shallow	LSR	0.3763 ± 0.0844	0.8227 ± 0.0956	0.4704 ± 0.1235	0.7184 ± 0.2072
	LSR-L1	0.3756 ± 0.0841	0.8221 ± 0.0960	0.4669 ± 0.1274	0.7163 ± 0.2188
	LSR-L2	0.3758 ± 0.0842	0.8223 ± 0.0959	0.4682 ± 0.1233	0.7181 ± 0.2076
	KNN	0.3708 ± 0.0958	0.8218 ± 0.1001	0.6225 ± 0.1469	0.5748 ± 0.2052
Deep	D-GEX	0.3204 ± 0.0879	0.8514 ± 0.0908	0.4393 ± 0.1239	0.7304 ± 0.2072
	SemiGAN	0.2997 ± 0.0869	0.8702 ± 0.0927	0.4223 ± 0.1266	0.7443 ± 0.2087
MTL	Deep-GO-MTL	0.2931 ± 0.0934	0.8717 ± 0.1075	0.4201 ± 0.1391	0.7434 ± 0.2153
	Deep-CNMTL	0.2946 ± 0.0928	0.8704 ± 0.1080	0.4199 ± 0.1393	0.7401 ± 0.2163
	Deep-AMTL	0.2942 ± 0.0936	0.8719 ± 0.1072	0.4238 ± 0.1388	0.7368 ± 0.2164
	Deep-AMTFL	0.2947 ± 0.0930	0.8703 ± 0.1081	0.4205 ± 0.1390	0.7428 ± 0.2154
Ours	DenseNet	0.2924 ± 0.0945	0.8727 ± 0.1070	0.4227 ± 0.1388	0.7416 ± 0.2156
	Deep-LSMTL	$\mathbf{0.2887 \pm 0.0949}$	$\mathbf{0.8753 \pm 0.1062}$	$\mathbf{0.4162 \pm 0.1390}$	$\mathbf{0.7510 \pm 0.2166}$

Implementation Details: In our model, we use a DenseNet structure with three hidden layers and $9,000$ hidden units on each layer. Leaky rectified linear unit [27] with leakiness ratio 0.2 is used as our activation function, and Adam algorithm [21] is employed as our optimization method. Moreover, we decrease our learning rates from 1×10^{-3} to 1×10^{-5} linearly from the first epoch to the maximum epoch 500. The batch size is set to 100. We also utilize batch normalization [16] as the layer normalization to speed up the convergence of training process. The parameters of all layers are initialized by Xavier approach [12]. We also select the dropout probability, λ, and number of hidden units in subspace layer from $dropout^{set} = \{0.05, 0.1, 0.25\}$, $\lambda^{set} = \{0.1, 1, 10\}$, and $units^{set} = \{500, 1000, 2000\}$ respectively based on the validation results. We use Pytorch toolbox for writing our code, and run the algorithm in a machine with one Titan X pascal GPU.

4.2   Performance Comparison

We compare the performance of Deep-LSMTL with other models on *GEO* and *GTEx* datasets. As shown in Table 1, the alternative models are grouped as the shallow regression models in the first part, the previous deep regression networks in the second part, the MTL algorithms applied on deep regression models in the third part, and our DenseNet baseline and Deep-LSMTL network in the forth part of the table. Regarding the MTL methods and our Deep-LSMTL network, we try to run the largest possible network with three hidden-layers on one GPU. The number of hidden-units for Deep-Go-MTL, Deep-CNMTL, Deep-AMTFL, Deep-AMTL and Deep-LSMTL are 8000, 4000, 5000, 7000 and 9000 respectively.

Table 2. Comparison of MTL algorithms for the gene expression inference problems on *GEO* and *GTEx* datasets. All of the models use a two hidden layers DenseNet as their structure, but have different number of hidden units in each part of the table. Better results correspond to lower MAE value or higher CC value.

Methods	GEO dataset		GTEx dataset		# params	# units
	MAE	CC	MAE	CC		
Deep-GO-MTL	0.3087 ± 0.0912	0.8602 ± 0.1120	0.4264 ± 0.1384	0.7347 ± 0.2179	8.08×10^7	3000
Deep-CNMTL	0.3070 ± 0.0912	0.8625 ± 0.1104	0.4263 ± 0.1390	0.7322 ± 0.2188	8.08×10^7	
Deep-AMTL	0.3073 ± 0.0912	0.8621 ± 0.1105	0.4265 ± 0.1385	0.7322 ± 0.0000	1.71×10^8	
Deep-AMTFL	0.3088 ± 0.0912	0.8599 ± 0.1121	0.4263 ± 0.1383	0.7346 ± 0.2180	1.47×10^8	
Deep-LSMTL	**0.3034 ± 0.0914**	**0.8626 ± 0.1153**	**0.4258 ± 0.1383**	**0.7377 ± 0.2188**	9.98×10^7	
Deep-GO-MTL	0.3014 ± 0.0922	0.8665 ± 0.1099	0.4267 ± 0.1388	0.7366 ± 0.2178	1.7×10^8	6000
Deep-CNMTL	0.2992 ± 0.0923	0.8696 ± 0.1079	0.4260 ± 0.1388	0.7345 ± 0.2179	1.7×10^8	
Deep-AMTL	0.2999 ± 0.0924	0.8688 ± 0.1085	0.4262 ± 0.1388	0.7351 ± 0.2175	2.61×10^8	
Deep-AMTFL	0.3016 ± 0.0922	0.8664 ± 0.1100	0.4265 ± 0.1387	0.7371 ± 0.2172	2.94×10^8	
Deep-LSMTL	**0.2951 ± 0.0927**	**0.8692 ± 0.1089**	**0.4234 ± 0.1391**	**0.7397 ± 0.2174**	1.89×10^8	
Deep-GO-MTL	0.2983 ± 0.0929	0.8693 ± 0.1089	0.4268 ± 0.1386	0.7376 ± 0.2167	2.78×10^8	9000
Deep-AMTL	0.2972 ± 0.0932	0.8713 ± 0.1077	0.4268 ± 0.1386	0.7367 ± 0.2170	3.69×10^8	
Deep-LSMTL	**0.2919 ± 0.0934**	**0.8717 ± 0.1080**	**0.4201 ± 0.1391**	**0.7439 ± 0.2170**	2.97×10^8	

Table 3. MAE comparison of D-GEX and Deep-LSMTL on *GEO* and *GTEx* datasets, when the number of hidden layers varies from 1 to 3, and the number of hidden units are 3000, 6000 or 9000. The structure of both models are based on the MLP network.

Methods	GEO dataset			GTEx dataset			# hidden layers
	# hidden units			# hidden units			
	3000	6000	9000	3000	6000	9000	
D-GEX	0.3421 ± 0.0858	0.3337 ± 0.0869	0.3300 ± 0.0874	0.4507 ± 0.1231	0.4428 ± 0.1246	0.4394 ± 0.1253	1
	0.3377 ± 0.0854	0.3280 ± 0.0869	0.3224 ± 0.0879	0.4586 ± 0.1194	0.4446 ± 0.1226	0.4393 ± 0.1239	2
	0.3362 ± 0.0850	0.3252 ± 0.0868	0.3204 ± 0.0879	0.5160 ± 0.1157	0.4595 ± 0.1186	0.4492 ± 0.1211	3
Deep-LSMTL	0.3179 ± 0.0901	0.3097 ± 0.0903	0.3054 ± 0.0903	0.4363 ± 0.1368	0.4349 ± 0.1369	0.4295 ± 0.1380	1
	0.3086 ± 0.0908	0.2985 ± 0.0915	0.2944 ± 0.0916	0.4338 ± 0.1374	0.4321 ± 0.1371	0.4289 ± 0.1379	2
	0.3067 ± 0.0913	0.2965 ± 0.0922	**0.2927 ± 0.0923**	0.4301 ± 0.1379	0.4286 ± 0.1373	**0.4253 ± 0.1383**	3

The MAE and CC results show that Deep-LSMTL significantly and consistently outperforms all of the alternative models on both *GEO* and *GTEx* datasets. As expected, Deep-LSMTL has large improvements against the shallow models, indicating the importance of deeper networks in capturing the complex nature of gene expression data. Deep-LSMTL also achieves better results than the existing deep inference models in the literature, proving the advantages of using the task interrelations in our MTL algorithm. Moreover, Deep-LSMTL not only shows better results compared to other MTL methods, but it also indicates the need for far less GPU memory than the other MTL methods.

Since the expressions of target genes are normalized, the direct comparisons of the errors may not be conclusive. In order to check if the improvement of Deep-LSMTL over the alternative models is statistically significant, we use the 5×2 cross validation method in [8]. In particular, we repeat 2-fold cross-validation of Deep-LSMTL and the best alternative model on *GEO* dataset (*i.e.* DenseNet) 5 times, and use a paired student's t-test on the MAE results. Based on the

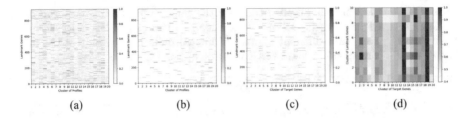

| (a) | (b) | (c) | (d) |

Fig. 2. Visualization of the relevance score calculated for each landmark gene on GEO dataset. (a) Relevance score of landmark genes *w.r.t.* cluster of profiles. We grouped the gene expression profiles into 20 clusters using K-means, and plot the contribution of each landmark gene to different clusters of profiles. (b) Cleaned version of landmark gene score. For each profile cluster, only the top 20 landmark genes in (a) are kept for clear visualization. (c) Relevance score of landmark genes *w.r.t.* cluster of target genes. We divide the 9520 target genes into 20 clusters via K-means, and demonstrate the contributions of cleaned landmark genes. (d) Relevance score of landmark gene clusters *w.r.t.* cluster of target genes. The landmark genes are clustered into 10 clusters, and their contributions in predicting of different clusters of target genes is plotted.

obtained p-values that is much less than 5%, we reject the null hypothesis that the results of the two models have the same distribution. Thus we can claim that Deep-LSMTL has statistically significant improvements compared to the other alternative models.

4.3   Ablation Study

While the previous experiments confirms the effectiveness of Deep-LSMTL in dealing with large-scale tasks by fitting a larger network on one GPU compared to other MTL methods, we design another experiment to compare the MTL methods with same structure. To do so, we consider the two-hidden-layer DenseNet architecture for all the MTL methods in three different settings with 3000, 6000, and 9000 hidden units. Table 2 shows the results of Deep-GO-MTL, Deep-CNMTL, Deep-AMTL, Deep AMTFL, and Deep-LSMTL on both *GEO* and *GTEx* Datasets. Note that there are still out-of-memory issues for Deep-CNMTL and Deep-AMTFL with 9000 hidden units. The results in Table 2 indicate better performance for Deep-LSMTL compared to the other MTL models on different architectures. Thus, Deep-LSMTL not only provides better scalable model in our inference problem, it also shows better performance even when the base network structure is similar.

In addition to investigate the effectiveness of Deep-LSMTL on the different base network than DenseNet, we compare Deep-LSMTL and D-GEX with MLP structure in Table 3. We report the results for the both models, where MLP network has one, two or three hidden layers and the hidden layers have 3000, 6000 or 9000 hidden units. Deep-LSMTL again outperforms D-GEX in all architectures consistently, and confirm its capability regardless of the base network structure.

4.4 Visualization

We perform a qualitative study on Deep-LSMTL to show the role of different landmark genes in the gene expression inference problem. In order to plot visualization figures, we adopt the Layer-wise Relevance Propagation (LRP) [3] method to calculate the importance of landmark genes that is learned in our model. Figure 2 shows the results of Deep-LSMTL with DenseNet structure (in Table 1) on *GEO* dataset. First, we divide the gene expression profiles into 20 clusters and then use LRP to calculate the relevance score of landmark genes *w.r.t.* each profile cluster in Fig. 2(a) and (b). These figures show that the landmark gene expression patterns are different for various profile groups, replicating the findings in previous cancer sub-type discovery and cancer landscape study that different group of samples usually exhibit different expression patterns [18,39].

Next, we analyze the relationship between landmark genes and target genes. We cluster the target genes into 20 groups and calculate the overall relevance score of landmark genes in the prediction of each target gene cluster in Fig. 2(c). For the sake of better visualization, we also group the landmark genes into 10 clusters and display the association between landmark gene clusters and target gene clusters in Fig. 2(d). We notice apparent difference in the relevance patterns for different target gene clusters, yet some similarity among certain clusters. This finding has also been validated by the previous gene cluster analysis [29], where genes cluster information is related to the structure of biosynthetic pathways and metabolites.

5 Conclusion

In this paper, we proposed a novel multi-task learning algorithm for training deep regression models on the gene expression inference problem. Our proposed method efficiently exploits the task interrelations to improve the generalizations of the predictors. We introduced a regularization on our learning framework that is easy to implement on deep models and scalable to large number of tasks. We validated our model on two gene expression datasets, and found consistent and significant improvements over all counterparts regardless of the base network architecture. Furthermore, we interpreted the role of landmark genes in the inference of target genes expression using visualization figures, providing insights into the information captured by our model.

References

1. Alipanahi, B., Delong, A., Weirauch, M.T., Frey, B.J.: Predicting the sequence specificities of DNA- and RNA-binding proteins by deep learning. Nat. Biotechnol. **33**(8), 831 (2015)
2. Argyriou, A., Evgeniou, T., Pontil, M.: Convex multi-task feature learning. Mach. Learn. **73**, 243–272 (2008). https://doi.org/10.1007/s10994-007-5040-8

3. Bach, S., Binder, A., Montavon, G., Klauschen, F., Müller, K.R., Samek, W.: On pixel-wise explanations for non-linear classifier decisions by layer-wise relevance propagation. PLoS ONE **10**(7), e0130140 (2015)
4. Bakker, B., Heskes, T.: Task clustering and gating for Bayesian multitask learning. J. Mach. Learn. Res. **4**, 83–99 (2003)
5. Brazma, A., et al.: ArrayExpress—a public repository for microarray gene expression data at the EBI. Nucleic Acids Res. **31**(1), 68–71 (2003)
6. Caruana, R.: Multitask learning. Mach. Learn. **28**, 41–75 (1997). https://doi.org/10.1023/A:1007379606734
7. Chen, Y., Li, Y., Narayan, R., Subramanian, A., Xie, X.: Gene expression inference with deep learning. Bioinformatics **32**(12), 1832–1839 (2016)
8. Dietterich, T.G.: Approximate statistical tests for comparing supervised classification learning algorithms. Neural Comput. **10**, 1895–1923 (1998)
9. Edgar, R., Domrachev, M., Lash, A.E.: Gene Expression Omnibus: NCBI gene expression and hybridization array data repository. Nucleic Acids Res. **30**(1), 207–210 (2002)
10. Evgeniou, T., Micchelli, C.A., Pontil, M.: Learning multiple tasks with kernel methods. J. Mach. Learn. Res. **6**, 615–637 (2005)
11. Ghasedi Dizaji, K., Wang, X., Huang, H.: Semi-supervised generative adversarial network for gene expression inference. In: Proceedings of the 24th ACM SIGKDD International Conference on Knowledge Discovery & Data Mining, pp. 1435–1444. ACM (2018)
12. Glorot, X., Bengio, Y.: Understanding the difficulty of training deep feedforward neural networks. In: Proceedings of the Thirteenth International Conference on Artificial Intelligence and Statistics, pp. 249–256 (2010)
13. Guo, X., Zhang, Y., Hu, W., Tan, H., Wang, X.: Inferring nonlinear gene regulatory networks from gene expression data based on distance correlation. PLoS ONE **9**(2), e87446 (2014)
14. Heimberg, G., Bhatnagar, R., El-Samad, H., Thomson, M.: Low dimensionality in gene expression data enables the accurate extraction of transcriptional programs from shallow sequencing. Cell Syst. **2**(4), 239–250 (2016)
15. Huang, G., Liu, Z., Weinberger, K.Q., van der Maaten, L.: Densely connected convolutional networks (2017)
16. Ioffe, S., Szegedy, C.: Batch normalization: accelerating deep network training by reducing internal covariate shift. In: International Conference on Machine Learning (ICML), pp. 448–456 (2015)
17. Jacob, L., Vert, J.P., Bach, F.R.: Clustered multi-task learning: a convex formulation. In: Advances in Neural Information Processing Systems (NIPS), pp. 745–752 (2009)
18. Kandoth, C., et al.: Mutational landscape and significance across 12 major cancer types. Nature **502**(7471), 333 (2013)
19. Kang, Z., Grauman, K., Sha, F.: Learning with whom to share in multi-task feature learning. In: International Conference on Machine Learning (ICML), pp. 521–528 (2011)
20. Keenan, A.B., et al.: The library of integrated network-based cellular signatures NIH program: system-level cataloging of human cells response to perturbations. Cell Syst. **6**(1), 13–24 (2017)
21. Kingma, D., Ba, J.: Adam: a method for stochastic optimization (2014)
22. Kishan, K., Li, R., Cui, F., Yu, Q., Haake, A.R.: GNE: a deep learning framework for gene network inference by aggregating biological information. BMC Syst. Biol. **13**(2), 38 (2019)

23. Kumar, A., Daumé III, H.: Learning task grouping and overlap in multi-task learning. In: Proceedings of the 29th International Conference on International Conference on Machine Learning (ICML), pp. 1723–1730. Omnipress (2012)

24. Lee, G., Yang, E., Hwang, S.: Asymmetric multi-task learning based on task relatedness and loss. In: International Conference on Machine Learning (ICML), pp. 230–238 (2016)

25. Lee, H., Yang, E., Hwang, S.J.: Deep asymmetric multi-task feature learning. In: Proceedings of the 35th International Conference on International Conference on Machine Learning (ICML) (2018)

26. Leung, M.K., Xiong, H.Y., Lee, L.J., Frey, B.J.: Deep learning of the tissue-regulated splicing code. Bioinformatics **30**(12), i121–i129 (2014)

27. Maas, A.L., Hannun, A.Y., Ng, A.Y.: Rectifier nonlinearities improve neural network acoustic models. In: International Conference on Machine Learning (ICML), vol. 30 (2013)

28. Maurer, A., Pontil, M., Romera-Paredes, B.: Sparse coding for multitask and transfer learning. In: International Conference on Machine Learning (ICML), pp. 343–351 (2013)

29. Medema, M.H., et al.: Minimum information about a biosynthetic gene cluster. Nat. Chem. Biol. **11**(9), 625 (2015)

30. Nelms, B.D., et al.: CellMapper: rapid and accurate inference of gene expression in difficult-to-isolate cell types. Genome Biol. **17**(1), 201 (2016)

31. Ntranos, V., Kamath, G.M., Zhang, J.M., Pachter, L., David, N.T.: Fast and accurate single-cell RNA-seq analysis by clustering of transcript-compatibility counts. Genome Biol. **17**(1), 112 (2016)

32. Peck, D., Crawford, E.D., Ross, K.N., Stegmaier, K., Golub, T.R., Lamb, J.: A method for high-throughput gene expression signature analysis, vol. 7, p. R61. BioMed Central (2006)

33. Rees, M.G., et al.: Correlating chemical sensitivity and basal gene expression reveals mechanism of action. Nat. Chem. Biol. **12**(2), 109 (2016)

34. Richiardi, J., et al.: Correlated gene expression supports synchronous activity in brain networks. Science **348**(6240), 1241–1244 (2015)

35. Ruder, S.: An overview of multi-task learning in deep neural networks (2017)

36. Ruder, S., Bingel, J., Augenstein, I., Søgaard, A.: Learning what to share between loosely related tasks (2017)

37. Shah, S., Lubeck, E., Zhou, W., Cai, L.: In situ transcription profiling of single cells reveals spatial organization of cells in the mouse hippocampus. Neuron **92**(2), 342–357 (2016)

38. Singh, R., Lanchantin, J., Robins, G., Qi, Y.: DeepChrome: deep-learning for predicting gene expression from histone modifications. Bioinformatics **32**(17), i639–i648 (2016)

39. Speicher, N.K., Pfeifer, N.: Integrating different data types by regularized unsupervised multiple kernel learning with application to cancer subtype discovery. Bioinformatics **31**(12), i268–i275 (2015)

40. Spencer, M., Eickholt, J., Cheng, J.: A deep learning network approach to *ab initio* protein secondary structure prediction. IEEE/ACM Trans. Comput. Biol. Bioinf. **12**(1), 103–112 (2015)

41. Stephens, P.J., et al.: The landscape of cancer genes and mutational processes in breast cancer. Nature **486**(7403), 400 (2012)

42. Thrun, S., O'Sullivan, J.: Discovering structure in multiple learning tasks: the TC algorithm. In: International Conference on Machine Learning (ICML), vol. 96, pp. 489–497 (1996)

43. Van't Veer, L.J., et al.: Gene expression profiling predicts clinical outcome of breast cancer. Nature **415**(6871), 530 (2002)
44. Wang, Z., He, Z., Shah, M., Zhang, T., Fan, D., Zhang, W.: Network-based multi-task learning models for biomarker selection and cancer outcome prediction. Bioinformatics (2019)
45. Yan, W., et al.: Transcriptional analysis of immune-related gene expression in p53-deficient mice with increased susceptibility to influenza A virus infection. BMC Med. Genomics **8**(1), 52 (2015). https://doi.org/10.1186/s12920-015-0127-8
46. Yang, Y., Hospedales, T.: Deep multi-task representation learning: a tensor factorisation approach. In: International Conference on Learning Representations (ICLR) (2017)
47. Yang, Y., Hospedales, T.M.: Trace norm regularised deep multi-task learning (2016)
48. Yıldırım, M.A., Goh, K.I., Cusick, M.E., Barabási, A.L., Vidal, M.: Drug-target network. Nat. Biotechnol. **25**(10), 1119–1126 (2007)
49. Zhou, J., Troyanskaya, O.G.: Predicting effects of noncoding variants with deep learning-based sequence model. Nat. Methods **12**(10), 931 (2015)

A Randomized Parallel Algorithm for Efficiently Finding Near-Optimal Universal Hitting Sets

Barış Ekim[1,2], Bonnie Berger[1,2]($\boxtimes$), and Yaron Orenstein[3]($\boxtimes$)

[1] Computer Science and Artificial Intelligence Laboratory,
Massachusetts Institute of Technology, Cambridge, MA 02139, USA
[2] Department of Mathematics, Massachusetts Institute of Technology,
Cambridge, MA 02139, USA
bab@mit.edu
[3] School of Electrical and Computer Engineering,
Ben-Gurion University of the Negev, 8410501 Beer-Sheva, Israel
yaronore@bgu.ac.il

Abstract. As the volume of next generation sequencing data increases, an urgent need for algorithms to efficiently process the data arises. *Universal hitting sets* (UHS) were recently introduced as an alternative to the central idea of minimizers in sequence analysis with the hopes that they could more efficiently address common tasks such as computing hash functions for read overlap, sparse suffix arrays, and Bloom filters. A UHS is a set of k-mers that hit every sequence of length L, and can thus serve as indices to L-long sequences. Unfortunately, methods for computing small UHSs are not yet practical for real-world sequencing instances due to their serial and deterministic nature, which leads to long runtimes and high memory demands when handling typical values of k (e.g. $k > 13$). To address this bottleneck, we present two algorithmic innovations to significantly decrease runtime while keeping memory usage low: (i) we leverage advanced theoretical and architectural techniques to parallelize and decrease memory usage in calculating k-mer hitting numbers; and (ii) we build upon techniques from randomized Set Cover to select universal k-mers much faster. We implemented these innovations in PASHA, the first randomized parallel algorithm for generating near-optimal UHSs, which newly handles $k > 13$. We demonstrate empirically that PASHA produces sets only slightly larger than those of serial deterministic algorithms; moreover, the set size is provably guaranteed to be within a small constant factor of the optimal size. PASHA's runtime and memory-usage improvements are orders of magnitude faster than the current best algorithms. We expect our newly-practical construction of UHSs to be adopted in many high-throughput sequence analysis pipelines.

Keywords: Universal hitting sets · Parallelization · Randomization

© Springer Nature Switzerland AG 2020
R. Schwartz (Ed.): RECOMB 2020, LNBI 12074, pp. 37–53, 2020.
https://doi.org/10.1007/978-3-030-45257-5_3

1 Introduction

The NIH Sequence Read Archive [8] currently contains over 26 petabases of sequence data. Increased use of sequence-based assays in research and clinical settings creates high computational processing burden; metagenomics studies generate even larger sequencing datasets [17,19]. New computational ideas are essential to manage and analyze these data. To this end, researchers have turned to k-mer-based approaches to more efficiently index datasets [7].

Minimizer techniques were introduced to select k-mers from a sequence to allow efficient binning of sequences such that some information about the sequence's identity is preserved [18]. Formally, given a sequence of length L and an integer k, its *minimizer* is the lexicographically smallest k-mer in it. The method has two key advantages: selected k-mers are close; and similar k-mers are selected from similar sequences. Minimizers were adopted for biological sequence analysis to design more efficient algorithms, both in terms of memory usage and runtime, by reducing the amount of information processed, while not losing much or any information [12]. The minimizer method has been applied in a large number of settings [4,6,20].

Orenstein and Pellow *et al.* [14,15] generalized and improved upon the minimizer idea by introducing the notion of a *universal hitting set* (UHS). For integers k and L, set $U_{k,L}$ is called a universal hitting set of k-mers if every possible sequence of length L contains at least one k-mer from $U_{k,L}$. Note that a UHS for any given k and L only needs to be computed once. Their heuristic DOCKS finds a small UHS in two steps: (i) remove a minimum-size set of vertices from a complete de Bruijn graph of order k to make it acyclic; and (ii) remove additional vertices to eliminate all $(L - k)$-long paths. The removed vertices comprise the UHS. The first step was solved optimally, while the second required a heuristic. The method is limited by runtime to $k \leq 13$, and thus applicable to only a small subset of minimizer scenarios. Recently, Marçais *et al.* [10] showed that there exists an algorithm to compute a set of k-mers that covers every path of length L in a de Bruijn graph of order k. This algorithm gives an asymptotically optimal solution for a value of k approaching L. Yet this condition is rarely the case for real applications where $10 \leq k \leq 30$ and $100 \leq L \leq 300$. The results of Marçais *et al.* show that for $k \leq 30$, the results are far from optimal for fixed L. A more recent method by DeBlasio *et al.* [3] can handle larger values of k, but with $L \leq 21$, which is impractical for real applications. Thus, it is still desirable to devise faster algorithms to generate small UHSs.

Here, we present PASHA (Parallel Algorithm for Small Hitting set Approximation), the first randomized parallel algorithm to efficiently generate near-optimal UHSs. Our novel algorithmic contributions are twofold. First, we improve upon the process of calculating vertex hitting numbers, i.e. the number of $(L - k)$-long paths they go through. Second, we build upon a randomized parallel algorithm for Set Cover to substantially speedup removal of k-mers for the UHS—the major time-limiting step—with a guaranteed approximation ratio on the k-mer set size. PASHA performs substantially better than current algorithms at finding a UHS in terms of runtime, with only a small increase in set size; it is consequently applicable to much larger values of k. Software and computed sets are available at: pasha.csail.mit.edu and github.com/ekimb/pasha.

2   Background and Preliminaries

Preliminary Definitions

For $k \geq 1$ and finite alphabet Σ, directed graph $B_k = (V, E)$ is a **de Bruijn graph** of order k if V and E represent k- and $(k + 1)$-long strings over Σ, respectively. An edge may exist from vertex u to vertex v if the $(k - 1)$-suffix of u is the $(k - 1)$-prefix of v. For any edge $(u, v) \in E$ with label $\mathcal{L}$, labels of vertices u and v are the prefix and suffix of length k of $\mathcal{L}$, respectively. If a de Bruijn graph contains all possible edges, it is *complete*, and the set of edges represents all possible $(k + 1)$-mers. An $\ell = (L - k)$-long path in the graph, i.e. a path of ℓ edges, represents an L-long sequence over Σ (for further details, see [1]).

For any L-long string s over Σ, k-mer set M **hits** s if there exists a k-mer in M that is a contiguous substring in s. Consequently, **universal hitting set** (UHS) $U_{k,L}$ is a set of k-mers that hits any L-long string over Σ. A trivial UHS is the set of all k-mers, but due to its size ($|\Sigma|^k$), it does not reduce the computational expense for practical use. Note that a UHS for any given k and L does not depend on a dataset, but rather needs to be computed only once.

Although the problem of computing a universal hitting set has no known hardness results, there are several NP-hard problems related to it. In particular, the problem of computing a universal hitting set is highly similar, although not identical, to the (k, L)-*hitting set* problem, which is the problem of finding a minimum-size k-mer set that hits an input set of L-long sequences. Orenstein and Pellow *et al.* [14, 15] proved that the (k, L)-*hitting set* problem is NP-hard, and consequently developed the near-optimal DOCKS heuristic. DOCKS relies on the Set Cover problem, which is the problem of finding a minimum-size collection of subsets $S_1, ..., S_k$ of finite set U whose union is U.

The DOCKS Heuristic

DOCKS first removes from a complete de Bruijn graph of order k a *decycling set*, turning the graph into a directed acyclic graph (DAG). This set of vertices represent a set of k-mers that hits all sequences of infinite length. A minimum-size decycling set can be found by Mykkelveit's algorithm [13] in $O(|\Sigma|^k)$ time. Even after all cycles, which represent sequences of infinite length, are removed from the graph, there may still be paths representing sequences of length L, which also need to be hit by the UHS. DOCKS removes an additional set of k-mers that hits all remaining sequences of length L, so that no path representing an L-long sequence, i.e. a path of length $\ell = L - k$, remains in the graph.

However, finding a minimum-size set of vertices to cover all paths of length ℓ in a directed acyclic graph (DAG) is NP-hard [16]. In order to find a small, but not necessarily minimum-size, set of vertices to cover all ℓ-long paths, Orenstein and Pellow *et al.* [14,15] introduced the notion of a *hitting number*, the number of ℓ-long paths containing vertex v, denoted by $T(v, \ell)$. DOCKS uses the hitting number to prioritize removal of vertices that are likely to cover a large number of paths in the graph. This, in fact, is an application of the greedy method for the Set Cover problem, thus guaranteeing an approximation ratio of $O(1 + \log(\max_v T(v, \ell)))$ on the removal of additional k-mers.

The hitting numbers for all vertices can be computed efficiently by dynamic programming: For any vertex v and $0 \leq i \leq \ell$, DOCKS calculates the number of i-long paths starting at v, $D(v, i)$, and the number of i-long paths ending at v, $F(v, i)$. Then, the hitting number is directly computable by

$$T(v, \ell) = \sum_{i=0}^{\ell} F(v, i) \cdot D(v, \ell - i) \tag{1}$$

and the dynamic programming calculation in graph $G = (V', E')$ is given by

$$\begin{aligned} &\forall v \in V', \ D(v, 0) = F(v, 0) = 1 \\ &D(v, i) = \sum_{(v,u) \in E'} D(u, i - 1) \\ &F(v, i) = \sum_{(u,v) \in E'} F(u, i - 1) \end{aligned} \tag{2}$$

Overall, DOCKS performs two main steps: First, it finds and removes a minimum-size decycling set, turning the graph into a DAG. Then, it iteratively removes vertex v with the largest hitting number $T(v, \ell)$ until there are no ℓ-long paths in the graph. DOCKS is sequential: In each iteration, one vertex with the largest hitting number is removed and added to the UHS output, and the hitting numbers are recalculated. Since the first phase of DOCKS is solved optimally in polynomial time, the bottleneck of the heuristic lies in the removal of the remaining set of k-mers to cover all paths of length $\ell = L - k$ in the graph, which represent all remaining sequences of length L.

As an additional heuristic, Orenstein and Pellow *et al.* [14,15] developed DOCKSany with a similar structure as DOCKS, but instead of removing the vertex that hits the most $(L-k)$-long paths, it removes a vertex that hits the most paths in each iteration. This reduces the runtime by a factor of L, as calculating the hitting number $T(v)$ for each vertex can be done in linear time with respect to the size of the graph. DOCKSanyX extends DOCKSany by removing X vertices with the largest hitting numbers in each iteration. DOCKSany and DOCKSanyX run faster compared to DOCKS, but the resulting hitting sets are larger.

3   Methods

Overview of the Algorithm. Similar to DOCKS, PASHA is run in two phases: First, a minimum-size decycling set is found and removed; then, an additional set of k-mers that hits remaining L-long sequences is removed. The removal of the decycling set is identical to that of DOCKS; however, in PASHA we introduce randomization and parallelization to efficiently remove the additional set of k-mers. We present two novel contributions to efficiently parallelize and randomize the second phase of DOCKS. The first contribution leads to a faster calculation of hitting numbers, thus reducing the runtime of each iteration. The second contribution leads to selecting multiple vertices for removal at each iteration, thus reducing the number of iterations to obtain a graph with no $(L - k)$-long paths. Together, the two contributions provide orthogonal improvements in runtime.

Improved Hitting Number Calculation

Memory Usage Improvements. We reduce memory usage through algorithmic and technical advances. Instead of storing the number of i-long paths for $0 \leq i \leq \ell$ in both F and D, we apply the following approach (Algorithm 1): We compute D for all $v \in V$ and $0 \leq i \leq \ell$. Then, while computing the hitting number, we calculate F for iteration i. For this aim, we define two arrays: F_{curr} and F_{prev}, to store only two instances of i-long path counts for each vertex: The current and previous iterations. Then, for some j, we compute F_{curr} based on F_{prev}, set $F_{prev} = F_{curr}$, and add $F_{curr}(v) \cdot D(v, \ell - j)$ to the hitting number sum. Lastly, we increase j, and repeat the procedure, adding the computed hitting numbers iteratively. This approach allows the reduction of matrix F, since in each iteration we are storing only two arrays, F_{curr} and F_{prev}, instead of the original F matrix consisting of $\ell + 1$ arrays. Therefore, we are able to reduce memory usage by close to half, with no change in runtime.

To further reduce memory usage, we use `float` variable type (of size 4 bytes) instead of `double` variable type (of size 8 bytes). The number of paths kept in F and D increase exponentially with i, the length of the paths. To be able to use the 8 bit exponent field, we initialize F and D to `float` minimum positive value. This does not disturb algorithm correctness, as path counting is only scaled to some arbitrary unit value, which may be 2^{-149}, the smallest positive value that can be represented by `float`. This is done in order to account for the high numbers that path counts can reach. The remaining main memory bottleneck is matrix D, whose size is $4 \cdot 4^k \cdot (\ell + 1)$ bytes.

Lastly, we utilized the property of a complete de Bruijn graph of order k being the line graph of a de Bruijn graph of order $k - 1$. While all k-mers are represented as the set of vertices in the graph of order k, they are represented as edges in the graph of order $k - 1$. If we remove edges of a de Bruijn graph of order $k - 1$, instead of vertices in a graph of order k, we can reduce memory usage by another factor of $|\Sigma|$. In our implementation we compute D and F for all vertices of a graph of order $k - 1$, and calculate hitting numbers for edges. Thus, the bottleneck of the memory usage is reduced to $4 \cdot 4^{k-1} \cdot (\ell + 1)$ bytes.

Runtime Reduction by Parallelization. We parallelize the calculation of the hitting numbers to achieve a constant factor reduction in runtime. The calculation of i-long paths through vertex v only depends on the previously calculated matrices for the $(i-1)$-long paths through all vertices adjacent to v (Eq. 2). Therefore, for some i, we can compute $D(v, i)$ and $F(v, i)$ for all vertices in V' in parallel, where V' is the set of vertices left after the removal of the decycling set. In addition, we can calculate the hitting number $T(v, \ell)$ for all vertices V' in parallel

---

Algorithm 1. Improved hitting number calculation. *Input:* $G = (V, E)$

1: $D \leftarrow [|V|][\ell + 1]$, with $[|V|][0]$ initialized to **1**
2: $F_{curr} \leftarrow [|V|]$
3: $F_{prev} \leftarrow [|V|]$ initialized to **1**
4: $T \leftarrow [|V|]$ initialized to **0**
5: **for** $1 \leq i \leq \ell$ **do:**
6:     **for** $v \in V$ **do:**
7:         **for** $(v, u) \in E$ **do:**
8:             $D[v][i] \mathrel{+}= D[u][i - 1]$
9: **for** $1 \leq i \leq \ell + 1$ **do:**
10:     **for** $v \in V$ **do:**
11:         $F_{curr}[v] = 0$
12:         **for** $(u, v) \in E$ **do:**
13:             $F_{curr}[v] \mathrel{+}= F_{prev}[u]$
14:         $T[v] \mathrel{+}= F_{prev}[v] \cdot D[v][\ell - i + 1]$
15:     $F_{prev} = F_{curr}$
16: **return** T

(similar to computing D and F), since the calculation does not depend on the hitting number of any other vertex (we call this parallel variant PDOCKS for the purpose of comparison with PASHA). We note that for DOCKSany and DOCK-SanyX, the calculations of hitting numbers for each vertex cannot be computed in parallel, since the number of paths starting and ending at each vertex both depend on those of the previous vertex in topological order.

Parallel Randomized k-mer Selection

Our goal is to find a minimum-size set of vertices that covers all ℓ-long paths. We can represent the remaining graph as an instance of the Set Cover problem. While the greedy algorithm for the second phase of DOCKS is serial, we will show that we can devise a parallel algorithm, which is close to the greedy algorithm in terms of performance guarantees, by picking a large set of vertices that cover nearly as many paths as the vertices that the greedy algorithm picks one by one.

In PASHA, instead of removing the vertex with the maximum hitting number in each iteration, we consider a set of vertices for removal with hitting numbers within an interval, and pick vertices in this set independently with constant probability. Considering vertices within an interval allows us to efficiently introduce randomization while still emulating the deterministic algorithm. Picking vertices independently in each iteration enables parallelization of the procedure. Our randomized parallel algorithm for the second phase of the UHS problem adapts that of Berger *et al.* [2] for the original Set Cover problem.

The UHS Selection Procedure. The input includes graph $G = (V, E)$ and randomization variables $0 < \varepsilon \leq \frac{1}{4}$, $0 < \delta \leq \frac{1}{\ell}$ (Algorithm 2). Let function calcHit() calculate the hitting numbers for all vertices, and return the maximum hitting number (line 2). We set $t = \lceil \log_{1+\varepsilon} T_{max} \rceil$ (line 3), and run a series of steps from t, iteratively decreasing t by 1. In step t, we first calculate the hitting numbers of all vertices (line 5); then, we define vertex set S to contain vertices with a hitting number between $(1+\varepsilon)^{t-1}$ and $(1+\varepsilon)^t$ for potential removal (lines 8–9).

Let P_S be the sum of all hitting numbers of the vertices in S, i.e. $P_S = \sum_{v \in S} T(v, \ell)$ (line 10). In each step, if the hitting number for vertex v is more than a δ^3 fraction of P_S, i.e. $T(v, \ell) \geq \delta^3 P_S$, we add v to the picked vertex set V_t (lines 11–13). For vertices with a hitting number smaller than $\delta^3 P_S$, we pairwise independently pick them with probability $\frac{\delta}{\ell}$. We test the vertices in pairs to impose pairwise independence: If an unpicked vertex u satisfies the probability $\frac{\delta}{\ell}$, we choose another unpicked vertex v and test the same probability $\frac{\delta}{\ell}$. If both are satisfied, we add both vertices to the picked vertex set V_t; if not, neither of them are added to the set (lines 14–16). This serves as a bound on the probability of picking a vertex. If the sum of hitting numbers of the vertices in set V_t is at least $|V_t|(1+\varepsilon)^t(1 - 4\delta - 2\varepsilon)$, we add the vertices to the output set, remove them from the graph, and decrease t by 1 (lines 17–20). The next iteration runs with decreased t. Otherwise, we rerun the selection procedure without decreasing t.

Algorithm 2. The selection procedure. *Input:* $G = (V, E), 0 < \varepsilon \leq \frac{1}{4}, 0 < \delta \leq \frac{1}{\ell}$

1: $R \leftarrow \{\}$
2: $T_{max} \leftarrow \text{calcHit}()$
3: $t \leftarrow \lceil \log_{1+\varepsilon} T_{max} \rceil$
4: **while** $t > 0$ **do**
5:     **if** calcHit() $== 0$ **then break**
6:     $S \leftarrow \{\}$
7:     $V_t \leftarrow \{\}$
8:     **for** $v \in V$ **do:**
9:         **if** $(1+\varepsilon)^{t-1} \leq T(v, \ell) \leq (1+\varepsilon)^t$ **then** $S \leftarrow S \cup \{v\}$
10:     $P_S \leftarrow \sum_{v \in S} T(v, \ell)$
11:     **for** $v \in S$ **do:**
12:         **if** $T(v, \ell) \geq \delta^3 P_S$ **then**
13:             $V_t \leftarrow V_t \cup \{v\}$
14:     **for** $u, v \in S$ **do:**
15:         **if** $u \notin V_t$ **and** unirand(0,1) $\leq \frac{\delta}{\ell}$ **and** $v \notin V_t$ **and** unirand(0,1) $\leq \frac{\delta}{\ell}$ **then**
16:             $V_t \leftarrow V_t \cup \{u, v\}$
17:     **if** $\sum_{v \in V_t} T(v, \ell) \geq |V_t| \cdot (1+\varepsilon)^t(1 - 4\delta - 2\varepsilon)$ **then**
18:         $R \leftarrow R \cup V_t$
19:         $G = G(V \setminus V_t, E)$
20:         $t \leftarrow t - 1$
21: **return** R

Performance Guarantees. At step t, we add the selected vertex set V_t to the output set if $\sum_{v \in V_t} T(v, \ell) \geq |V_t|(1 + \varepsilon)^t(1 - 4\delta - 2\varepsilon)$. Otherwise, we rerun the selection procedure with the same value of t. We show in Appendix A that with high probability, $\sum_{v \in V_t} T(v, \ell) \geq |V_t|(1 + \varepsilon)^t(1 - 4\delta - 2\varepsilon)$. We also show that PASHA produces a cover $\alpha(1 + \log T_{max})$ times the optimal size, where $\alpha = 1/(1 - 4\delta - 2\varepsilon)$. In Appendix B, we give the asymptotic number of the selection steps and prove the average runtime complexity of the algorithm. Performance summaries in terms of theoretical runtime and approximation ratio are in Table 1.

Table 1. Summary of theoretical results for the second phase of different algorithms for generating a set of k-mers hitting all L-long sequences. PDOCKS is DOCKS with the improved hitting number calculation, i.e. greedy removal of one vertex at each iteration. p_D, p_{DA} denote the total number of picked vertices for DOCKS/PDOCKS and DOCKSany, respectively. m denotes the number of parallel threads used, T_{max} the maximum vertex hitting number, and ϵ and δ PASHA's randomization parameters.

Algorithm	DOCKS	PDOCKS	DOCKSany	PASHA
Theoretical runtime	$O((1 + p_D)|\Sigma|^{k+1} \cdot L)$	$O((1 + p_D)|\Sigma|^{k+1} \cdot L/m)$	$O((1 + p_{DA})|\Sigma|^{k+1})$	$O((L^2 \cdot |\Sigma|^{k+1} \cdot \log^2(|\Sigma|^k))/(\varepsilon\delta^3 m))$
Approximation ratio	$1 + \log T_{max}$	$1 + \log T_{max}$	N/A	$(1 + \log T_{max})/(1 - 4\delta - 2\varepsilon)$

4   Results

PASHA Outperforms Extant Algorithms for $k \leq 13$

We compared PASHA and PDOCKS to extant methods on several combinations of k and L. We ran DOCKS, DOCKSany, PDOCKS, and PASHA over $5 \leq k \leq 10$, DOCKSanyX, PDOCKS, and PASHA for $k = 11$ and $X = 10$, and PASHA and DOCKSanyX for $X = 100, 1000$ for $k = 12, 13$ respectively, for $20 \leq L \leq 200$. We say that an algorithm is *limited by runtime* if for some value of $k \leq 13$ and for $L = 100$, its runtime exceeds 1 day (86400 s), in which case we stopped the operation and excluded the method from the results for the corresponding value of k. While running PASHA, we set $\delta = 1/\ell$, and $1 - 4\delta - 2\varepsilon = 1/2$ to set an emulation ratio $\alpha = 2$ (see Sect. 3 and Appendix A). The methods were benchmarked on a 24-CPU Intel Xeon Gold (2.10 GHz) with 754 GB of RAM. We ran all tests using all available cores ($m = 24$ in Table 1).

Comparing Runtimes and UHS Sizes. We ran DOCKS, PDOCKS, DOCKSany, and PASHA for $k = 10$ and $20 \leq L \leq 200$. As seen in Fig. 1A, DOCKS has a significantly higher runtime than the parallel variant PDOCKS, while producing identical sets (Fig. 1B). For small values of L, DOCKSany produces the largest UHSs compared to other methods, and as L increases, the differences in both runtime and UHS size for all methods decrease, since there are fewer k-mers to add to the removed decycling set to produce a UHS.

We ran PDOCKS, DOCKSany10, and PASHA for $k = 11$ and $20 \leq L \leq 200$. As seen in Fig. 1C, for small values of L, both PDOCKS and DOCKSany10 have significantly higher runtimes than PASHA; while for larger L, DOCKSany10 and PASHA are comparable in their runtimes (with PASHA being negligibly slower). In Fig. 1D, we observe that PDOCKS computes the smallest sets for all values of L. Indeed, its guaranteed approximation ratio is the smallest among all three benchmarked methods. While the set sizes for all methods converge to the same value for larger L, DOCKSany10 produces the largest UHSs for small values of L, in which case PASHA and PDOCKS are preferable.

PASHA's runtime behaves differently than that of other methods. For all methods but PASHA, runtime decreases as L increases. Instead of gradually decreasing with L, PASHA's runtime gradually decreases up to $L = 70$, at which it starts to increase at a much slower rate. This is explained by the asymptotic complexity of PASHA (Table 1). Since computing a UHS for small L requires a larger number of vertices to be removed, the decrease in runtime with increasing L up to $L = 70$ is significant; however, due to PASHA's asymptotic complexity being quadratic with respect to L, we see a small increase from $L = 70$ to $L = 200$. All other methods depend linearly on the number of removed vertices, which decreases as L increases.

Despite the significant decrease in runtime in PDOCKS compared to DOCKS, PDOCKS was still limited by runtime to $k \leq 12$. Therefore, we ran DOCKSany100 and PASHA for $k = 12$ and $20 \leq L \leq 200$. As seen in Figs. 1E and F, both methods follow a similar trend as in $k = 11$, with DOCKSany100 being significantly slower and generating significantly larger UHSs for small values of L. For larger values of L, DOCKSany100 is slightly faster, while PASHA produces sets that are slightly smaller.

At $k = 13$ we observed the superior performance of PASHA over DOCKSany1000 in both runtime and set size for all values of L. We ran DOCKSany1000 and PASHA for $k = 13$ and $20 \leq L \leq 200$. As seen in Figs. 1G and H, DOCKSany1000 produces larger sets and is significantly slower compared to PASHA for all values of L. This result demonstrates that the slow increase in runtime for PASHA compared to other algorithms for $k < 13$ does not have a significant effect on runtime for larger values of k.

PASHA Enables UHS for $k = 14, 15, 16$

Since all existing algorithms and PDOCKS are limited by runtime to $k \leq 13$, we report the first UHSs for $14 \leq k \leq 16$ and $L = 100$ computed using PASHA, run on a 24-CPU Intel Xeon Gold (2.10 GHz) with 754 GB of RAM using all 24 cores. Figure 2 shows runtimes and sizes of the sets computed by PASHA.

Density Comparisons for the Different Methods

In addition to runtimes and UHS sizes, we report values of another measure of UHS performance known as *density*. The *density* of the minimizers scheme $d(M, S, k)$ is the fraction of selected k-mers' positions over the number of k-mers in the sequence. Formally, the density of scheme M over sequence S is defined

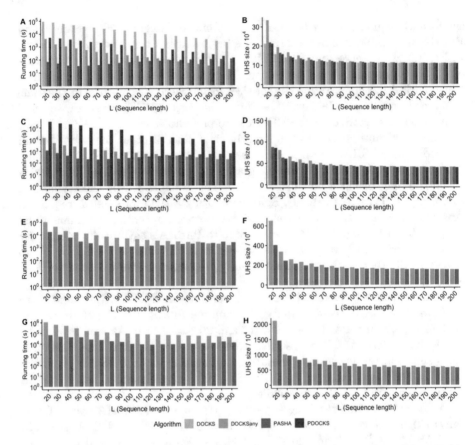

Fig. 1. Runtimes (left) and UHS sizes (divided by 10^4, right) for values of $k = 10$ (A, B), 11 (C, D), 12 (E, F), and 13 (G, H) and $20 \leq L \leq 200$ for the different methods. Note that the y-axes for runtimes are in logarithmic scale.

as

$$d(M, S, k) = \frac{|M(S, k)|}{|S| - k + 1} \tag{3}$$

where $M(S, k)$ is the set of positions of the k-mers selected over sequence S.

We calculate densities for a UHS by selecting the lexicographically smallest k-mer that is in the UHS within each window of $L - k + 1$ consecutive k-mers, since at least one k-mer is guaranteed to be in each such window. Marçais et al. [11] showed that using UHSs for k-mer selection in this manner yields smaller densities than lexicographic or random minimizer selection schemes. Therefore, we do not report comparisons between UHSs and minimizer schemes, but rather comparisons among UHSs constructed by different methods.

Marçais et al. [11] also showed that the expected density of a minimizers scheme for any k and window size $L - k + 1$ is equal to the density of the minimizers scheme on a de Bruijn sequence of order L. This allows for exact

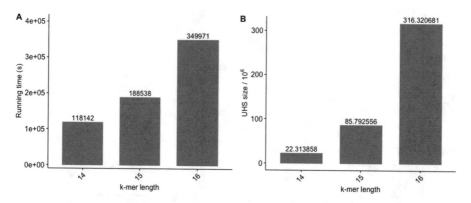

Fig. 2. Runtimes (A) and UHS sizes (divided by 10^6) (B) for $14 \leq k \leq 16$ and $L = 100$ for PASHA. Note that the y-axis for runtime is in logarithmic scale.

calculation of expected density for any k-mer selection procedure. However, for $14 \leq k \leq 16$ we calculated UHSs only for $L = 100$, and iterating over a de Bruijn sequence of order 100 is infeasible. Therefore, we computed the approximate expected density on long random sequences, since the computed expected density on these sequences converges to the expected density [11]. In addition, we computed the density of different methods on the entire human reference genome (GRCh38).

We computed the density values of UHSs generated by PDOCKS, DOCK-Sany, and PASHA over 10 random sequences of length 10^6, and the entire human reference genome (GRCh38), for $5 \leq k \leq 16$ and $L = 100$, when a UHS was available for such (k, L) combination.

As seen in Fig. 3, the differences in both approximate expected density and density computed on the human reference genome are negligible when comparing UHSs generated by the different methods. For most values of k, DOCKS yields the smallest approximate expected density and human genome density values, while DOCKSany generally yields lower human genome density values, but higher expected density values than PASHA. For $k \leq 6$, the UHS is only the decycling set; therefore, density values for these values of k are identical for the different methods.

Since there is no significant difference in the density of the UHSs generated by the different methods, other criteria, such as runtime and set size, are relevant when evaluating the performance of the methods: As k increases, PASHA produces sets that are only slightly smaller or larger in density, but significantly smaller in size and significantly faster than extant methods.

5   Discussion

We presented an efficient randomized parallel algorithm for generating a small set of k-mers that hits every possible sequence of length L and produces a set that

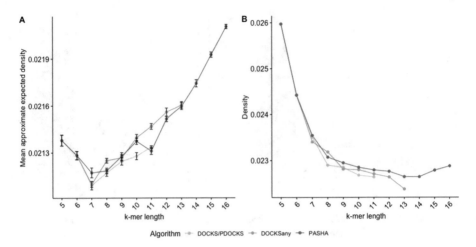

Fig. 3. Mean approximate expected density (A), and density on the human reference genome (B) for different methods, for $5 \leq k \leq 16$ and $L = 100$. Error bars represent one standard deviation from the mean across 10 random sequences of length 10^6. Density is the fraction of selected k-mer positions over the number of k-mers in the sequence.

is a small guaranteed factor away from the optimal set size. Since the runtime of DOCKS variants and PASHA depend exponentially on k, these greedy heuristics are eventually limited by runtime. However, using these heuristics in conjunction with parallelization, we are newly able to compute UHSs for values of k and L large enough for most biological applications.

The improvements in runtime for the hitting number calculation are due to parallelization of the dynamic programming phase, which is the bottleneck in sequential DOCKS variants. A minimum-size set that hits all infinite-length sequences is optimally and rapidly removed; however, the remaining sequences of length L are calculated and removed in time polynomial in the output size. We show that a constant factor reduction is beneficial in mitigating this bottleneck for practical use. In addition, we reduce the memory usage of this phase by theoretical and technical advancements. Last, we build on a randomized parallel algorithm for Set Cover to significantly speed up vertex selection. The randomized algorithm can be derandomized, while preserving the same approximation ratio, since it requires only pairwise independence of the random variables [2].

One main open problem still remains from this work. Although the randomized approximation algorithm enables us to generate a UHS more efficiently, the hitting numbers still need to be calculated at each iteration. The task of computing hitting numbers remains as the bottleneck in computing a UHS. Is there a more efficient way of calculating hitting numbers than the dynamic programming calculation done in DOCKS and PASHA? A more efficient calculation of hitting numbers will enable PASHA to run over $k > 16$ in a reasonable time.

As for long reads, which are becoming more popular for genome assembly tasks, a k-mer set that hits all infinite long sequences, as computed optimally by Mykkelveit's algorithm [13], is enough due to the length of these long read sequences. Still, due to the inaccuracies and high cost of long read sequencing compared to short read sequencing, the latter is still the prevailing method to produce sequencing data, and is expected to remain so for the near future.

We expect the efficient calculation of UHSs to lead to improvements in sequence analysis and construction of space-efficient data structures. Unfortunately, previous methods were limited to small values of k, thus allowing application to only a small subset of sequence analysis tasks. As there is an inherent exponential dependency on k in terms of both runtime and memory, efficiency in calculating these sets is crucial. We expect that the UHSs newly-enabled by PASHA for $k > 13$ will be useful in improving various applications in genomics.

6   Conclusion

We developed a novel randomized parallel algorithm PASHA to compute a small set of k-mers which together hit every sequence of length L. It is based on two algorithmic innovations: (i) improved calculation of hitting numbers through paralleization and memory reduction; and (ii) randomized parallel selection of additional k-mers to remove. We demonstrated the scalability of PASHA to larger values of k up to 16. Notably, the universal hitting sets need to be computed only once, and can then be used in many sequence analysis applications. We expect our algorithms to be an essential part of the sequence analysis toolkit.

Acknowledgments. This work was supported by NIH grant R01GM081871 to B.B. B.E. was supported by the MISTI MIT-Israel program at MIT and Ben-Gurion University of the Negev. We gratefully acknowledge the support of Intel Corporation for giving access to the Intel®AI DevCloud platform used for part of this work.

A   Emulating the Greedy Algorithm

The greedy Set Cover algorithm was developed independently by Johnson and Lovász for unweighted vertices [5,9]. Lovász [9] proved:

Theorem 1. *The greedy algorithm for Set Cover outputs cover R with $|R| \leq (1 + \log T_{max})|OPT|$, where T_{max} is the maximum cardinality of a set.*

We adapt a definition for an algorithm emulating the greedy algorithm for the Set Cover problem to the second phase of DOCKS [2]. We say that an algorithm for the second phase of DOCKS α-**emulates** the greedy algorithm if it outputs a set of vertices serially, during which it selects a vertex set A such that

$$\frac{|A|}{|P_A|} \leq \frac{\alpha}{T_{max}},$$

where P_A is the set of ℓ-long paths covered by A. Using this definition, we come up with a near-optimal approximation by the following theorem:

Theorem 2. *An algorithm for the second phase of DOCKS that α-emulates the greedy algorithm produces cover $R \subseteq V$ with $|R| \le \alpha(1 + \log T_{max})|OPT|$, where OPT is the optimal cover.*

Proof. We define the *cost* of covering path p as $\mathcal{C}(p) = \frac{|S|}{|P_S|}$, where S is the set of vertices selected in the selection step in which p was covered, and P_S the set of ℓ-long paths covered by S. Then, $\sum_{p \in P_S} \mathcal{C}(p) = |S|$.

Let P_ℓ be the set of all ℓ-long paths in G. A **fractional cover** of graph $G = (V, E)$ is function $\mathcal{F} : V \to \{0, 1\}$ s.t. for all $p \in P_\ell$, $\sum_{v \in p} \mathcal{F}(v) \ge 1$. The optimal cover $\mathcal{F}_{OPT}$ has minimum $\sum_{v \in V} \mathcal{F}_{OPT}(v)$.

Let $\mathcal{F}$ be such an optimal fractional cover. The size of the cover produced is

$$|R| = \sum_{p \in P_\ell} \mathcal{C}(p) \le \sum_{v \in V} \left(\mathcal{F}(v) \sum_{p \in P_v} \mathcal{C}(p) \right)$$

where P_v is the set of all ℓ-long paths through vertex v.

Lemma 1. *There are at most $\frac{\alpha}{k}$ paths $p \in P_v$ such that $\mathcal{C}(p) \ge k$ for any v, k.*

Proof. Assume the contrary: Before such a path p is covered, $T(v, \ell) > \frac{\alpha}{k}$. Thus,

$$\frac{|S|}{|P_S|} \ge k > \alpha/T(v, \ell) \ge \alpha/T_{max},$$

contradicting the definition.

Suppose we rank the $T(v, \ell)$ paths $p \in P_v$ by decreasing order of $\mathcal{C}(p)$. From the above remark, if the ith path has cost k, then $i \le \alpha/k$. Then, we can write

$$\sum_{p \in P_v} \mathcal{C}(p) \le \sum_{i=1}^{T(v,\ell)} \alpha/i \le \alpha \sum_{i=1}^{T(v,\ell)} 1/i \le \alpha(1 + \log T(v, \ell)) \le \alpha(1 + \log T_{max})$$

Then,

$$\sum_{p \in P_\ell} \mathcal{C}(p) \le \sum_{v \in V} \mathcal{F}(v)\alpha(1 + \log T_{max})$$

and finally

$$|R| \le \alpha(1 + \log T_{max})|OPT|.$$

In PASHA, we ensure that in step t, the sum of vertex hitting numbers of selected vertex set V_t is at least $|V_t|(1 + \varepsilon)^t(1 - 4\delta - 2\varepsilon)$. We now show that this is satisfied with high probability in each step.

Theorem 3. *With probability at least $1/2$, the sum of vertex hitting numbers of selected vertex set V_t at step t is at least $|V_t|(1 + \varepsilon)^t(1 - 4\delta - 2\varepsilon)$.*

Proof. For any vertex v in selected vertex set V_t at step t, let X_v be an indicator variable for the random event that vertex v is picked, and $f(X) = \sum_{v \in V_t} X_v$.

Note that $\mathrm{Var}[f(X)] \leq |V_t| \cdot \delta/\ell$, and $|V_t| \geq \ell/\delta^3$, since we are given that no vertex covers a δ^3 fraction of the ℓ-long paths covered by the vertices in V_t. By Chebyshev's inequality, for any $k \geq 0$,

$$\Pr[|f(X) - \mathrm{E}[f(X)]| \geq k(|V_t| \cdot \delta/\ell)] \leq \frac{1}{k^2}$$

and with probability $3/4$,

$$(f(X) - \mathrm{E}[f(X)])^2 \leq 4|V_t|^2 \cdot \frac{\delta^4}{\ell^2}$$

and

$$|f(X) - \mathrm{E}[f(X)]| \leq 2|V_t| \cdot \frac{\delta^2}{\ell}.$$

Let P_{V_t} denote the set of ℓ-long paths covered by vertex set V_t. Then,

$$|P_{V_t}| \geq \sum_{u \in V_t} T(u, \ell)X_u - \sum_{p \in P_{V_t}} \sum_{u,v \in p} X_u X_v$$

We know that $\sum_{u \in V_t} T(u, \ell)X_u \geq |V_t|(1+\varepsilon)^{t-1}$, which is bounded below by $((\delta - 2\delta^2) \cdot |V_t|(1+\varepsilon)^{t-1})/\ell$. Let $g(X) = \sum_{p \in P_{V_t}} \sum_{u,v \in p} X_u X_v$. Then,

$$\mathrm{E}[g(X)] = \sum_{p \in P_{V_t}} \mathrm{E}[\sum_{u,v \in p} X_u X_v] = \sum_{p \in P_{V_t}} \binom{\ell}{2}(\delta/\ell)^2 = \sum_{p \in P_{V_t}} \frac{(\ell-1) \cdot \delta^2}{2\ell} \leq \sum_{p \in P_{V_t}} \frac{\delta^2}{2}.$$

Hence, with probability at least $3/4$,

$$g(X) \leq 4\mathrm{E}[g(X)] \leq 2\delta^2 \cdot |V_t|(1+\varepsilon)^t$$

Both events hold with probability at least $1/2$, and the sum of vertex hitting numbers is at least

$$((\delta - 2\delta^2) \cdot |V_t|(1+\varepsilon)^{t-1}) \cdot \ell - 2\delta^2 \cdot |V_t|(1+\varepsilon)^t \geq |V_t|(1+\varepsilon)^{t-1}(\delta\ell - 2\delta^2\ell - 2\delta^2 - 2\delta^2\varepsilon)$$
$$\geq |V_t|(1+\varepsilon)^t(\delta\ell - 2\delta^2\ell - 2\delta^2 - 2\delta^2\varepsilon)/(1+\varepsilon)$$
$$\geq |V_t|(1+\varepsilon)^t(1 - 4\delta - 2\varepsilon).$$

B   Runtime Analysis

Here, we show the number of the selection steps and the average-time asymptotic complexity of PASHA.

Lemma 2. *The number of selection steps is $O(\log|V| \log|P_\ell|/(\varepsilon\delta^3 m))$.*

Proof. The number of steps is $O(\log|V|/\varepsilon)$, and within each step, there are $O(\log|P_S|/(\delta^3 m))$ selection steps (where P_S is the sum of vertex hitting numbers of the vertex set S for that step and m the number of threads used), since we are guaranteed to remove at least δ^3 fraction of the paths during that step. Overall, there are $O(\log|V|\log|P_\ell|/(\varepsilon\delta^3 m))$ selection steps.

Theorem 4. *For $\varphi < 1$, there is an approximation algorithm for the second phase of DOCKS that runs in $O((L^2 \cdot |\Sigma|^{k+1} \cdot \log^2(|\Sigma|^k))/(\varepsilon\delta^3 m))$ average time, where m is the number of threads used, and produces a cover of size at most $(1+\varphi)(1+\log T_{max})$ times the optimal size, where $1+\varphi = 1/(1-4\delta-2\epsilon)$.*

Proof. Follows immediately from Theorem 2 and Lemma 2.

References

1. Berger, B., Peng, J., Singh, M.: Computational solutions for omics data. Nat. Rev. Genet. **14**(5), 333 (2013)
2. Berger, B., Rompel, J., Shor, P.W.: Efficient NC algorithms for set cover with applications to learning and geometry. J. Comput. Syst. Sci. **49**(3), 454–477 (1994)
3. DeBlasio, D., Gbosibo, F., Kingsford, C., Marçais, G.: Practical universal k-mer sets for minimizer schemes. In: Proceedings of the 10th ACM International Conference on Bioinformatics, Computational Biology and Health Informatics, pp. 167–176. ACM (2019)
4. Deorowicz, S., Kokot, M., Grabowski, S., Debudaj-Grabysz, A.: KMC 2: fast and resource-frugal k-mer counting. Bioinformatics **31**(10), 1569–1576 (2015)
5. Johnson, D.S.: Approximation algorithms for combinatorial problems. J. Comput. Syst. Sci. **9**(3), 256–278 (1974)
6. Kawulok, J., Deorowicz, S.: CoMeta: classification of metagenomes using k-mers. PLoS ONE **10**(4), e0121453 (2015)
7. Kucherov, G.: Evolution of biosequence search algorithms: a brief survey. Bioinformatics **35**(19), 3547–3552 (2019)
8. Leinonen, R., Sugawara, H., Shumway, M., Collaboration, I.N.S.D.: The sequence read archive. Nucleic Acids Res. **39**, D19–D21 (2010)
9. Lovász, L.: On the ratio of optimal integral and fractional covers. Discret. Math. **13**(4), 383–390 (1975)
10. Marçais, G., DeBlasio, D., Kingsford, C.: Asymptotically optimal minimizers schemes. Bioinformatics **34**(13), i13–i22 (2018)
11. Marçais, G., Pellow, D., Bork, D., Orenstein, Y., Shamir, R., Kingsford, C.: Improving the performance of minimizers and winnowing schemes. Bioinformatics **33**(14), i110–i117 (2017)
12. Marçais, G., Solomon, B., Patro, R., Kingsford, C.: Sketching and sublinear data structures in genomics. Ann. Rev. Biomed. Data Sci. **2**, 93–118 (2019)
13. Mykkeltveit, J.: A proof of Golomb's conjecture for the de Bruijn graph. J. Comb. Theory **13**(1), 40–45 (1972)
14. Orenstein, Y., Pellow, D., Marçais, G., Shamir, R., Kingsford, C.: Compact universal k-mer hitting sets. In: Frith, M., Storm Pedersen, C.N. (eds.) WABI 2016. LNCS, vol. 9838, pp. 257–268. Springer, Cham (2016). https://doi.org/10.1007/978-3-319-43681-4_21

15. Orenstein, Y., Pellow, D., Marçais, G., Shamir, R., Kingsford, C.: Designing small universal k-mer hitting sets for improved analysis of high-throughput sequencing. PLoS Comput. Biol. **13**(10), e1005777 (2017)
16. Paindavoine, M., Vialla, B.: Minimizing the number of bootstrappings in fully homomorphic encryption. In: Dunkelman, O., Keliher, L. (eds.) SAC 2015. LNCS, vol. 9566, pp. 25–43. Springer, Cham (2016). https://doi.org/10.1007/978-3-319-31301-6_2
17. Qin, J., et al.: A human gut microbial gene catalogue established by metagenomic sequencing. Nature **464**(7285), 59 (2010)
18. Roberts, M., Hayes, W., Hunt, B.R., Mount, S.M., Yorke, J.A.: Reducing storage requirements for biological sequence comparison. Bioinformatics **20**(18), 3363–3369 (2004)
19. Turnbaugh, P.J., Ley, R.E., Hamady, M., Fraser-Liggett, C.M., Knight, R., Gordon, J.I.: The human microbiome project. Nature **449**(7164), 804 (2007)
20. Ye, C., Ma, Z.S., Cannon, C.H., Pop, M., Douglas, W.Y.: Exploiting sparseness in de novo genome assembly. BMC Bioinform. **13**(6), S1 (2012)

Multiple Competition-Based FDR Control and Its Application to Peptide Detection

Kristen Emery[1], Syamand Hasam[1], William Stafford Noble[2] ⬤,
and Uri Keich[1]([✉]) ⬤

[1] School of Mathematics and Statistics F07, University of Sydney, Sydney, Australia
{keme6477,uri.keich}@sydney.edu.au, shas8442@uni.sydney.edu.au
[2] Department of Genome Sciences and of Computer Science and Engineering,
University of Washington, Seattle, USA
william-noble@uw.edu

Abstract. Competition-based FDR control has been commonly used for over a decade in the computational mass spectrometry community [5]. Recently, the approach has gained significant popularity in other fields after Barber and Candés laid its theoretical foundation in a more general setting that included the feature selection problem [1]. In both cases, the competition is based on a head-to-head comparison between an observed score and a corresponding decoy/knockoff. We recently demonstrated some advantages of using multiple rather than a single decoy when addressing the problem of assigning peptide sequences to observed mass spectra [17]. In this work, we consider a related problem—detecting peptides based on a collection of mass spectra—and we develop a new framework for competition-based FDR control using multiple null scores. Within this framework, we offer several methods, all of which are based on a novel procedure that rigorously controls the FDR in the finite sample setting. Using real data to study the peptide detection problem we show that, relative to existing single-decoy methods, our approach can increase the number of discovered peptides by up to 50% at small FDR thresholds.

Keywords: Multiple hypothesis testing · Peptide detection · Tandem mass spectrometry · False discovery rate

1 Introduction

Proteins are the primary functional molecules in living cells, and tandem mass spectrometry (MS/MS) currently provides the most efficient means of studying proteins in a high-throughput fashion. Knowledge of the protein complement in a cellular population provides insight into the functional state of the cells. Thus, MS/MS can be used to functionally characterize cell types, differentiation

---

Supplementary Material: https://bitbucket.org/noblelab/multi-competition-fdr/.

© Springer Nature Switzerland AG 2020
R. Schwartz (Ed.): RECOMB 2020, LNBI 12074, pp. 54–71, 2020.
https://doi.org/10.1007/978-3-030-45257-5_4

stages, disease states, or species-specific differences. For this reason, MS/MS is the driving technology for much of the rapidly growing field of proteomics.

Paradoxically, MS/MS does not measure proteins directly. Because proteins themselves are difficult to separate and manipulate biochemically, an MS/MS experiment involves first digesting proteins into smaller pieces, called "peptides." The peptides are then measured directly. A typical MS/MS experiment generates ~10 observations ("spectra") per second, so a single 30-min MS/MS experiment will generate approximately 18,000 spectra. Canonically, each observed spectrum is generated by a single peptide. Thus, the first goal of the downstream analysis is to infer which peptide was responsible for generating each observed spectrum. The resulting set of detected peptides can then be used, in a second analysis stage, to infer what proteins are present in the sample.

In this work, we are interested in the first problem—peptide detection. This problem is important not only as a stepping stone toward the downstream goal of detecting proteins; in many proteomics studies, the peptides themselves are of primary interest. For example, MS/MS is being increasingly applied to complex samples, ranging from the microbiome in the human gut [21] to microbial communities in environmental samples such as soil or ocean water [30]. In these settings, the genome sequences of the species in the community are only partially characterized, so protein inference is problematic. Nonetheless, observation of a particular peptide can often be used to infer the presence of a group of closely related species (a taxonomic clade) or closely related proteins (a homology group). Peptide detection is also of primary interest in studies that aim to detect so-called "proteoforms"—variants of the same protein that arise due to differential splicing of the mature RNA or due to post-translational modifications of the translated protein. Identifying proteoforms can be critically important, for example, in the study of diseases like Alzheimer's or Parkinson's disease, in which the disease is hypothesized to arise in part due to the presence of deviant proteoforms [23,28,37].

Specifically, we focus on the task of assigning confidence estimates to peptides that have been identified by MS/MS. As is common in many molecular biology contexts, these confidence estimates are typically reported in terms of the false discovery rate (FDR), i.e., the expected value of the proportion of false discoveries among a set of detected peptides. For reasons that will be explained below, rather than relying on standard methods for control of the FDR such as the Benjamini-Hochberg (BH) procedure [2] the proteomics field employs a strategy known as "target-decoy competition" (TDC) to control the FDR in the reported list of detected peptides [5]. TDC works by comparing the list of peptides detected with a list of artificial peptides, called "decoys," detected using the same spectra set. The decoys are created by reversing or randomly shuffling the letters of the real ("target") peptides. The TDC protocol, which is described in detail in Sect. 2.2, estimates the FDR by counting the number of detected decoy peptides and using this count as an estimate for the number of incorrectly detected target peptides.

One clear deficiency of TDC is its reliance on a single set of decoy peptides to estimate the FDR. Thus, with ever increasing computational resources one can ask whether we can gain something by exploiting multiple randomly drawn decoys for each target peptide. We recently described such a procedure, called "average target-decoy competition" (aTDC), that, in the context of the related spectrum identification problem (described in Sect. 2.1), reduces the variability associated with TDC and can provide a modest boost in power [17,18].

In this paper we propose a new approach to using multiple decoy scores. The proposed procedure relies on a direct competition between the target and its corresponding decoy scores, rather than on averaging single competitions. We formulate our approach in the following more general setting. Suppose that we can compute a test statistic Z_i for each null hypothesis H_i, so that the larger Z_i is, the less likely is the null. However, departing from the standard multiple hypotheses setup, we further assume that we cannot compute p-values for the observed scores. Instead, we can only generate a *small* sample of independent decoys or competing null scores for each hypothesis H_i: $\tilde{Z}_i^j$ $j = 1, \ldots, d$ (Definition 2). Note that the case $d = 1$ corresponds to the TDC setup described above. We will show using both simulated and real data that the novel method we propose yields more power (more discoveries) than our aforementioned averaging procedure.

In addition to the peptide detection problem, our proposed procedure is applicable in several other bioinformatics applications. For example, the procedure could be used when analyzing a large number of motifs reported by a motif finder, e.g., [12], where creating competing null scores can require the time consuming task of running the finder on randomized versions of the input sets, e.g., [25]. In addition, our procedure is applicable to controlling the FDR in selecting differentially expressed genes in microarray experiments where a small number of permutations is used to generate competing null scores [36].

Our proposed method can also be viewed as a generalization of Barber and Candés' "knockoff" procedure [1], which is a competition-based FDR control method that was initially developed for feature selection in a classical linear regression model. The procedure has gained a lot of interest in the statistical and machine learning communities, where it has been applied to various applications in biomedical research [10,29,38] and has been extended to work in conjunction with deep neural networks [22] and with time series data [9]. Despite the different terminology, both knockoffs and decoys serve the same purpose in competition-based FDR control; hence, for the ideas presented in this paper the two are interchangeable. A significant part of Barber and Candés' work is the sophisticated construction of their knockoff scores; controlling the FDR then follows exactly the same competition that TDC uses. Indeed, their Selective SeqStep+ (SSS+) procedure rigorously formalizes in a much more general setting the same procedure described above in the context of TDC. Note that Barber and Candés suggested that using multiple knockoffs could improve the power of their procedure so the methods we propose here could provide a stepping stone toward that. However, we would still need to figure out how to generalize their construction from one to multiple knockoffs.

2   Background

2.1   Shotgun Proteomics and Spectrum Identification

In a "shotgun proteomics" MS/MS experiment, proteins in a complex biological sample are extracted and digested into peptides, each with an associated charge. These charged peptides, called "precursors," are measured by the mass spectrometer, and a subset is then selected for further fragmentation into charged ions, which are detected and recorded by a second round of mass spectrometry [26,27]. The recorded tandem fragmentation spectra, or spectra for short, are then subjected to computational analysis.

This analysis typically begins with the spectrum identification problem, which involves inferring which peptide was responsible for generating each observed fragmentation spectrum. The most common solution to this problem is peptide database search. Pioneered by SEQUEST [8], the search engine extracts from the peptide database all "candidate peptides," defined by having their mass lie within a pre-specified tolerance of the measured precursor mass. The quality of the match between each one of these candidate peptides and the observed fragmentation spectrum is then evaluated using a score function. Finally, the optimal peptide-spectrum match (PSM) for the given spectrum is reported, along with its score [24].

In practice, many expected fragment ions will fail to be observed for any given spectrum, and the spectrum is also likely to contain a variety of additional, unexplained peaks [26]. Hence, sometimes the reported PSM is correct—the peptide assigned to the spectrum was present in the mass spectrometer when the spectrum was generated—and sometimes the PSM is incorrect. Therefore, we report a thresholded list of top-scoring PSMs, together with the critical estimate of the fraction of incorrect PSMs in our reported list.

2.2   False Discovery Rate Control in Spectrum Identification

The general problem of controlling the proportion of false discoveries has been studied extensively in the context of multiple hypotheses testing (MHT). We briefly review this setup in Supplementary Sect. 6.1; however, these techniques cannot be applied directly to the spectrum identification problem. A major reason for that is the presence in any shotgun proteomics dataset of both "native spectra" (those for which their generating peptide is in the target database) and "foreign spectra" (those for which it is not). These create different types of false positives, implying that we typically cannot apply FDR controlling procedures that were designed for the general MHT context to the spectrum identification problem [16].

Instead, the mass spectrometry community uses TDC to control the FDR in the reported list of PSMs [3,5,6,15]. TDC works by comparing searches against a target peptide database with searches against a decoy database of peptides. More precisely, let Z_i be the score of the optimal match (PSM) to the ith spectrum in the target database, and let $\tilde{Z}_i$ be the corresponding optimal match in the decoy database. Each decoy score $\tilde{Z}_i$ directly competes with its corresponding target score Z_i for determining the reported list of discoveries. Specifically, for each score threshold T we only report target PSMs that won their competition: $Z_i > \max\{T, \tilde{Z}_i\}$. Subsequently, the number of decoy wins ($\tilde{Z}_i > \max\{T, Z_i\}$) is used to estimate the number of false discoveries in the list of target wins. Thus, the ratio between that estimate and the number of target wins yields an estimate of the FDR among the target wins. To control the FDR at level α we choose the smallest threshold $T = T(\alpha)$ for which the estimated FDR is still $\leq \alpha$. It was recently shown that, assuming that incorrect PSMs are independently equally likely to come from a target or a decoy match and provided we add 1 to the number of decoy wins before dividing by the number of target wins, this procedure rigorously controls the FDR [13, 20].

2.3   The Peptide Detection Problem

The spectrum identification is largely used as the first step in addressing the peptide identification problem that motivates the research presented here. Indeed, to identify the peptides we begin, just like we do in spectrum identification, by assigning each spectrum to the unique target/decoy peptide which offers the best match to this spectrum in the corresponding database. We then assign to each target peptide a score Z_j which is the maximum of all PSM scores of spectra that were assigned to this peptide in the first phase. Similarly, we assign to the corresponding decoy peptide a score $\tilde{Z}_j$, which again is the maximum of all PSM scores involving spectra that were assigned to that decoy peptide. The rest continues using the same TDC protocol we outlined above for the spectrum identification problem [11, 31].

3   Controlling the FDR Using Multiple Decoys

3.1   Why Do We Need a New Approach?

A key feature of our problem is that due to computational costs the number of decoys, d, is small. Indeed, if we are able to generate a large number of independent decoys for each hypothesis, then we can simply apply the standard FDR controlling procedures (Supplementary Sect. 6.1) to the empirical p-values. These p-values are estimated from the empirical null distributions, which are constructed for each hypothesis H_i using its corresponding decoys. Specifically, these empirical p-values take values of the form $(d_1 - r_i + 1)/d_1$, where $d_1 = d + 1$, and $r_i \in \{1, \ldots, d_1\}$ is the rank of the originally observed score ("original score" for short) Z_i in the combined list of d_1 scores: $\left(\tilde{Z}_i^0 = Z_i, \tilde{Z}_i^1, \ldots, \tilde{Z}_i^d\right)$ ($r_i = 1$ is the lowest rank). Using these p-values the BH procedure [2] rigorously controls the FDR, and Storey's method [32] will asymptotically control the FDR as the number of hypotheses $m \to \infty$.

Unfortunately, because d is small, applying those standard FDR control procedures to the rather coarse empirical p-values may yield very low power. For example, if $d = 1$, each empirical p-value is either $1/2$ or 1, and therefore for many practical examples both methods will not be able to make any discoveries at usable FDR thresholds.

Alternatively, one might consider pooling all the decoys regardless of which hypothesis generated them. The pooled empirical p-values attain values of the form $i/(m \cdot d + 1)$ for $i = 1, \ldots, md + 1$; hence, particularly when m is large, the p-values generally no longer suffer from being too coarse. However, other significant problems arise when pooling the decoys. These issues — discussed in Supplementary Sect. 6.2 — imply that in general, applying BH or Storey's procedure to p-values that are estimated by pooling the competing null scores can be problematic both in terms of power and control of the FDR.

3.2 A Novel Meta-procedure for FDR Control Using Multiple Decoys

The main technical contribution of this paper is the introduction of several procedures that effectively control the FDR in our multiple competition-based setup and that rely on the following meta-procedure.

Input: an original/target score Z_i and d competing null scores $\tilde{Z}_i^j$ for each null hypothesis H_i.

Parameters: an FDR threshold $\alpha \in (0, 1)$, two tuning parameters $c = i_c/d_1$ ($d_1 = d+1$), the "original/target win" threshold, and $\lambda = i_\lambda/d_1$, the "decoy win" threshold where $i_\lambda, i_c \in \{1, \ldots, d\}$ and $c \leq \lambda$, as well as a (possibly randomized) mapping function $\varphi : \{1, \ldots, d_1 - i_\lambda\} \mapsto \{d_1 - i_c + 1, \ldots, d_1\}$.

Procedure:

1. Each hypothesis H_i is assigned an original/decoy win label:

$$
L_i = \begin{cases} 1 & r_i \geq d_1 - i_c + 1 \quad \text{(original win)} \\ 0 & r_i \in (d_1 - i_\lambda, d_1 - i_c + 1) \quad \text{(ignored hypothesis)}, \\ -1 & r_i \leq d_1 - i_\lambda \quad \text{(decoy win)} \end{cases} \quad (1)
$$

where $r_i \in \{1, \ldots, d_1\}$ is the rank of the original score when added to the list of its d decoy scores.

2. Each hypothesis H_i is assigned a score $W_i = \tilde{Z}_i^{(s_i)}$, where $\tilde{Z}_i^{(j)}$ is the jth order statistic or the jth largest score among $\left(\tilde{Z}_i^0 = Z_i, \tilde{Z}_i^1, \ldots, \tilde{Z}_i^d\right)$, and the "selected rank", s_i, is defined as

$$
s_i = \begin{cases} r_i & L_i = 1 \text{ (so } W_i = Z_i \text{ in an original win)} \\ u_i & L_i = 0 \text{ (where } u_i \text{ is randomly chosen uniformly in } \{d_1 - i_c + 1, \ldots, d_1\}) \\ \varphi(r_i) & L_i = -1 \text{ (so } W_i \text{ coincides with a decoy score in a decoy win)} \end{cases}
$$

$$(2)$$

3. The hypotheses are reordered so that W_i are decreasing, and the list of discoveries is defined as the subset of original wins $D(\alpha, c, \lambda) := \{i : i \le i_{\alpha c \lambda}, L_i = 1\}$, where

$$i_{\alpha c \lambda} := \max \left\{ i : \frac{1 + \#\{j \le i : L_j = -1\}}{\#\{j \le i : L_j = 1\} \vee 1} \cdot \frac{c}{1 - \lambda} \le \alpha \right\}. \tag{3}$$

We assume above that all ties in determining the ranks r_i, as well as the order of W_i, are broken randomly, although other ways to handle ties are possible (e.g., Sect. 8.3 in our technical report [7]).

Note that the hypotheses for which $L_i = 0$ can effectively be ignored as they cannot be considered discoveries nor do they factor in the numerator of (3).

Our procedures vary in how they define the (generally randomized) mapping function φ (and hence s_i in (2)), as well as in how they set the tuning parameters c, λ. For example, in the case $d = 1$ setting $c = \lambda = 1/2$ and $\varphi(1) := 2$ our meta-procedure coincides with TDC. For $d > 1$ we have increasing flexibility with d, but one obvious generalization of TDC is to set $c = \lambda = 1/d_1$. In this case $L_i = 1$ if the original score is larger than all its competing decoys and otherwise $L_i = -1$. Thus, by definition, φ is constrained to the constant value d_1 so $s_i \equiv d_1$ and W_i is always set to $Z_i^{(d_1)} = \max \left\{ \tilde{Z}_i^0, \ldots, \tilde{Z}_i^d \right\}$. Hence we refer to this as the "max method." As we will see, the max method controls the FDR, but this does not hold for any choice of c, λ and φ. The following section specifies a sufficient condition on c, λ and φ that guarantees FDR control.

3.3   Null Labels Conditional Probabilities Property

Definition 1. *Let N be the indices of all true null hypotheses. We say the null labels conditional probabilities property (NLCP) is satisfied if conditional on all the scores $W = (W_1, \ldots W_m)$ the random labels $\{L_i : i \in N\}$ are (i) independent and identically distributed (iid) with $P(L_i = 1 \mid W) = c$ and $P(L_i = -1 \mid W) = 1 - \lambda$, and (ii) independent of the false null labels $\{L_i : i \notin N\}$.*

Note that in claiming that TDC controls the FDR we implicitly assume that a false match is equally likely to arise from a target win as it is from a decoy win independently of all other scores [13]. This property coincides with the NLCP with $d = 1$ and $c = \lambda = 1/2$. Our next theorem shows that the NLCP generally guarantees the FDR control of our meta-procedure. Specifically, we argue that with NLCP established step 3 of our meta-procedure can be viewed as a special case of Barber and Candés' SSS+ procedure [1] and its extension by Lei and Fithian's Adaptive SeqStep (AS) [19]. Both procedures are designed for sequential hypothesis testing where the order of the hypotheses is pre-determined – by the scores W_i in our case.

Theorem 1. *If the NLCP holds then our meta-procedure controls the FDR in a finite-sample setting, that is, $E(|D(\alpha, c, \lambda) \cap N|/|D(\alpha, c, \lambda)|) \le \alpha$, where the expectation is taken with respect to all the decoy draws.*

Why does Theorem 1 make sense? If the NLCP holds then a true null H_i is an original win ($L_i = 1$) with probability c and is a decoy win with probability $1 - \lambda$. Hence, the factor $\frac{c}{1-\lambda}$ that appears in (3) adjusts the observed number of decoy wins, $\#\{j \leq i : L_j = -1\}$, to estimate the number of (unobserved) false original wins (those for which the corresponding H_i is a true null). Ignoring the $+1$ correction, the adjusted ratio of (3) therefore estimates the FDR in the list of the first i original wins. The procedure simply takes the largest such list for which the estimated FDR is $\leq \alpha$.

Proof. To see the connection with SSS+ and AS we assign each hypothesis H_i a p-value $p_i := P(L_i \geq l)$. Clearly, if the NLCP holds then

$$p_i = \begin{cases} c & l = 1 \\ \lambda & l = 0 \\ 1 & l = -1 \end{cases}. \tag{4}$$

Moreover, the NLCP further implies that for any $u \in (0, 1)$ and $i \in N$, $P(p_i \leq u \mid \mathcal{W}) \leq u$, and that the true null labels L_i, and hence the true null p-values, p_i, are independent conditionally on $\mathcal{W}$. It follows that, even after sorting the hypotheses by the decreasing order of the scores W_i, the p-values of the true null hypotheses are still iid valid p-values that are independent from the false nulls. Hence our result follows from Theorem 3 (SSS+) of [1] for $c = \lambda$, and more generally for $c \leq \lambda$ from Theorem 1 (AS) of [19] (with $s = c$). $\square$

Remark 1. With the risk of stating the obvious we note that one cannot simply apply SSS+ or AS by selecting $W_i = Z_i$ for all i with the corresponding empirical p-values $(d_1 - r_i + 1)/d_1$. Indeed, in this case the order of the hypotheses (by W_i) is not independent of the true null p-values.

3.4   When Does the NLCP Hold for Our Meta-procedure?

To further analyze the NLCP we make the following assumption on our decoys.

Definition 2 (formalizing the multiple-decoy problem). *If the d_1 (original and decoy) scores corresponding to each true null hypothesis are iid independently of all other scores then we say we have "iid decoys".*

It is clear that if we have iid decoys then for each fixed $i \in N$ the rank r_i is uniformly distributed on $1, \ldots, d_1$, and hence $P(L_i = 1) = c$ and $P(L_i = -1) = 1 - \lambda$. However, to determine whether or not r_i is still uniformly distributed when conditioning on $\mathcal{W}$ we need to look at the mapping function φ as well.

More specifically, in the iid decoys case the conditional distribution of $\{L_i : i \in N\}$ given $\mathcal{W}$ clearly factors into the product of the conditional distribution of each true null L_i given W_i: a true null's L_i is independent of all $\{L_j, W_j : j \neq i\}$. Thus, it suffices to show that L_i is independent of W_i for each $i \in N$. Moreover, because W_i is determined in terms of s_i and the *set* of scores $\{\tilde{Z}_i^0, \ldots, \tilde{Z}_i^d\}$, and because a true null's label L_i and s_i are independent of the last set (a set is unordered), it suffices to show that L_i is independent of s_i. Of course, s_i is determined by φ as specified in (2).

For example, consider the max method where $s_i \equiv d_1$ (equivalently $\varphi \equiv d_1$): in this case, L_i is trivially independent of s_i and hence by the above discussion the method controls the FDR. In contrast, assuming d_1 is even and choosing $\varphi \equiv d_1$ with $c = \lambda = 1/2$ we see that the scores $\{W_i : i \in N, L_i = -1\}$ will generally be larger than the corresponding $\{W_i : i \in N, L_i = 1\}$. Indeed, when $L_i = -1$ we always choose the maximal score $W_i = Z_i^{(d_1)}$, whereas W_i is one of the top half scores when $L_i = 1$. Hence, $P(L_i = -1 \mid \text{higher } W_i) > 1/2$.

So how can we guarantee that the NLCP holds for pre-determined values of $c = i_c/d_1$ and $\lambda = i_\lambda/d_1$? The next theorem provides a sufficient condition on φ (equivalently on s_i) to ensure the property holds.

Theorem 2. *If the iid decoys assumption holds, and if for any $i \in N$ and $j \in \{d_1 - i_c + 1, \ldots, d_1\}$*

$$P\left(s_i = j, r_i \le d_1 - i_\lambda\right) = P\left(s_i = j, L_i = -1\right) = \frac{d_1 - i_\lambda}{d_1 \cdot i_c}, \tag{5}$$

then the NLCP holds and hence our meta-procedure with those values of c, λ and φ controls the FDR.

Proof. By (5), for any $i \in N$ and $j \in \{d_1 - i_c + 1, \ldots, d_1\}$,

$$P\left(L_i = 1 \mid s_i = j\right) = \frac{P\left(s_i = j, L_i = 1\right)}{\sum_{l \in \{-1,0,1\}} P\left(s_i = j, L_i = l\right)}$$

$$= \frac{1/d_1}{(d_1 - i_\lambda)/(d_1 \cdot i_c) + (i_\lambda - i_c)/d_1 \cdot 1/i_c + 1/d_1} = \frac{i_c}{d_1} = c,$$

$$P\left(L_i = -1 \mid s_i = j\right) = \frac{(d_1 - i_\lambda)/(d_1 \cdot i_c)}{(d_1 - i_\lambda)/(d_1 \cdot i_c) + (i_\lambda - i_c)/d_1 \cdot 1/i_c + 1/d_1} = \frac{d_1 - i_\lambda}{d_1} = 1 - \lambda.$$

At the same time $P\left(L_i = 1 \mid s_i = j\right) = 1$ for $j \in \{1, \ldots, i_c\}$ always holds; therefore, L_i is independent of s_i and by the above discussion the NLCP holds. Theorem 1 completes the proof.

For any fixed values of c, λ we can readily define a randomized $\varphi = \varphi_u$ so that the NLCP holds: randomly and uniformly map $\{1, \ldots, d_1 - i_\lambda\}$ onto $\{d_1 - i_c + 1, \ldots, d_1\}$. Indeed, in this case (5) holds:

$$P\left(s_i = j, s_i \ne r_i\right) = P\left(r_i \in \{1, \ldots, d_1 - i_\lambda\}\right) \cdot P\left(s_i = j \mid r_i \in \{1, \ldots, d_1 - i_\lambda\}\right) = \frac{d_1 - i_\lambda}{d_1} \cdot \frac{1}{i_c}. \tag{6}$$

3.5   Mirroring and Mirandom

Using the above randomized uniform map φ_u we have a way to define an FDR-controlling variant of our meta-procedure for any pre-determined c, λ. However, we can design more powerful procedures using alternative definitions of φ (for the same values of c, λ).

For example, with $c = \lambda = 1/2$ and an even d_1 we can consider, in addition to φ_u, the mirror map: $\varphi_m(j) := d_1 - j + 1$. It is easy to see that under the conditions of Theorem 2, $P(s_i = j, r_i \leq d_1 - i_\lambda) = 1/d_1$ hence (5) holds and the resulting method, which we refer to as the "mirror method" (because when $L_i = -1$, s_i is the rank symmetrically across the median to r_i), controls the FDR. Similarly, we can choose to use a shift map φ_s: $\varphi_s(j) = j + d_1/2$, which will result in a third FDR-controlling variant of our meta-procedure for $c = \lambda = 1/2$.

Comparing the shift and the mirror maps we note that when $L_i = -1$, φ_s replaces middling target scores with high decoy scores, whereas φ_m replaces low target scores with high decoy scores. Of course, the high decoy scores are the ones more likely to appear in the numerator of (3), and generally we expect the density of the target scores to monotonically decrease with the quality of the score. Taken together, it follows that the estimated FDR will generally be higher when using φ_s than when using φ_m, and hence the variant that uses φ_s will be weaker than the mirror. By extension the randomized φ_u will fall somewhere between the other two maps, as can be partly verified by the comparison of the power using φ_m and φ_u in panel A of Supplementary Fig. 1.

We can readily extend the mirroring principle to other values of c and λ where i_c divides $d_1 - i_\lambda$, however when $i_c \nmid d_1 - i_\lambda$ we need to introduce some randomization into the map. Basically, we accomplish this by respecting the mirror principle as much as we can while using the randomization to ensure that (5) holds—hence the name *mirandom* for this map/procedure. It is best described by an example.

Suppose $d = 7$. Then for $i_c = 3$ ($c = 3/8$) and $i_\lambda = 4$ ($\lambda = 1/2$) the mirandom map, φ_{md}, is defined as

$$
\varphi_{md}(j) = \begin{cases}
8 & j = 1 \\
8 \text{ (with probability } 1/3\text{), or } 7 \text{ (with probability } 2/3\text{)} & j = 2 \\
7 \text{ (with probability } 2/3\text{), or } 6 \text{ (with probability } 1/3\text{)} & j = 3 \\
6 & j = 4
\end{cases}
$$

Note the uniform coverage $(4/3)$ of each value in the range, implying that if j is randomly and uniformly chosen in the domain then $\varphi_{md}(j)$ is uniformly distributed over $\{6, 7, 8\}$.

More generally the mirandom map φ_{md} for a given $c \leq \lambda$ is defined in two steps. In the first step it defines a sequence of $d_1 - i_\lambda$ distributions $F_1, \ldots, F_{d_1 - i_\lambda}$ on the range $\{d_1 - i_c + 1, \ldots, d_1\}$ so that

- each F_l is defined on a contiguous sequence of natural numbers, and
- if $j < l$ then F_j stochastically dominates F_l and $\min \text{support}\{F_j\} \geq \max \text{support}\{F_l\}$.

In practice, it is straightforward to construct this sequence of distributions and to see that, when combined, they necessarily satisfy the following equal coverage property: for each $j \in \{d_1 - i_c + 1, \ldots, d_1\}$, $\sum_{l=1}^{d_1 - i_\lambda} F_l(j) = \frac{d_1 - i_\lambda}{i_c}$. In the second step, mirandom defines s_i for any i with $r_i \in \{1, \ldots, d_1 - \tilde{i}_c\}$ by randomly drawing a number from F_{r_i} (independently of everything else).

It follows from the equal coverage property that for any $i \in N$ and $j \in \{d_1 - i_c + 1, \ldots, d_1\}$ (5) holds for φ_{md} for essentially the same reason it held for φ_u in (6). Hence, the mirandom map allows us to controls the FDR for any pre-determined values of c, λ.

3.6   Data-Driven Setting of the Tuning Parameters c, λ

All the procedures we consider henceforth are based on the mirandom map. Where they differ is in how they set c, λ. For example, choosing $c = \lambda = 1/2$ gives us the mirror (assuming d_1 is even), $c = \lambda = 1/d_1$ yields the max, while choosing $\lambda = 1/2$ and $c = \alpha \leq 1/2$ coincides with Lei and Fithian's recommendation in the related context of sequential hypothesis testing (technically we set $c = \lfloor \alpha \cdot d_1 \rfloor / d_1$ and refer to this method as "LF"). All of these seem plausible; however, our extensive simulations (Supplementary Sect. 6.3) show that none dominates the others with substantial power to be gained/lost for any particular problem (Supplementary Fig. 1, panels B-D). As the optimal values of c, λ seem to vary in a non-trivial way with the nature of the data, as well as with d and α, we turned to data-driven approaches to setting c, λ.

Lei and Fithian pointed out the connection between the (c, λ) (they refer to c as s) parameters of their AS procedure and the corresponding parameters in Storey's procedure. Specifically, AS's λ is analogous to the parameter λ of [33] that determines the interval $(\lambda, 1]$ from which π_0, the fraction of true null hypotheses, is estimated, and AS's c is Storey's rejection threshold (Supplementary Sect. 6.4).

We take this analogy one step further and essentially use the procedure of [33] to determine c by applying it to the empirical p-values, $\tilde{p}_i := (d_1 - r_i + 1)/d_1$. However, to do that, we first need to determine λ.

We could have determined λ by applying the bootstrap approach of [33] to $\tilde{p}_i$. However, in practice we found that using the bootstrap option of the qvalue package [35] in our setup can significantly compromise our FDR control. Therefore, instead we devised an alternative approach inspired by the spline-based method of [34] for estimating π_0, where we look for the flattening of the tail of the p-value histogram as we approach 1. Because our p-values, $\tilde{p}_i$, lie on the lattice i/d_1 for $i = 1, \ldots, d_1$, instead of threading a spline as in [34], we repeatedly test whether the number of p-values in the first half of the considered tail interval $(\lambda, 1]$ is significantly larger than their number in the second half of this interval (Supplementary Sect. 6.5).

Our finite-decoy Storey (FDS) procedure starts with determining λ as above then essentially applies the methodology of [33] to $\tilde{p}_i$ to set $c = t_\alpha$ before applying mirandom with the chosen c, λ (Supplementary Sect. 6.6). We defined FDS as close as possible to Storey, Taylor and Siegmund's recommended procedure for guaranteed FDR control in the finite setting [33]. Indeed, as we argue in Supplementary Sect. 6.7, FDS converges to a variant of Storey's procedure once we let $d \longrightarrow \infty$ (the mirror and mirandom maps in general have an interesting limit in that setup). However, we found that a variant of FDS that we denote as FDS_1 (Supplementary Sect. 6.6), often yields better power in our setting, so we considered both variants.

FDS and FDS_1 peek at the data to set c, λ hence they no longer fall under mirandom's guaranteed FDR control. Still, our extensive simulations show they essentially control the FDR: their empirical violations of FDR control are roughly in line with that of the max method, which provably controls the FDR (Supplementary Fig. 2). Importantly, FDS_1 seems to deliver overall more power than the mirror, max, LF, FDS and TDC, and often substantially more (Supplementary Fig. 3). We note, however, that at times FDS_1 has 10–20% less power than the optimal method, and we observe similar issues with the examples mentioned in Supplementary Sect. 6.2 where BH has no power (Supplementary Sect. 6.10). These issues motivate our next procedure.

3.7 A Bootstrap Procedure for Selecting an Optimal Method

Our final, and ultimately our recommended multi-decoy procedure, uses a novel resampling approach to choose the optimal procedure among several of the above candidates. Our optimization strategy is indirect: rather than using the resamples to choose the method that maximizes the number of discoveries, we use the resamples to advise us whether or not such a direct maximization approach is likely to control the FDR.

Clearly, a direct maximization would have been ideal had we been able to sample more instances of the data. In reality, that is rarely possible all the more so with our underlying assumption that the decoys are given and that it is forbiddingly expensive to generate additional ones. Hence, when a hypothesis is resampled it comes with its original, as well as its decoy scores, thus further limiting the variability of our resamples. In particular, direct maximization will occasionally fail to control the FDR. Our Labeled Bootstrap monitored Maximization (LBM) procedure tries to identify those cases.

In order to gauge the rate of false discoveries we need labeled samples. To this end, we propose a segmented resampling procedure that makes informed guesses (described below) about which of the hypotheses are false nulls before resampling the indices. The scores $\left\{ \tilde{Z}_i^j \right\}_{j=0}^d$ associated with each resampled conjectured *true null* index are then randomly permuted, which effectively boils down to randomly sampling $j \in \{0, 1, \ldots, d\}$ and swapping the corresponding original score $\tilde{Z}_i^0 = Z_i$ with $\tilde{Z}_i^j$.

The effectiveness of our resampling scheme hinges on how informed are our guesses of the false nulls. To try and increase the overlap between our guesses and the true false nulls we introduced two modifications to the naive approach of estimating the number of false nulls in our sample and then uniformly drawing that many conjectured false nulls. First, we consider increasing sets of hypotheses $\mathcal{H}_j \subset \mathcal{H}_{j+1}$ and verify that the number of conjectured false nulls we draw from each $\mathcal{H}_j$ agrees with our estimate of the number of false nulls in $\mathcal{H}_j$. Second, rather than being uniform, our draws within each set $\mathcal{H}_j$ are weighted according to the empirical p-values so that hypotheses with more significant empirical p-values are more likely to be drawn as conjectured false nulls. Our segmented resampling procedure is described in detail in Supplementary Sect. 6.8.

In summary, LBM relies on the labeled resamples of our segmented resampling approach to estimate whether we are likely to control the FDR when using direct maximization (we chose FDS, mirror, and FDS_1 as the candidate methods). If so, then LBM uses the maximizing method; otherwise, LBM chooses a pre-determined fall-back method (here we consistently use FDS_1, see Supplementary Sect. 6.9 for details).

Our simulations suggest that LBM's control of the FDR is on-par with that of the, provably FDR-controlling, max: the overall maximal observed violation is 5.0% for LBM while it is 6.7% for max, and the number of curves (out of 1200) in which the maximal violation exceeds 2% is 21 for LBM, and 24 for the max (panels A and D, Supplementary Fig. 2). Power-wise LBM arguably offers the best balance among our proposed procedures, offering substantially more power in many of the experiments while never giving up too much power when it is not optimal (Supplementary Fig. 4). Finally, going back to the examples where BH and Storey's procedure applied to the empirical p-values fail we find that all our methods, including LBM, essentially control the FDR where Storey's procedure substantially failed to do so, and similarly that LBM delivers substantial power where BH had none (Supplementary Sect. 6.10).

4   The Peptide Detection Problem

Our peptide detection procedure starts with a generalization of the WOTE procedure of [11]. We use Tide [4] to find for each spectrum its best matching peptide in the target database as well as in the d decoy peptide databases. We then assign to the ith target peptide the observed score, Z_i, which is the maximum of all the PSM scores that were optimally matched to this peptide. We similarly define the maximal scores of each of that peptide's d randomly shuffled copies as the corresponding decoy scores: $\tilde{Z}_i^1, \ldots, \tilde{Z}_i^d$. If no spectrum was optimally matched to a peptide then that peptide's score is $-\infty$.

We then applied to the above scores TDC ($d = 1$, with the $+1$ finite sample correction)—representing a peptide-level analogue of the picked target-decoy strategy of [31]—as well as the mirror, LBM and the averaging-based $aTDC^1$ each using $d \in \{3, 5, 7, 9\}$. Note that to ameliorate the effect of decoy-induced variability on our comparative analysis we report the average of our analysis over 100 applications of each method using that many randomly drawn decoy sets (Supplementary Sect. 6.11).

We used three datasets in our analysis: "human", "yeast" and "ISB18" (Supplementary Sect. 6.11). Panel D of Supplementary Fig. 5 suggests that when

---

[1] We used the version named $aTDC_1^+$, which was empirically shown to control the FDR even for small thresholds/datasets [18].

applied to the ISB18 dataset all our procedures seem to control the FDR:[2] the empirically estimated FDR is always below the selected threshold. In terms of power, again we see that LBM is the overall winner: it typically delivers the largest number of discoveries, and even in the couple of cases where it fails to do so it is only marginally behind the top method (panels A–C). In contrast, each of the other methods has some cases where it delivers noticeably fewer discoveries.

More specifically, for $\alpha = 0.01$ LBM's average of 142.0 ISB18 discoveries ($d = 3$) represents an 8.0% increase over TDC's average of 131.5 ISB18 discoveries, and we see a 9.4% increase over TDC when using $d = 5$ (143.3 discoveries). In the human dataset and for the same $\alpha = 0.01$ we see a 2.8% increase in power going from TDC to LBM with $d = 3$ (532.4 vs. 547.1 discoveries), and a 4.2% increase when using LBM with $d = 5$ (555.0 discoveries). LBM offers the biggest gains in the yeast dataset where we see (again $\alpha = 0.01$) a 45.5% increase in power going from TDC to LBM with $d = 3$ (76.3 vs. 111.0 discoveries), and a 46.7% increase when using LBM with $d = 5$ (111.9 discoveries). Moreover, we note that for this $\alpha = 0.01$ TDC reported 0 yeast discoveries in 33 of the 100 runs (each using a different decoy database), whereas LBM reported a positive number of discoveries in all 100 runs for each $d > 1$ we considered.

At the higher FDR thresholds of 0.05 and 0.1 LBM offers a much smaller power advantage over TDC and is marginally behind for $\alpha = 0.1$ and $d = 3$ in the human and yeast datasets. Also, consistent with our simulations, we find that the mirror lags behind LBM, and in fact in these real datasets it is roughly on par with TDC.

Finally, although aTDC was designed for the spectrum identification problem and in practice was never applied to the peptide detection problem, it was instructive to add aTDC to this comparison. LBM consistently delivered more detected peptides than aTDC did, although in some cases the difference is marginal. Still, in the human dataset for $\alpha = 0.01$ with $d = 3$ we see a 4.4% increase in power going from aTDC to LBM (524.2 vs. 547.1 discoveries), and with $d = 5$ a 4.6% increase when using LBM (530.8 vs. 555.0 discoveries). Similarly, in the ISB18 dataset for $\alpha = 0.01$ with $d = 3$ we see a 7.3% increase in power going from aTDC to LBM (132.3 vs. 142.0 discoveries), and with $d = 5$ a 6.4% increase when using LBM (134.7 vs. 143.3 discoveries).

In Supplementary Sect. 6.12 we discuss a further analysis where we added two more spectra runs to the yeast dataset representing a higher budget experiment. In this case at 1% FDR the average number of TDC discoveries was 275.9 and for LBM using $d = 5$ decoys it was 294. Subsequent Gene Ontology enrichment test of the 54 proteins imputed from the peptide discovered by LBM yielded two overrepresented biological process terms that were not present in the 50 proteins imputed from TDC. The two missing terms—"cellular protein localization" and "cellular macromolecule localization"—are closely related and imply that the sample under investigation is enriched for proteins responsible in shuttling or maintaining other proteins in their proper cellular compartments. Critically, an

---

[2] Being a controlled experiment, the ISB18 dataset allows us to empirically gauge the FDR.

analysis based solely on the traditional TDC approach would entirely miss this property of the sample being analyzed.

5   Discussion

We consider a new perspective on the peptide detection problem which can be framed more broadly as multiple-competition based FDR control. The problem we pose and the tools we offer can be viewed as bridging the gap between the canonical FDR controlling procedures of BH and Storey and the single-decoy approach of the popular TDC used in spectrum identification (ID). Indeed, our proposed FDS converges to Storey's method as the number of decoys $d \longrightarrow \infty$ (Supplementary Sect. 6.7).

The methods we propose here rely on our novel mirandom procedure, which guarantees FDR control in the finite sample case for any pre-determined values of the tuning parameters c, λ. Our extensive simulations show that which of our methods delivers the maximal power varies with the properties of the experiment, as well as with the FDR threshold α. This variation motivates our introduction of LBM. LBM relies on a novel labeled resampling technique, which allows it to select its preferred method after testing whether a direct maximization approach seems to control the FDR. Our simulations, as well as our analysis of peptide detection using real datasets, suggest that LBM largely controls the FDR and seems to offer the best balance among our multi-decoy methods as well as a significant power advantage over the single-decoy TDC.

Finally, as mentioned, our approach is applicable beyond peptide detection. Moreover, while we stated our results assuming iid decoys, the results hold in a more general setting of *"conditional null exchangeability"* (Supplementary Sect. 6.13). This exchangeability is particularly relevant for future work on generalizing the construction of [1] to multiple knockoffs, where the iid decoys assumption is unlikely to hold.

Related work. We recently developed aTDC in the context of spectrum ID. The goal of aTDC was to reduce the decoy-induced variability associated with TDC by averaging a number of single-decoy competitions [17,18]. As such, aTDC fundamentally differs from the methods of this paper which simultaneously use all the decoys in a single competition; hence, the methods proposed here can deliver a significant power advantage over aTDC (panel F, Supplementary Fig. 4 and Supplementary Fig. 5). Our new methods are designed for the iid (or exchangeable) decoys case, which is a reasonable assumption for the peptide detection problem studied here but does not hold for the spectrum ID for which aTDC was devised. Indeed, as pointed out in [16], due to the different nature of native/foreign false discoveries, the spectrum ID problem fundamentally differs from the setup of this paper and even the above, weaker, null exchangeability property does not hold in this case. Thus, LBM cannot replace aTDC entirely; indeed, LBM is too liberal in the context of the spectrum ID problem. Note that in practice aTDC has not previously been applied to the peptide detection problem.

While working on this manuscript we became aware of a related Arxiv submission [14]. The initial version of that paper had just the mirror method, which as we show is quite limited in power. A later version that essentially showed up simultaneously with the submission of our technical report [7] extended their approach to a more general case; however, the method still consists of a subset of our independently developed research in that: (a) they do not consider the λ tuning parameter, (b) they use the uniform random map φ_u which, as we show, is inferior to mirandom, and (c) they do not offer either a general deterministic (FDS) or bootstrap based (LBM) data-driven selection of the tuning parameter(s), relying instead on a method that works only in the limited case-control scenario they consider.

Acknowledgement. This work was supported by National Institutes of Health award R01GM121818.

References

1. Barber, R.F., Candès, E.J.: Controlling the false discovery rate via knockoffs. Ann. Stat. **43**(5), 2055–2085 (2015)
2. Benjamini, Y., Hochberg, Y.: Controlling the false discovery rate: a practical and powerful approach to multiple testing. J. Roy. Stat. Soc. Ser. B **57**, 289–300 (1995)
3. Cerqueira, F.R., Graber, A., Schwikowski, B., Baumgartner, C.: MUDE: a new approach for optimizing sensitivity in the target-decoy search strategy for large-scale peptide/protein identification. J. Proteome Res. **9**(5), 2265–2277 (2010)
4. Diament, B., Noble, W.S.: Faster SEQUEST searching for peptide identification from tandem mass spectra. J. Proteome Res. **10**(9), 3871–3879 (2011)
5. Elias, J.E., Gygi, S.P.: Target-decoy search strategy for increased confidence in large-scale protein identifications by mass spectrometry. Nat. Methods 4(3), 207–214 (2007)
6. Elias, J.E., Gygi, S.P.: Target-decoy search strategy for mass spectrometry-based proteomics. Methods Mol. Biol. **604**, 55–71 (2010). https://doi.org/10.1007/978-1-60761-444-9_5
7. Emery, K., Hasam, S., Noble, W.S., Keich, U.: Multiple competition based FDR control. arXiv (2019). arXiv:1907.01458
8. Eng, J.K., McCormack, A.L., Yates, J.R.: An approach to correlate tandem mass spectral data of peptides with amino acid sequences in a protein database. J. Am. Soc. Mass Spectrom. **5**(11), 976–989 (1994). https://doi.org/10.1016/1044-0305(94)80016-2
9. Fan, Y., Lv, J., Sharifvaghefi, M., Uematsu, Y.: IPAD: stable interpretable forecasting with knockoffs inference. Available at SSRN 3245137 (2018)
10. Gao, C., et al.: Model-based and model-free machine learning techniques for diagnostic prediction and classification of clinical outcomes in parkinson's disease. Sci. Rep. **8**(1), 7129 (2018)
11. Granholm, V., Navarro, J.F., Noble, W.S., Käll, L.: Determining the calibration of confidence estimation procedures for unique peptides in shotgun proteomics. J. Proteomics **80**(27), 123–131 (2013)
12. Harbison, C.T., et al.: Transcriptional regulatory code of a eukaryotic genome. Nature **431**, 99–104 (2004)

13. He, K., et al.: A theoretical foundation of the target-decoy search strategy for false discovery rate control in proteomics. arXiv (2015). https://arxiv.org/abs/1501.00537
14. He, K., Li, M., Fu, Y., Gong, F., Sun, X.: A direct approach to false discovery rates by decoy permutations (2018). arXiv preprint arXiv:1804.08222
15. Jeong, K., Kim, S., Bandeira, N.: False discovery rates in spectral identification. BMC Bioinform. **13**(Suppl. 16), S2 (2012)
16. Keich, U., Noble, W.S.: Controlling the FDR in imperfect database matches applied to tandem mass spectrum identification. J. Am. Stat. Assoc. (2017). https://doi.org/10.1080/01621459.2017.1375931
17. Keich, U., Noble, W.S.: Progressive calibration and averaging for tandem mass spectrometry statistical confidence estimation: why settle for a single decoy? In: Sahinalp, S.C. (ed.) RECOMB 2017. LNCS, vol. 10229, pp. 99–116. Springer, Cham (2017). https://doi.org/10.1007/978-3-319-56970-3_7
18. Keich, U., Tamura, K., Noble, W.S.: Averaging strategy to reduce variability in target-decoy estimates of false discovery rate. J. Proteome Res. **18**(2), 585–593 (2018)
19. Lei, L., Fithian, W.: Power of ordered hypothesis testing. In: International Conference on Machine Learning, pp. 2924–2932 (2016)
20. Levitsky, L.I., Ivanov, M.V., Lobas, A.A., Gorshkov, M.V.: Unbiased false discovery rate estimation for shotgun proteomics based on the target-decoy approach. J. Proteome Res. **16**(2), 393–397 (2017)
21. Lin, H., He, Q.Y., Shi, L., Sleeman, M., Baker, M.S., Nice, E.C.: Proteomics and the microbiome: pitfalls and potential. Exp. Rev. Proteomics **16**(6), 501–511 (2019)
22. Lu, Y.Y., Fan, Y., Lv, J., Noble, W.S.: DeepPINK: reproducible feature selection in deep neural networks. In: Advances in Neural Information Processing Systems (2018)
23. Morris, M., Knudsen, G.M., Maeda, S., Trinidad, J.C., Ioanoviciu, A., Burlingame, A.L., Mucke, L.: Tau post-translational modifications in wild-type and human amyloid precursor protein transgenic mice. Nat. Neurosci. **18**, 1183–1189 (2015)
24. Nesvizhskii, A.I.: A survey of computational methods and error rate estimation procedures for peptide and protein identification in shotgun proteomics. J. Proteomics **73**(11), 2092–2123 (2010)
25. Ng, P., Keich, U.: Gimsan: a gibbs motif finder with significance analysis. Bioinformatics **24**(19), 2256–2257 (2008)
26. Noble, W.S., MacCoss, M.J.: Computational and statistical analysis of protein mass spectrometry data. PLOS Comput. Biol. **8**(1), e1002296 (2012)
27. Hernandez, P., Muller, M., Appel, R.D.: Automated protein identification by tandem mass spectrometry: issues and strategies. Mass Spectrom. Rev. **25**, 235–254 (2006)
28. Ping, L., et al.: Global quantitative analysis of the human brain proteome in Alzheimer's and Parkinson's disease. Sci. Data **5**, 180036 (2018)
29. Read, D.F., Cook, K., Lu, Y.Y., Le Roch, K., Noble, W.S.: Predicting gene expression in the human malaria parasite plasmodium falciparum. J. Proteome Res. **15**(9), e1007329 (2019)
30. Saito, M.A., et al.: Progress and challenges in ocean metaproteomics and proposed best practices for data sharing. J. Proteome Res. **18**(4), 1461–1476 (2019)
31. Savitski, M.M., Wilhelm, M., Hahne, H., Kuster, B., Bantscheff, M.: A scalable approach for protein false discovery rate estimation in large proteomic data sets. Mol. Cell. Proteomics **14**(9), 2394–2404 (2015)

32. Storey, J.D.: A direct approach to false discovery rates. J. Roy. Stat. Soc. Ser. B **64**, 479–498 (2002)
33. Storey, J.D., Taylor, J.E., Siegmund, D.: Strong control, conservative point estimation, and simultaneous conservative consistency of false discovery rates: a unified approach. J. Roy. Stat. Soc. Ser. B **66**, 187–205 (2004)
34. Storey, J.D., Tibshirani, R.: Statistical significance for genome-wide studies. Proc. Nat. Acad. Sci. US Am. **100**, 9440–9445 (2003)
35. Storey, J.D., Bass, A.J., Dabney, A., Robinson, D.: qvalue: Q-value estimation for false discovery rate control (2019). http://github.com/jdstorey/qvalue, r package version 2.14.1
36. Tusher, V.G., Tibshirani, R., Chu, G.: Significance analysis of microarrays applied to the ionizing radiation response. Proc. Nat. Acad. Sci. US Am. **98**, 5116–5121 (2001). https://doi.org/10.1073/pnas.091062498
37. Wildburger, N.C., et al.: Diversity of amyloid-beta proteoforms in the Alzheimer's disease brain. Sci. Rep. **7**, 9520 (2017)
38. Xiao, Y., Angulo, M.T., Friedman, J., Waldor, M.K., WeissT, S.T., Liu, Y.Y.: Mapping the ecological networks of microbial communities. Nat. Commun. **8**(1), 2042 (2017)

Supervised Adversarial Alignment of Single-Cell RNA-seq Data

Songwei Ge[1(✉)], Haohan Wang[2], Amir Alavi[1], Eric Xing[2,3],
and Ziv Bar-Joseph[1,3(✉)]

[1] Computational Biology Department, Carnegie Mellon University,
Pittsburgh 15213, USA
{songweig,aalavi,zivbj}@cs.cmu.edu
[2] Language Technologies Institute, Carnegie Mellon University,
Pittsburgh 15213, USA
{haohanw,epxing}@cs.cmu.edu
[3] Machine Learning Department, Carnegie Mellon University,
Pittsburgh 15213, USA

Abstract. Dimensionality reduction is an important first step in the analysis of single cell RNA-seq (scRNA-seq) data. In addition to enabling the visualization of the profiled cells, such representations are used by many downstream analyses methods ranging from pseudo-time reconstruction to clustering to alignment of scRNA-seq data from different experiments, platforms, and labs. Both supervised and unsupervised methods have been proposed to reduce the dimension of scRNA-seq. However, all methods to date are sensitive to batch effects. When batches correlate with cell types, as is often the case, their impact can lead to representations that are batch rather than cell type specific. To overcome this we developed a domain adversarial neural network model for learning a reduced dimension representation of scRNA-seq data. The adversarial model tries to simultaneously optimize two objectives. The first is the accuracy of cell type assignment and the second is the inability to distinguish the batch (domain). We tested the method by using the resulting representation to align several different datasets. As we show, by overcoming batch effects our method was able to correctly separate cell types, improving on several prior methods suggested for this task. Analysis of the top features used by the network indicates that by taking the batch impact into account, the reduced representation is much better able to focus on key genes for each cell type.

Keywords: Dimensionality reduction · Single-cell RNA-seq · Batch effect removal · Domain adversarial training · Data integration

1 Introduction

Single-cell RNA sequencing (scRNA-seq) has revolutionized the study of gene expression programs [16,27]. The ability to profile genes at the single-cell level

© Springer Nature Switzerland AG 2020
R. Schwartz (Ed.): RECOMB 2020, LNBI 12074, pp. 72–87, 2020.
https://doi.org/10.1007/978-3-030-45257-5_5

has revealed novel specific interactions and pathways within cells [43], differences in the proportions of cell types between samples [17,44], and the identity and characterization of new cell types [39]. Several biological tissues, systems, and processes have recently been studied using this technology [17,43,44].

While studies using scRNA-seq provide many insights, they also raise new computational challenges. One of the major challenges involves the ability to integrate and compare results from multiple scRNA-seq studies. There are several different commercial platforms for performing such experiments, each with their own biases. Furthermore, similar to other high throughput genomic assays, scRNA-seq suffers from batch effects which can make cells profiled in one lab look very different from the same cells profiled at another lab [37,38]. Moreover, other types of high throughput transcriptomics profiling, including microscopy-based techniques, are also generating single cell expression datasets [8,40]. The goal of fully utilizing these spatial datasets motivates the development of methods that can combine them with scRNA-seq when studying specific biological tissues and processes.

A number of recent methods have attempted to address this challenge by developing methods for aligning scRNA-seq data from multiple studies of the same biological system. Many of these methods rely on identifying nearest neighbors between the different datasets and using them as anchors. Methods that use this approach include Mutual Nearest Neighbors (MNN) [13] and Seurat [36]. Others including scVI and scAlign first embed all datasets into a common lower dimensional space. scVI encodes the scRNA-seq data with a deep generative model conditioning on the batch identifiers [24] while scAlign regularizes the representation between two datasets by minimizing the random walk probability differences between the original and embedding spaces. While these methods were successful for some datasets, here we show that they are not always able to correctly match all cell types. A key problem with these methods is the fact that they are unsupervised and rely on the assumption that cell types profiled by the different studies overlap. While this works for some datasets, it may fail for studies in which cells do not fully overlap or for those containing rare cell types. Unsupervised methods tend to group rare types with the larger types making it hard to identify them in a joint space.

Recent machine learning work has focused on a related problem termed "domain adaptation/generalization". Methods developed for these problems attempt to learn representations of diverse data that are invariant to technical confounders [5,25,42]. These methods have been used for multiple applications such as machine translation for domain specific corpus [4] and face detection [28]. Several methods proposed for domain adaptation rely on the use of adversarial methods [5,10,21,41], which has been proved effective to align latent distributions. In addition to the original task such as classification, these methods apply a domain classifier upon the learned representations. The encoder network is used for both improving accurate classification while at the same time reducing the impact of the domain (by "fooling" a domain classifier). This is achieved by learning encoder weights that simultaneously perform gradient *descent* on the label classification task and gradient *ascent* on the domain classification task.

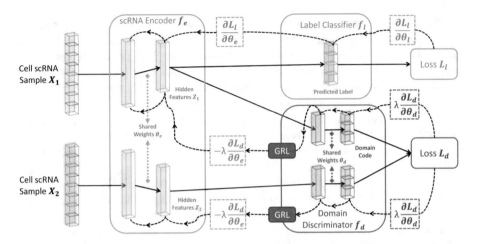

Fig. 1. Architecture of scDGN. The network includes three modules: scRNA encoder f_e (blue), label classifier f_l (orange) and domain discriminator f_d (red). Note that the red and orange networks use the same encoding as input. Solid lines represent the forward direction of the neural network while the dashed lines represent the backpropagation direction with the corresponding gradient it passes. Gradient Reversal Layers (GRL) have no effect in forward propagation, but flip the sign of the gradients that flow through them during backpropagation. This allows the combined network to simultaneously optimize label classification and attempt to "fool" the domain discriminator. Thus, the encoder leads to representations that are invariant to the different domains while still distinguishing cell types.

Here we extend these approaches, coupling them with Siamese network learning [20] for overcoming batch effects in scRNA-seq analysis. We define a "domain" in this paper as a standalone dataset profiled at a single lab using a single platform. We define "label" as the cell type for each cell in the dataset. Considering the specificity of the cell types in the scRNA-seq datasets, we propose a conditional pair sampling strategy that constrains input pair selection when training the adversarial network. We discuss how to formulate a domain adaptation network for scRNA-seq data, how to learn the parameters for the network, and how to train it using available data.

We tested our method on several datasets ranging in size from 10 to 39 cell types and from 4 to 155 batches. As we show, for all of the datasets our domain adversarial method improves on previous methods, in some cases significantly. Visualization of the learned representation from several different methods helps highlight the advantages of the domain adversarial framework. As we show, the framework is able to accurately mitigate the batch effects while maintaining the grouping of cells from the same type across different batches. Biological analysis of the resulting model identifies key genes that can correctly distinguish between cell types across different experiments. Such batch invariant genes are promising candidates for a cell type specific signature that can be used across different studies to annotate cells.

2   Methods

2.1   Problem Formulation

To formulate the problem we start with a few notation definitions. We assume that the single cell RNA-seq data are drawn from the input space $\mathbf{X} \in \mathbb{R}^p$ where each sample (a cell) $\mathbf{x}$ has p features corresponding to the gene expression values. Cells are also associated with the label $y \in \mathbf{Y} = \{1, 2, ..., K\}$ which represents their cell types. We associate each sample with a specific domain/batch $d \in \mathcal{D}$ that represents any standalone dataset profiled at a single lab using a single platform. Note that we will use domain and batch interchangeably in this paper for convenience. The data are divided into a training set and a test set that are drawn from multiple studies. The domains used to collect training data are not used for the test set and so batch effects can vary between the training and test data. In practice, each of the domains only contains a small subset of the cell types. This means that the distribution of cell types is correlated with the distribution of domains. Thus, the methods that naively learn cell types based on expression profile [3,18,22] may instead fit domain information and not generalize well to the unobserved studies.

2.2   Domain Adversarial Training with Siamese Network

To overcome this problem and remove the domain impact when learning a cell type representation we propose a neural network framework which includes three modules as shown in Fig. 1: scRNA encoder, label classifier, and domain discriminator. The encoder module $f_e(\mathbf{x}; \theta_e)$ is used to reduce the dimensions of the data and contains fully connected layers which produce the hidden features, where θ_e represents the parameters in these layers. The label classifier $f_l(f_e; \theta_l)$ attempts to predict the label of input $\mathbf{x_1}$ whereas the goal of the domain discriminator $f_d(f_e; \theta_d)$ is to determine whether a pair of inputs $\mathbf{x_1}$ and $\mathbf{x_2}$ are from the same domain or not. Past work for classifying scRNA-seq data only attempted to minimize the loss function for the label classifier $\mathcal{L}_l(f_l(f_e; \theta_l))$ [3,23]. Here, we extend these methods by adding a regularization term based on the adversarial loss of the domain discriminator $\mathcal{L}_d(f_d(f_e; \theta_d))$ which we will elaborate later. The overall loss E on a pair of samples $\mathbf{x_1}$ and $\mathbf{x_2}$ is denoted by:

$$E(\theta_e, \theta_l, \theta_d) = \mathcal{L}_l\big(f_l(f_e(\mathbf{x_1}; \theta_e); \theta_l)\big) - \lambda \mathcal{L}_d\big(f_d(f_e(\mathbf{x_1}; \theta_e); \theta_d), f_d(f_e(\mathbf{x_2}; \theta_e); \theta_d)\big),$$

where λ can control the trade-off between the goals of domain invariance and higher classification accuracy. For convenience, we use $\mathbf{z_1}$ and $\mathbf{z_2}$ to denote the hidden representations of $\mathbf{x_1}$ and $\mathbf{x_2}$ calculated from $f_e(\mathbf{x}; \theta_e)$. Inspired by Siamese networks [20], we implement our domain discriminator by adopting a contrastive loss [12]:

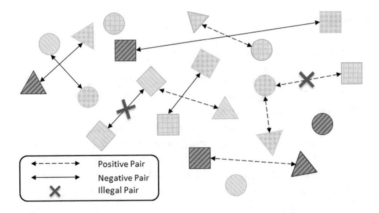

Fig. 2. Conditional domain generalization strategy: Shapes represent different labels and colors (or patterns) represent different domains. For negative pairs from different domains, we only select those samples with the same label. For positive pairs from the same domain, we only select the samples with different labels.

$$\mathcal{L}_d\big(f_d(\mathbf{z_1};\theta_d),\ f_d(\mathbf{z_2};\theta_d)\big) = U\frac{1}{2}D\big(f_d(\mathbf{z_1}),f_d(\mathbf{z_2})\big)^2$$
$$+\ (1-U)\frac{1}{2}\big(\max\{0,m-D(f_d(\mathbf{z_1}),f_d(\mathbf{z_2}))\}\big)^2,$$

where $U = 1$ indicates that two samples are from the same domain d and $U = 0$ indicates that they are not, $D(\cdot)$ is the euclidean distance, and m is the margin that indicates the prediction boundary. The domain discriminator parameters, θ_d, are updated using back propagation to *maximize* the total loss E while the encoder and classifier parameters, θ_e and θ_l, are updated to *minimize* E. To allow all three modules to be updated together end-to-end, we use a Gradient Reversal Layer (Fig. 1) [10,29]. Specifically, Gradient Reversal Layers (GRL) have no effect in forward propagation, but flip the sign of the gradients that flow through them during backpropagation. The following provides the overall optimization problems solved for the network parameters:

$$(\widehat{\theta}_e,\widehat{\theta}_l) = \arg\min_{\theta_e,\theta_l} E(\theta_e,\theta_l,\widehat{\theta}_d)$$
$$(\widehat{\theta}_d) = \arg\max_{\theta_d} E(\widehat{\theta}_e,\widehat{\theta}_l,\theta_d)$$

In other words, the goal of the domain discriminator is to tell if two samples are drawn from the same or different batches. By optimizing the scRNA encoder adversarially against the domain discriminator, we attempt to make sure that the network representation cannot be used to classify based on domain knowledge. During the training, the maximization and minimization tasks compete with each other, which is achieved by adjusting the representations to improve the accuracy of the label classifier and simultaneously fool the domain discriminator.

2.3 Conditional Domain Generalization Strategy

Most prior domain adaption or generalization methods focused on the cases where the distribution of labels is independent of the domains [5,25]. In contrast, as we show in Results, for scRNA-seq experiments different studies tend to focus on certain cell types [17,43,44]. Consequently, it is not reasonable to completely merge the scRNA-seq data from different batches. To be specific, aligning the scRNA-seq data from two batches with different sets of cell types would sacrifice its biological significance and prevent the cell classifier from predicting effectively. To overcome this issue, instead of arbitrarily choosing positive pairs (samples from the same domain) and negative pairs (samples from different domains), we constrain the selection as follows: (1) for positive pairs, only the samples with different labels from the same domain are selected. (2) for negative pairs, only the samples with the same label from different domains are selected. Figure 2 provides a visual interpretation of this strategy. Formally, letting y_i and z_i represent the i-th sample's cell-type label and domain label respectively, we have the following equations to define the value of U for sample pairs:

$$U = \begin{cases} 0, & z_1 \neq z_2 \text{ and } y_1 = y_2 \\ 1, & z_1 = z_2 \text{ and } y_1 \neq y_2 \end{cases}$$

This strategy prevents the domain adversarial training from aligning samples with different labels or separating samples with same labels. For example, in order to fool the discriminator with a positive pair, the encoder must implicitly increase the distance of two samples with different cell types. Therefore, combining this strategy with domain adversarial training allows the network to learn cell type specific, focused representations. We term our model Single Cell Domain Generalization Network (scDGN).

3 Results

3.1 Experiment Setups

Datasets. To test our method and to compare it to previous methods for aligning and classifying scRNA-seq data, we used several recent datasets. These datasets contain between 6,000 and 45,000 cells, and all include cells profiled in multiple experiments by different labs and on different platforms.

scQuery: We use a subset of the dataset provided by scQuery, which includes uniformly processed data from hundreds of studies [3][1]. The dataset we use contains 44,490 samples from 155 different experiments. scQuery assigns cells to 39 types spanning diverse categories ranging from immune cells to neurons to organ specific cells. We use 99 of the 155 batches for training, 26 for validation, and 30 for testing. We provide a list of the studies used for each set in Appendix

---

[1] https://scquery.cs.cmu.edu/processed_data/.

Table 1. Basic statistics for scQuery, Suerat pancreas, and PBMC datasets

	scQuery			Seurat pancreas			Seurat pbmc		
	Data	Cell type	Domain	Data	Cell type	Domain	Data	Cell type	Domain
Training	37697	39	99	6321	13	3	25977	10	8
Validation	3023	19	26	–	–	–	–	–	–
Test	3770	23	30	638	13	1	2992	10	1

A.1 [11]. Statistics for the different datasets are shown in Table 1. RPKM normalization is applied to the 20,499 genes in each sample. Note that while there are 39 cell-types in the training set, only 19 and 23 of them are included in the validation and test set. This mimics the application of the methods to future studies that may not profile all types of cells.

PBMC: The Peripheral Blood Mononuclear Cells (PBMC) dataset contains 28,969 cells assigned to 10 blood cell types. The data are profiled in 9 batches (from 8 different sequencing technologies) [6]. We use the data from the *10xChromiumv2A* platform as test data and the rest as training data. Following the provided tutorial [2], we use the top 3000 variable genes for the analysis.

Seurat Pancreas: The Seurat pancreas dataset is designed for evaluating single cell alignment algorithms and contains 6321 scRNA-seq samples of human pancreatic islet cell produced by four studies. We use the smallest study for the test data and the other three for training as shown in Table 1. Thirteen canonical labels of the pancreatic islet cell are assigned to cells in each study. Similar to the Seurat PBMC dataset, we only used the 3000 most variable genes. To further simulate the correlation between cell types and domains for this dataset we randomly remove the data for 6 of the 13 cell types for each of the training domains. As a result, we construct 6 synthetic datasets based this strategy to evaluate the alignment performance of different methods under a high label-domain correlation setting. The specific cell type information of each dataset is listed in Appendix A.3 [11].

Model Configurations. We used the network of Lin et al. [23] as the components for the encoder and the label classifier in our model. The encoder contains two hidden layers with 1136 and 100 units. The label classifier is directly connected to the 100 unit layer and makes predictions based on these values. The domain discriminator contains an additional hidden layer with 64 units and is also connected to the 100 unit layer of the encoder (Fig. 1). For each layer, $tanh()$ is used as the non-linear activation function. We test several other possible configurations but did not observe improvement in performance. As is commonly done, we use a validation set to tune the hyperparameters for learning including learning rates, decay, momentum, and the adversarial weight and margin parameters λ and m. Generally, our analysis indicates that for larger datasets a

Table 2. Overall performances of different methods. *MI* represents the mutual information between batch and cell type in the corresponding dataset. The highest test accuracy for each dataset is bolded.

Experiments	MI	NN [23]	CaSTLe [22]	MNN [13]	scVI [24]	Seurat [36]	scDGN
scQuery	3.025	0.255	0.156	0.200	0.257	0.144	**0.286**
PBMC	0.112	0.861	0.865	0.859	0.808	0.830	**0.868**
Pancreas 1	0.902	0.720	0.705	0.591	0.855	0.812	**0.856**
Pancreas 2	0.733	0.891	0.764	0.764	0.852	0.825	**0.918**
Pancreas 3	0.931	0.545	0.722	0.722	0.651	**0.751**	0.663
Pancreas 4	0.458	0.927	0.914	0.914	**0.925**	0.881	**0.925**
Pancreas 5	0.849	0.928	0.882	**0.932**	0.895	0.865	0.923
Pancreas 6	0.670	0.944	0.917	0.946	0.893	0.907	**0.950**
Average	–	0.826	0.817	0.842	0.845	0.840	**0.872**

lower weight λ and larger margin m for the adversarial training is preferred and vice versa. More details about the hyperparameters and training are provided in Appendix A.2 [11].

Baselines. We compared scDGN to several prior methods for classifying and aligning scRNA-seq data. These included the neural network (NN) model of Lin et al. [23] which is developed for classifying scRNA-seq data, CaSTLe [22] which performs cell type classification based on transfer learning, and several state-of-the-art alignment methods. For alignment, we compared to MNN [13] which utilizes mutual nearest neighbors to align data from different batches, scVI [24] which trains a deep generative model on the scRNA-seq data and uses an explicit batch identifier to retain conditional independence property of the representation, and Seurat [36] which first identifies the anchors among different batches and then projects different datasets using a correction vector based on the order defined by hierarchical clustering with pairwise distances. Our comparisons include both visual projection of the learned alignment (Figs. 4 and 5) and quantitative analysis of the accuracy of the predicted test cell types (Table 2). For the latter, to enable comparisons of the supervised and unsupervised methods, we used the resulting aligned data from the unsupervised methods to train a neural network that has the same configuration as Lin et al. [23]. For scVI, which results in a much lower dimensional representation, we used a smaller input vector and a smaller hidden layer. Note that these alignment methods actually use the scRNA-seq test data to determine the final dimensionality reduction function while our method does not utilize the test data for any model decision or parameter learning. To effectively apply Seurat to scQuery, we remove the batches which have <100 samples. Also, for those datasets that the assumption of overlapped cell types is not guaranteed such as scQuery, we find that the performance of MNN highly depends on the order of alignment. Therefore, for

MNN on the scQuery dataset, we use 10 random permutations of batch orders and report the average accuracy.

3.2 Overall Performance

As mentioned above, we use the validation set to select the best model when using the scQuery dataset. For the smaller datasets, we use the model obtained after 250 epochs (all models converged after this number of epochs). Test accuracy for the different methods is presented in Table 2. We show both mean and standard deviation of the accuracy for 10 randomly initialized experiments. We also report the performances on different cell types in Appendix B [11]. In addition, Table 2 presents the Mutual Information (MI) between labels and domains which corresponds to the difficulty of the dataset. A larger MI indicates that models that do not account for the domain are likely to fit the domain information rather than the cell type. For the scQuery dataset, we find the accuracy is low for all methods indicating that this dataset is relatively difficult. This is corroborated by the large MI value. For such data we see a clear advantage for the scDGN: scDGN improves by over 10% over all other methods ($p = 5.069 \times 10^{-5}$ based on Student's t-test when compared to the NN baseline which is tied for second best). The improvements over other single cell alignment methods are even more significant. scDGN also achieves the best performance on the second largest dataset, the PBMC dataset. However, given the very low MI for this dataset the performance of the other methods, including the baseline NN, is almost as good as the performance of scDGN. The third dataset we test on is the Seurat pancreas dataset. This is the smallest dataset and so it has the least number of training samples. Still, of the 6 settings we tested (which differed in the subset of cells that were excluded from training), we find that scDGN is the top performer in 4 of them, comparable to the top performer for another 1 and in only one setting (Pancreas 3, with the highest MI) is significantly outperformed by Seurat. Note that even for the Pancreas 3 data the domain adversarial training helps: using this the scDGN is able to improve by more than 20% over the baseline NN used for the label classifier.

3.3 Visualization of the Representation Learned by Alignment and Classification Methods

To further explore the effectiveness of the batch removal provided by our proposed domain adversarial training with conditional domain generalization strategy, we visualize the 100-dimensional hidden representations learned by NN and scDGN: Fig. 3 presents both PCA and t-SNE plots for several different cell types across the three datasets. Points are colored using their batch IDs in order to evaluate batch effects. As can be seen, using scDGN we obtain results that are much better at mixing cells from the different batches when compared to the baseline NN model. The impact is larger for the pancreas datasets which have larger MI compared to the PBMC dataset, which helps explain the large increase in performance for these two datasets.

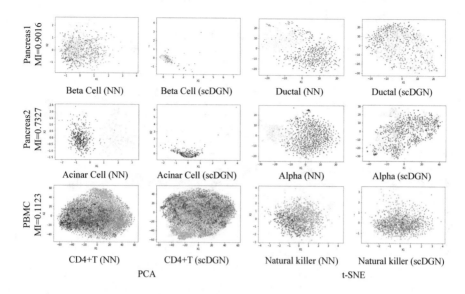

Fig. 3. Visualization of learned representations for NN and scDGN: using PCA and t-SNE Rows: The three datasets we tested the method on. Columns: Methods and cell types. For each row, data from different batches are distinguished using different colors.

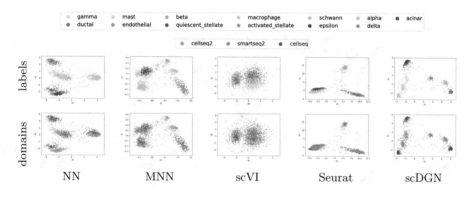

Fig. 4. PCA visualizations of the representations learned by different models on the full Pancreas2 dataset. Colors for different cell types and domains are shown in the legend at the top.

We next extended this comparison and visualized the learned (aligned) representations for all methods using data from both the Pancreas2 and scQuery datasets (Figs. 4 and 5). For the Pancreas2 dataset, we visualize the entire dataset. For scQuery, given the large number of cell types and domains, we present PCA visualization of a subset of cell types and domains. As can be seen, in addition to scDGN, Seurat is also able to successfully mix the data from different batches. However, as the results in Table 2 indicate this may come at the

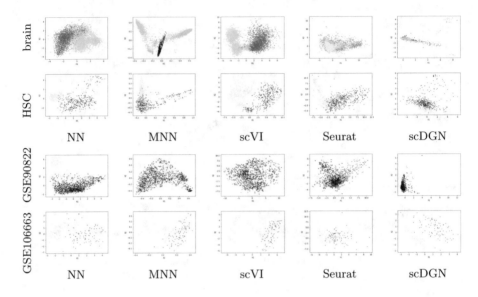

Fig. 5. PCA visualizations of the representations of certain cell types and batches by different models for the scQuery dataset. Top two rows: Cell types. Colors represent different batches. HSC = hematopoietic stem cell. Bottom two rows: Batches. Colors represent different cell types.

expense of not correctly separating cell types. MNN and scVI are not always effective at removing batch effects for the cell types. In contrast, scDGN is able to do both domain mixing and cell type assignment, leading to its better performance overall. For example, for the acinar and alpha cell types in the pancreas dataset (Fig. 4), only scDGN , MNN, and Seurat are able to align the data from different domains. However, MNN and Seurat over-correct the representation by aligning different cell types from different domains, mixing acinar and gamma cells. Additional visualizations for other cell types and domains can be found in Appendix C [11], where the same advantages of scDGN over other methods can be consistently observed.

3.4   Analysis of Key Genes

While NNs are often treated as black boxes, recent methods provide useful directions for making them more interpretable [31]. Here we use activation maximization, which relies on the gradient of the correct category logit with respect to the input vector to select the key inputs for each of the models [9,33,34]. Formally, given a particular cell type i and a trained neural network ϕ, activation maximization looks for important input genes x' by solving the following optimization problem:

$$x' = \max_x(\phi(x) \cdot e_i),$$

Table 3. GO analysis results for top 100 scQuery liver genes in the NN method.

term_name	term_id	p_{adj}	$-\log_{10} p_{adj}$
symbiotic process	GO:0044403	1.16E-08	7.935246875
interspecies interaction between organisms	GO:0044419	3.14E-08	7.503093471
viral process	GO:0016032	3.69E-08	7.433145019
immune response	GO:0006955	2.5491E-06	5.593613105
multi-organism process	GO:0051704	1.40837E-05	4.851282542
immune effector process	GO:0002252	4.53533E-05	4.34339136
response to stress	GO:0006950	5.56335E-05	4.254663785
defense response	GO:0006952	6.18759E-05	4.208478308

Table 4. GO analysis results for top 100 scQuery liver genes in the scDGN method.

term_name	term_id	p_{adj}	$-\log_{10} p_{adj}$
chylomicron remodeling	GO:0034371	3.04042E-05	4.517066786
positive reg. of cholesterol esterification	GO:0010873	3.04042E-05	4.517066786
negative reg. of cellular component organization	GO:0051129	3.94437E-05	4.404022507
protein-lipid complex remodeling	GO:0034368	7.34551E-05	4.133978335
plasma lipoprotein particle remodeling	GO:0034369	7.34551E-05	4.133978335
protein-containing complex remodeling	GO:0034367	8.8522E-05	4.052948555

where e_i is the natural basis vector associated with the i-th category. This can be solved through backpropagation, where the gradient of $\phi(x)$ with respect to x, which can be viewed as the weight of the first-order Taylor expansion of the neural network, are calculated to iteratively update the input. We follow a previous method [33] and initialize the optimization with a zero vector. Given this setting, we ran the optimization for 100 iterations with learning rate set to 1. The important genes are selected as those inputs leading to the largest changes compared with the initialization values. To compare scDGN and NN for certain cell types, we select the top k genes with the largest changes and perform GO analysis on these selected genes.

As an example, consider the genes identified for the liver cell type using the scQuery dataset. We select the top 100 genes for this cell type from NN and scDGN and present the enriched GO categories on Biological Process with adjusted p-value $<1.0 \times 10^{-4}$ in Tables 3 and 4. We also list these genes by order in Appendix A.3 [11]. As can be seen, while a number of significant GO categories are identified for the top 100 NN genes, these are generic and not liver specific. They include general terms related to interactions between organs and immune response categories that are active in multiple organs and cell types. In sharp contrast, the categories identified for scDGN are much more specific and highlight key pathways that are mainly utilized in the liver. For example, the top category for the scDGN genes, "chylomicron remodeling", refers to the main physiological purpose of chylomicron remnants: to facilitate the return of bile lipoproteins and cholesterol to the liver [30]. Specifically, in this pathway chylomicrons (lipoproteins) are broken down (remodeled via hydrolysis) and converted to a form called "chylomicron remnant" that is taken up by specific receptors that exist primarily on the surface of liver cells [14]. The second term, "pos. regulation of cholesterol esterification" refers to cholesterol esterification, a critical step in reverse cholesterol transport, the process in which excess cholesterol is sent to the liver to be removed from the body [1, 26]. Furthermore, Cholesteryl Ester Transfer Protein (CETP) is a key enzyme involved in this process and is highly expressed in liver cells, and variants of CETP are associated with increased risk of atherosclerosis [1, 32]. The fifth most significant term, "lipoprotein remodeling" is part of the two aforementioned processes. The top 100 genes identified by the scDGN include *apoa1* (main protein component of High-Density Lipoprotein cholesterol), *apoa2*, and *apoc1*, all of which encode lipoproteins that are primarily expressed in the liver [7, 19]. These genes were not included in the top 100 genes by the NN. We present the GO analysis results comparison for several additional cell types in Appendix D.2 [11].

4   Discussion

Single cell computational methods that do not account for batch effects are likely to fit the noise introduced by the batches. Several recent methods have been proposed for aligning scRNA-seq from multiple studies of the same tissues or processes. Most of these methods are unsupervised and assume that the cell types among different batches overlap. However, we show that these methods would fail on the studies in which cell types do not fully overlap, which is often the case when dealing with multiple datasets. To overcome this problem we extend a supervised scRNA-seq cell type assignment method based on NN and regularize its prediction to be invariant to batch effects.

Our method is based on the ideas of domain adversarial training. In such training, two competing tasks are used to optimize the representation of scRNA-seq data. The first focuses on the traditional goal of cell type identification while the second attempts to construct representations that are not affected by specific batch or experimental artifacts. This is accomplished by jointly minimizing a loss function that takes into account both goals, accounting for the weight of each of the goals using a gradient reversal layer. We also proposed a conditional strategy to avoid over-correction. We presented efficient learning methods for this setting and tested it on three large scale scRNA-seq datasets containing experiments from several different platforms for partially overlapping cell types.

As we show, our scDGN method is able to correctly identify cell types in the test datasets. For the largest dataset we tested on which contained close to 40 different cell types, scDGN significantly outperformed all prior methods. It also ranked first for the 2nd largest dataset and for all but 1 of the 6 tests on the third dataset. Importantly, it always outperformed the supervised learning based method indicating that batch effects should be addressed when designing such methods for cell type assignments. In addition to accurately assigning cell types, further analysis of significant genes indicates that by overcoming batch effects scDGN is better able to focus on relevant sets of genes when compared to prior supervised methods, explaining its improvement in accuracy.

While scDGN performed best on the data we analyzed, there are a number of possible issues with this approach. First, it learns a large number of parameters which require large input datasets. However, as we showed, scDGN is able to perform well even for datasets with a few thousand cells which matches current sizes of scRNA-seq datasets. Second, scDGN is based on NNs which are often seen as a black box, making it hard to interpret the resulting model and its biological relevance. Recent work provides a number of directions that can be used to overcome this issue. As we showed, using activation maximization we were able to identify several relevant cell type specific genes in the learned network. Future work would include using additional NN interpretation methods, including LIME [31] or ROAR and KAR [15], to further identify the set of genes that play the largest role in the decisions the network makes. Third, as shown in Third, as shown in Appendix C.13 [11], scDGN sometimes does not mix up the representations from different batches for all cell types. Considering the visualization results for NN in Appendix C.18 [11] and its competitive performance in Table 2 together, it may indicate that it is not always necessary to remove batch effects for the model to achieve high test accuracy. Therefore, it is worthwhile to further study when the alignment is imperative. Finally, unlike prior scRNA-seq alignment methods scDGN is supervised. While this is an advantage when it comes to accuracy, as we have shown, it may be a problem for the new data. We believe that as more scRNA-seq and other high throughput single cell data accumulate, we would have labeled data for most cell types which would enable training an scDGN for even more cell types. As we have shown with the scQuery dataset, for which scDGN significantly outperformed all other methods, when such data exists scDGN is able to correctly align experiments and platforms not seen in the training set.

scDGN is implemented in Python with the PyTorch API [35] and users can obtain the code and sampled data from https://github.com/SongweiGe/scDGN.

Acknowledgements. This work was partially supported by National Institute of Health grants 1R01GM122096 and OT2OD026682 to Z.B.J. and by a Scholars Award in Studying Complex Systems from the James S. McDonnell Foundation to Z.B.J. HW was supported by the National Institutes of Health grants R01-GM093156 and P30-DA035778.

References

1. Inazu, A.: Plasma cholesteryl ester transfer protein (CETP) in relation to human pathophysiology (Chap. 3). In: Komoda, T. (ed.) The HDL Handbook, pp. 35–59. Academic Press, Boston (2010)
2. Integration and label transfer - standard workflow, October 2019. https://satijalab.org/seurat/v3.1/integration.html#standard-workflow

3. Alavi, A., Ruffalo, M., Parvangada, A., Huang, Z., Bar-Joseph, Z.: A web server for comparative analysis of single-cell RNA-seq data. Nat. Commun. **9**(1), 4768 (2018)
4. Chu, C., Wang, R.: A survey of domain adaptation for neural machine translation. arXiv preprint arXiv:1806.00258 (2018)
5. Csurka, G.: Domain adaptation for visual applications: a comprehensive survey. arXiv preprint arXiv:1702.05374 (2017)
6. Ding, J., et al.: Systematic comparative analysis of single cell RNA-seq methods. BioRxiv, p. 632216 (2019)
7. Domingo-Espín, J., Nilsson, O., Bernfur, K., Giudice, R.D., Lagerstedt, J.O.: Site-specific glycations of apolipoprotein A-I lead to differentiated functional effects on lipid-binding and on glucose metabolism. Biochimica et Biophysica Acta (BBA) Mol. Basis Dis. **1864**(9, Part B), 2822–2834 (2018)
8. Eng, C.H.L., et al.: Transcriptome-scale super-resolved imaging in tissues by RNA seqFISH+. Nature **568**(7751), 235 (2019)
9. Erhan, D., Bengio, Y., Courville, A., Vincent, P.: Visualizing higher-layer features of a deep network. Univ. Montreal **1341**(3), 1 (2009)
10. Ganin, Y., et al.: Domain-adversarial training of neural networks. J. Mach. Learn. Res. **17**(1), 1–35 (2016)
11. Ge, S., Wang, H., Alavi, A., Xing, E., Bar-Joseph, Z.: Supporting information for: Supervised adversarial alignment of scRNA-seq data. bioRxiv (2020). https://doi.org/10.1101/2020.01.06.896621v1.full.pdf
12. Hadsell, R., Chopra, S., LeCun, Y.: Dimensionality reduction by learning an invariant mapping. In: 2006 IEEE Computer Society Conference on Computer Vision and Pattern Recognition (CVPR 2006), vol. 2, pp. 1735–1742. IEEE (2006)
13. Haghverdi, L., Lun, A.T., Morgan, M.D., Marioni, J.C.: Batch effects in single-cell RNA-sequencing data are corrected by matching mutual nearest neighbors. Nat. Biotechnol. **36**(5), 421 (2018)
14. Hara, T., Tan, Y., Huang, L.: In vivo gene delivery to the liver using reconstituted chylomicron remnants as a novel nonviral vector. Proc. Natl. Acad. Sci. **94**(26), 14547–14552 (1997)
15. Hooker, S., Erhan, D., Kindermans, P.J., Kim, B.: Evaluating feature importance estimates. arXiv preprint arXiv:1806.10758 (2018)
16. Hwang, B., Lee, J.H., Bang, D.: Single-cell RNA sequencing technologies and bioinformatics pipelines. Exp. Mol. Med. **50**(8), 1–14 (2018)
17. Jaitin, D.A., et al.: Massively parallel single-cell RNA-seq for marker-free decomposition of tissues into cell types. Science **343**(6172), 776–779 (2014)
18. Kiselev, V.Y., Yiu, A., Hemberg, M.: scmap: projection of single-cell RNA-seq data across data sets. Nat. Methods **15**(5), 359 (2018)
19. Ko, H.L., Wang, Y.S., Fong, W.L., Chi, M.S., Chi, K.H., Kao, S.J.: Apolipoprotein C1 (APOC 1) as a novel diagnostic and prognostic biomarker for lung cancer: a marker phase I trial. Thorac. Cancer **5**(6), 500–508 (2014)
20. Koch, G., Zemel, R., Salakhutdinov, R.: Siamese neural networks for one-shot image recognition. In: ICML Deep Learning Workshop, vol. 2 (2015)
21. Li, H., Pan, S.J., Wang, S., Kot, A.C.: Domain generalization with adversarial feature learning. In: CVPR (2018)
22. Lieberman, Y., Rokach, L., Shay, T.: Castle-classification of single cells by transfer learning: harnessing the power of publicly available single cell RNA sequencing experiments to annotate new experiments. PLoS One **13**(10), e0205499 (2018)
23. Lin, C., Jain, S., Kim, H., Bar-Joseph, Z.: Using neural networks for reducing the dimensions of single-cell RNA-seq data. Nucleic Acids Res. **45**(17), e156 (2017)

24. Lopez, R., Regier, J., Cole, M.B., Jordan, M.I., Yosef, N.: Deep generative modeling for single-cell transcriptomics. Nat. Methods **15**(12), 1053 (2018)
25. Motiian, S., Piccirilli, M., Adjeroh, D.A., Doretto, G.: Unified deep supervised domain adaptation and generalization. In: ICCV, vol. 2, p. 3 (2017)
26. Murakami, T., et al.: Triglycerides are major determinants of cholesterol esterification/transfer and HDL remodeling in human plasma. Arterioscler. Thromb. Vasc. Biol. **15**(11), 1819–1828 (1995)
27. Papalexi, E., Satija, R.: Single-cell RNA sequencing to explore immune cell heterogeneity. Nat. Rev. Immunol. **18**(1), 35 (2018)
28. Patel, V.M., Gopalan, R., Li, R., Chellappa, R.: Visual domain adaptation: a survey of recent advances. IEEE Signal Process. Mag. **32**(3), 53–69 (2015)
29. Pei, Z., Cao, Z., Long, M., Wang, J.: Multi-adversarial domain adaptation. In: AAAI Conference on Artificial Intelligence (2018)
30. Redgrave, T.: Chylomicron metabolism. Biochem. Soc. Trans. **32**(1), 79–82 (2004). https://doi.org/10.1042/bst0320079
31. Ribeiro, M.T., Singh, S., Guestrin, C.: Why should i trust you?: explaining the predictions of any classifier. In: SIGKDD, pp. 1135–1144. ACM (2016)
32. Seidman, M.A., Mitchell, R.N., Stone, J.R.: Pathophysiology of atherosclerosis (Chap. 12). In: Willis, M.S., Homeister, J.W., Stone, J.R. (eds.) Cellular and Molecular Pathobiology of Cardiovascular Disease, pp. 221–237. Academic Press, San Diego (2014)
33. Simonyan, K., Vedaldi, A., Zisserman, A.: Deep inside convolutional networks: Visualising image classification models and saliency maps. arXiv preprint arXiv:1312.6034 (2013)
34. Springenberg, J.T., Dosovitskiy, A., Brox, T., Riedmiller, M.: Striving for simplicity: the all convolutional net. arXiv preprint arXiv:1412.6806 (2014)
35. Steiner, B., et al.: Pytorch: An imperative style, high-performance deep learning library. In: NeurIPS, vol. 32 (2019)
36. Stuart, T., et al.: Comprehensive integration of single-cell data. Cell **177**, 1888–1902 (2019)
37. Stuart, T., Satija, R.: Integrative single-cell analysis. Nat. Rev. Genet. **20**, 257–272 (2019)
38. Tung, P.Y., et al.: Batch effects and the effective design of single-cell gene expression studies. Sci. Rep. **7**, 39921 (2017)
39. Villani, A.C., et al.: Single-cell RNA-seq reveals new types of human blood dendritic cells, monocytes, and progenitors. Science **356**(6335), eaah4573 (2017)
40. Wang, G., Moffitt, J.R., Zhuang, X.: Multiplexed imaging of high-density libraries of RNAs with MERFISH and expansion microscopy. Sci. Rep. **8**(1), 4847 (2018)
41. Wang, H., Ge, S., Xing, E.P., Lipton, Z.C.: Learning robust global representations by penalizing local predictive power. arXiv preprint arXiv:1905.13549 (2019)
42. Wang, H., He, Z., Lipton, Z.C., Xing, E.P.: Learning robust representations by projecting superficial statistics out. arXiv preprint arXiv:1903.06256 (2019)
43. Yu, Y., et al.: Single-cell RNA-seq identifies a PD-1 hi ILC progenitor and defines its development pathway. Nature **539**(7627), 102 (2016)
44. Zeisel, A., et al.: Cell types in the mouse cortex and hippocampus revealed by single-cell RNA-seq. Science **347**(6226), 1138–1142 (2015)

Bagging MSA Learning: Enhancing Low-Quality PSSM with Deep Learning for Accurate Protein Structure Property Prediction

Yuzhi Guo[1,2], Jiaxiang Wu[2], Hehuan Ma[1], Sheng Wang[1], and Junzhou Huang[1,2(✉)]

[1] University of Texas at Arlington, Arlington, TX 76019, USA
jzhuang75@gmail.com
[2] Tencent AI Lab, Shenzhen 518057, China

Abstract. Accurate predictions of protein structure properties, *e.g.* secondary structure and solvent accessibility, are essential in analyzing the structure and function of a protein. PSSM (Position-Specific Scoring Matrix) features are widely used in the structure property prediction. However, some proteins may have low-quality PSSM features due to insufficient homologous sequences, leading to limited prediction accuracy. To address this limitation, we propose an enhancing scheme for PSSM features. We introduce the "Bagging MSA" method to calculate PSSM features used to train our model, and adopt a convolutional network to capture local context features and bidirectional-LSTM for long-term dependencies, and integrate them under an unsupervised framework. Structure property prediction models are then built upon such enhanced PSSM features for more accurate predictions. Empirical evaluation of CB513, CASP11, and CASP12 datasets indicate that our unsupervised enhancing scheme indeed generates more informative PSSM features for structure property prediction.

Keywords: Deep learning · Unsupervised learning · Enhancing PSSM · Protein secondary structure prediction

1 Introduction

The function of a protein is closely related to its structure, which is largely determined by the amino-acid sequence. However, predicting one protein's structure based on its amino-acid sequence alone remains an open and challenging problem. An alternative approach is to firstly predict structure properties, including secondary structure, solvent accessibility, and backbone dihedral angles [1]. Those predictions are combined eventually to help the final prediction of protein structure.

PSSM (Position-Specific Scoring Matrix) features [3], which reflect per-residue evolution patterns in the sequence profile, are commonly used in the

© Springer Nature Switzerland AG 2020
R. Schwartz (Ed.): RECOMB 2020, LNBI 12074, pp. 88–103, 2020.
https://doi.org/10.1007/978-3-030-45257-5_6

structure property prediction [4,5]. The quality of PSSM features is basically determined by the underlying multiple sequence alignments (MSA) [6]. MSA requires searching the query amino-acid sequence through a large-scale sequence database, *e.g.* UniRef [18] and UniClust [19]. The MSA quality of the protein can be evaluated by counting the number of homologous proteins, or the non-redundant sequence homologs (Meff [2]) retrieved from the database. However, for those proteins with a limited number of high-quality homologous sequences, the prediction quality is often limited due to less informative PSSM features [7]. One possible solution is to develop more efficient and accurate MSA search algorithm, such as SABERTOOTH [8], hhblits [9], jackhmmer [10], and HBLAST [11]. These algorithms have achieved certain performance improvement by speeding up the searching process, as well as find more accurate homologous protein sequences in the database. However, if the database did not contain enough homologous protein sequences for the target protein, it is still inaccessible to obtain sufficient quantity or high quality of the MSA, yet the corresponding high-quality PSSM features.

In this paper, we propose an unsupervised deep learning method to enhance the low-quality PSSM features of proteins. To be specific, during the training of our model, we randomly sample the MSA of each protein in a certain proportion in each learning iteration, which we called "Bagging MSA". Then, we use the "Weak PSSMs" calculated by these bags and the "Original PSSM" calculated by all MSA to train our network. In this way, our network can learn how to generate high-quality PSSM from a protein that has low-quality PSSM features.

The most commonly predicted one-dimensional structural property of a protein is the secondary structure. Therefore, in order to evaluate our method on different prediction networks, we use two widely used deep learning techniques in the protein secondary prediction area, which are CNN and bi-LSTM models [26,27,33]. The knowledge of the secondary structure of proteins and the network of validation of our method are described in Sects. 2 and 3.

The technical contributions of this paper are summarized as: (1) Our method is the first attempt to enhance low quality PSSMs of proteins. According to the experimental results, our method significantly improve the secondary structure prediction task of proteins with weak PSSM. (2) In the unsupervised module, our method calculate PSSM features by randomly sampling 10% to 20% MSA in each training iteration as the input data, and use the original PSSM features as unsupervised labels. This approach not only increases the diversity of the data, but also make the network more flexible to learn different PSSM quality differences so as to give full play to unsupervised learning. (3) Our method is generalizable since it is capable for any prediction model with PSSM as the input other than just secondary protein prediction task. (4) The unsupervised part of our method is independent, so the output could be used as the input directly for the inference phase of any prediction network, which is more flexible and efficient.

2   Related Work

2.1   Position-Specific Scoring Matrix

MSA. A multiple sequence alignment (MSA) is a sequence alignment of multiple homologous protein sequences for the target protein [6]. See Fig. 1 for an example of MSA. MSA is an important step in comparative analyses and property predicting of biological sequences, since a lot of information *e.g.* evolution and co-evolution clusters, are displayed on the MSA and can be mapped to the target sequence of choice or on the protein structure [12]. Almost all existing approaches to studying proteins utilize MSAs indirectly, that is, they convert MSAs into a position-specific scoring matrix (PSSM) that represents the distribution of amino acid types on each column [13].

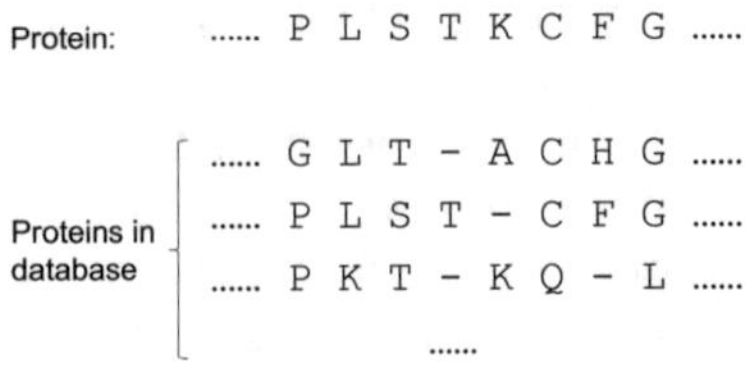

Protein: ...... P L S T K C F G ......

Proteins in   ...... G L T - A C H G ......
database   ...... P L S T - C F G ......
                  ...... P K T - K Q - L ......
                               ......

Fig. 1. An example of MSA.

PSSMs Calculation. PSSM scores are generally expressed as positive or negative integers. A positive score indicates that the frequency of substitutions in a given amino acid sequence is higher than expected, while a negative score indicates that the frequency of substitutions is lower than expected [14,15].

We extract the PSSM features of size $n \times 21$ based on Eqs. (1) and (2), where, n is the protein sequence length, 21 is the sum of twenty known amino acids appeared in the genetic code and one unknown amino acid marker. *Frequency* is the count of occurrences of residue j ($j = 1, 2, 3, \ldots, 21$) in column i ($i = 1, 2, 3, \ldots, n$), 20 represents the known amino acids. A simple procedure called pseudo-counts assigns minimal scores to residues which do not appear at a certain position of the alignment according to the following Eq. (1), where we set the *Pseudocount* equal to 1. N is the number of sequences in the multiple alignments. The *Background frequency* in Eq. (2) is the frequency of each residue appearing in the entire MSA of the protein.

$$score_{i,j} = \frac{Frequency + Pseudocount}{N + 20 Pseudocount} \tag{1}$$

$$PSSM_{i,j} = \log(score/Background frequency) \tag{2}$$

2.2 Scoring Criteria for PSSM

Count Score. The number of sequence homologs is recorded as the Count score. As we mentioned before, PSSM is a matrix calculated from the MSA, and the quality of the MSA directly determines the quality of the PSSM. We can use the number of homologous proteins of the MSA to evaluate the quality of the PSSM, which is represented as Count score. The larger Count score leads to more reliable PSSM. Thus, the Count score is one important criteria to evaluate the quality of the PSSM features.

Meff Score. We introduce the Meff score as the number of non-redundant sequence homologs. As in [7], homologous sequence in MSA of proteins have some redundancy, so we use Meff score as another criteria for PSSM to demonstrates the superiority and stability of our model under various evaluation standards.

The calculation formula of Meff score is shown in Eq. (3). where both i and j go over all the sequence homologs, $S_{i,j}$ is a binary number which describes the similarity of two proteins. We use the hamming distance to compute the similarity of two sequence homologs [2]: $S_{i,j}$ is 1 if the normalized hamming distance is less than 0.3; otherwise $S_{i,j}$ is set to 0.

$$Meff = \sum_i \frac{1}{\sum_j S_{i,j}} \qquad (3)$$

2.3 Protein Secondary Structure Prediction

The sequence space of proteins is vast, with perhaps 20 residues at each position, and evolution has been sampling it over billions of years. One of the most important sub-problems in protein studies is the secondary structure prediction. Protein secondary structure refers to the local conformation of the polypeptide backbone of proteins. There are two regular SS states: alpha-helix (H) and beta-strand (E), and one irregular SS type: coil region (C) [16]. The other way is a DSSP algorithm [17] to classify SS into 8 fine-grained states. In particular, the algorithm assigns 3 types for helix (G, H and I), 2 types for strand (E and B), and 3 types for coil (T, S and L). Overall, many computational methods have been developed to predict both 3-state secondary structure and a few to predict 8-state secondary structure. Meanwhile, since a chain of 8-state secondary structures contains more precise structural information for a variety of applications [23,35], the focus of secondary structure prediction has been shifted from 3-state secondary structure (Q3) prediction to the prediction of 8-state secondary structures (Q8). Because the Q8 problem is much more complicated than the Q3 problem, deep learning methods would be more suitable for addressing the Q8 problem.

3 Method

3.1 Framework Overview

Our method consists of two stages: enhancing PSSM and secondary structure prediction. The workflow of the inference phase is shown in Fig. 2. We input the low-quality PSSM into the trained unsupervised model with the protein sequence features to generate enhanced PSSM features. Then the enhanced PSSM features with sequence features are concatenated as the input of the inference phase for the prediction network. Finally, the results of the enhanced PSSM and the original PSSM on the prediction model are compared for evaluation.

3.2 Unsupervised Learning to Enhance PSSM

The architecture of our unsupervised learning method is shown in Fig. 3, which mainly contains four parts: Bagging MSA module, Local contexts feature encoding module, Long-distance interdependencies feature encoding module and Generation module. For each amino acid in a protein sequence, its input features are concatenated by its sequence features and PSSM features, which form a $2l$ ($l = 21$) dimensional vector. We denote the size of the entire input features as $N \times 2l$, and the size of the output from unsupervised learning network is $N \times l$, where N is the length of the protein sequence. The details regarding input features are explained in the experiments section.

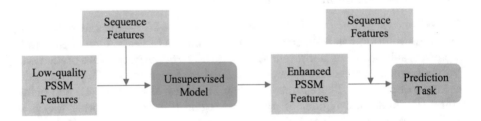

Fig. 2. Framework overview.

Bagging MSA. The main purpose of our enhancing PSSM module is to generate higher-quality PSSM features from low-quality PSSM features calculated from MSA with fewer rows or lower quality. Here we introduce the concept of 'Bagging MSA': As shown in Fig. 3, we randomly sample a small part of MSA for a protein and repeat this operation in each training iteration and for each protein. We bring in a hyper-parameter R to determine the proportion of selected homologous proteins in MSA randomly per training iteration, e.g. when R = [10%, 20%], a number greater than 10% and less than 20% would be randomly selected for each batch, and the homologous proteins in MSA would be randomly sampled according to this proportion. In this way, we are able to get many MSA bags, and each MSA bag would calculate a so-called 'Weak PSSM'. We used the weak PSSM calculated by these bags as a part of the input unsupervised data, and the original PSSM calculated by the complete MSA as the unsupervised labels. This module is ideal for unsupervised learning due to the size of the PSSM matrix is always the same for the same protein, even though the MSA size of each bag and label is different.

Local Contexts Feature Encoding Module. We introduce a fully convolutional architecture as the local contexts feature encoding module. Recently, CNN has been successfully used in the seq2seq model [21] and machine translation [22], as well as applied in several protein studies, which achieved remarkable successes [23,24]. This one-dimensional convolution operation is usually used to process sequence data, such as emotional analysis and sequence structure prediction [25,26], so CNN would be a good fit for our prediction task.

In our method, the local contexts feature encoding module exploits the One-dimensional convolution to extract the local hidden patterns and features of adjacent amino-acid residues from the input matrix. This module contains three 1-d convolutional layers with the ReLU activation function, and the window size is equal to three for each layer, as shown in Appendix A.1.

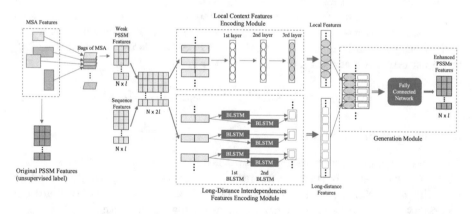

Fig. 3. Unsupervised learning model. (1) Bagging MSA module has two outputs: "Original PSSM" calculated by all MSA are used as the unsupervised labels; "Weak PSSM" calculated via the bags of MSA are fed into the two encoding networks. (2) The outputs of the two encoding networks are local features and long-distance features respectively. (3) The output of the generation module is the "Enhanced PSSM", which is used to calculate the loss from the "Original PSSM" to adjust the networks.

Long-Distance Interdependencies Feature Encoding Module. As we mentioned before, CNNs have the ability to capture local relationships of spatial or temporal structures, but we can not capture sufficient long-range sequence information by increasing the window size and network depth infinitely. However, long-distance interdependencies [27] of amino-acid residues are also critical for amino acid sequence information. Inspired by the success of some methods which use a combination of multiple neural networks, for example, coupling residual two-dimensional bidirectional long short-term memory with convolutional neural networks [28], ACLSTM [29] and CRRNNs [30], our method not only uses convolutional neural network with a few layers but also another network to catch Long-distance interdependencies feature.

RNN-based model has achieved remarkable results in sequence modeling; however, the gradient vector may grow or degrade exponentially over a long sequence during the training process. Thus LSTM neural networks are designed to avoid this problem by introducing the gate structures, which is good at capturing the long-range relations (from the first atom to the last one).

In our method, the long-distance interdependencies feature encoding module includes two stacked bidirectional LSTM neural networks. As shown in Appendix A.1, the input data are fed into the feature encoding model by its original order as well as the reverse order, and then the two outputs are concatenated together as the final features representation.

Generation Module. Our method has one fully connected hidden layer in the generation module. Moreover, in order to get the complete information of protein sequence, as shown in Fig. 3, we directly concatenate the outputs of the previous two modules and feed them into the fully connected (FC) layer with the ReLU activation function to generate the enhanced PSSMs. We use the MSE loss [20] to adjust our unsupervised network, as shown in Eq. 4.

$$Loss_{unsup} = MSE(PSSM_{Enhanced}, PSSM_{Full}) \tag{4}$$

3.3   Prediction Network

Since our unsupervised learning method is an independent enhancing PSSM network, we are able to use any deep learning network for the prediction module to verify the generalization of our method. In this study, we use two protein secondary structure prediction networks to evaluate our method: CNN-based network and LSTM-based network, which are two widely used deep learning prediction networks. For CNN-based method, we use five CNN layers [26], and fix the window size to 11 since the average length of an alpha-helix is around eleven residues [31] and that of a beta-strand is around six [32]. For LSTM-based method, we use two stacked bidirectional LSTM neural networks [33] and a fully connected (FC) layer.

The input data for the prediction network is the same as the input for the unsupervised learning model, which is the concatenation of sequence information and PSSM features calculated by the complete MSA of the protein. The protein secondary structure is used as the label. Based on the validation results, we select the best model as the secondary structure predictor, then feed the enhanced

PSSM features generated by our unsupervised network and the original PSSM into the predictor respectively. Last, the prediction performances of the two PSSM features are compared to evaluate the effectiveness of our enhanced PSSM model.

4   Experiments

4.1   Experiments Set up

Dataset. We use four publicly available datasets: CullPDB [34] of 5926 proteins, CB513 [35] of 513 proteins, CASP11 of 85 proteins, and CASP12 of 40 proteins. CASP11 and CASP12 datasets are downloaded from the official CASP website [36]. 53 duplicated proteins observed in the CullPDB are removed and 591 proteins are randomly sampled for validation, then the remaining proteins are used for training. The other three datasets are used as the test dataset. We generate the position specific scoring matrix (PSSM) by searching the Uniref50 [37] database. And the labels used for the prediction network are 8-state protein secondary structures which are generated by DSSP [17,41].

Input Features. The input features for the encoding networks of our method are described in [35]. We extract the MSA from Uniref50 databases using Jackhmmer [10], and set the parameters refer to their guide [38], details are listed in Appendix A.3. We randomly sample 10% to 20% (R = [10%, 20%]) of the MSA for each protein within each learning iteration (Bagging MSA), and then we calculate PSSM using Eqs. (1) and (2). We transform those PSSMs by the Sigmoid function $1/(1 + \exp(-x))$ where x is a PSSM entry to map each PSSM value in between 0 and 1. As shown in Fig. 3, the input features of the two encoding modules is a $N \times 2l$ matrix, where N is the length of the input sequence and $2l$ is the dimension of the concatenated vectors. In our method, the sequence feature vectors are sparse one-hot vectors of 21 elements ($l = 21$) since there might be some unknown amino acids in a protein sequence. Therefore, there are 42 input features in total for each residue, 21 from PSSM features and the other 21 from sequence feature.

For the prediction part, there are 42 input features for each residue too, 21 of them are from PSSM features and the others are from sequence feature. We compare the testing results of the enhanced input features with the original input features to evaluate the effectiveness of our unsupervised model.

Neural Network Structure and Learning Hyper-parameters. The framework of our unsupervised learning method is very flexible in the network structure selection.

In the long-distance interdependencies feature encoding module, we can set different hidden layers and hidden dimensions (with different layers and layer hidden sizes). Moreover, different types of network can be chosen in addition to the bi-LSTM network, such as LSTM [39]. Due to the space limitation of this paper, two stacked bi-LSTM with 512 hidden units are used for all experiments. Then, we use 1d-CNN of 3 hidden layers, and 100 neurons for each layer in the local contexts feature encoding module. The window size at each layer is set to 3.

For optimization, we use multi-step LR (learning rate) descent with [30, 100, 200] for epoch indices. The multiplicative factor of learning rate decay is 0.1. We use Adam [40] as the optimizer of our method. The initial learning rate for all training models is 0.0001.

For the protein secondary structure prediction task, we have two kinds of networks. For CNN network, we use five 1-dim CNN layers with window size 11, and neurons size 100 for each layer. For LSTM network, we use two stacked bi-LSTM with 512 hidden units and one fully connected (FC) layer.

Evaluation Metric. For the unsupervised learning, we calculate the RMSE of the Enhanced PSSM and the Original PSSM in the input feature as the evaluation matrix. Q8 accuracy is the criterion of the prediction module.

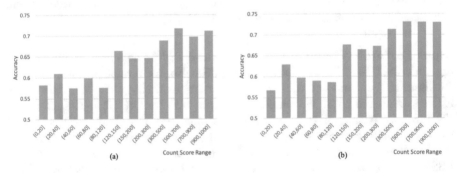

Fig. 4. The average accuracy of proteins within Count score ranges (a) CNN-based prediction model; (b) LSTM-based prediction model.

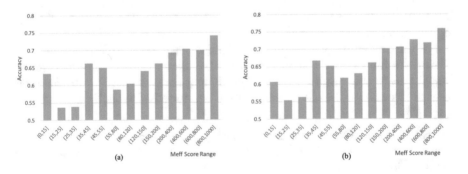

Fig. 5. The average accuracy of proteins within Meff score ranges (a) CNN-based prediction model; (b) LSTM-based prediction model.

4.2   Results

Relationship Between PSSM Quality and Performance. As we mentioned before, we use two methods to score the quality of the protein PSSM, higher score represents better quality. Figures 4 and 5 show the relationship between the quality of PSSM and the corresponding performance on the prediction networks on CB513 dataset. Figure 4 shows the average accuracy obtained by using Count score as the evaluation standard on the prediction network of CNN and LSTM respectively, and Fig. 5 for the Meff score. We can find that proteins with high-quality PSSM performs better than proteins with low-quality PSSM both CNN-based and LSTM-based prediction network, as well as under all evaluations including Count score or Meff score. Tables 1 and 2 show the data distribution within the ranges Count and Meff Scores. Thus, our method aims at improving the prediction performance for those proteins with original low-quality PSSM by enhancing their PSSM features. See the gray-scale images in Appendix A.2, which show the difference between "before" and "after" PSSM enhancement.

Table 1. Number of proteins in certain Count score ranges.

range	(0,20]	(20,40]	(40,60]	(60,80]	(80,120]	(120,150]	(150,200]	(200,300]	(300,500]	(500,700]	(700,900]	(900,1000]
num	2	16	18	19	29	11	23	27	45	26	26	271

Table 2. Number of proteins in certain Meff score ranges.

range	(0,15]	(15,25]	(25,35]	(35,45]	(45,55]	(55,80]	(80,120]	(120,150]	(150,200]	(200,400]	(400,600]	(600,800]	(800,1000]
num	12	23	18	9	16	18	19	15	23	68	89	89	114

Enhancement on Low-Quality PSSM Protein. Our method is used to enhance the performance of proteins with low-quality PSSM in secondary structure prediction task. However, while improving the low-quality PSSM, noise might have been added to the high-quality PSSM, which would end up with a lower accuracy score. Therefore, we need to find a standard to determine the definition of low-quality proteins for our method, which would be the thresholds of the Count score and the Meff score. As shown in Fig. 6, our method increase or decrease the accuracy of prediction tasks under certain ranges. Greater than 0 means that the average accuracy of our method has improved under the threshold, while less than 0 means that it has decreased. Based on the accuracy results, we are able to find a consistent trend for both CNN-based and LSTM-based models: our method shows significant superiority for proteins with a Count score less than 60 and a Meff score less than 35.

In addition, in order to verify the threshold we selected is suitable for other datasets, we also report the results of casp11 and casp12, which are shown

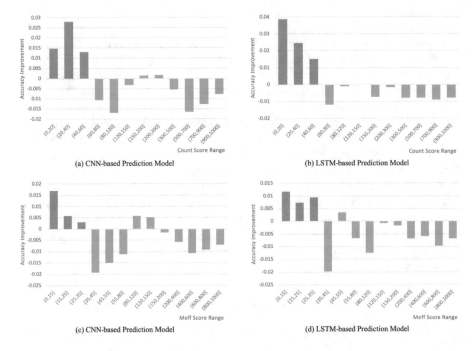

(a) CNN-based Prediction Model

(b) LSTM-based Prediction Model

(c) CNN-based Prediction Model

(d) LSTM-based Prediction Model

Fig. 6. Our method has achieved significant improvement in all prediction tasks (CNN-based and LSTM-based) when the Count score is less than 60 (a, b), and the Meff score is less than 35 (c, d). These figures are the results on CB513 dataset.

Table 3. Comparison results (Q8 accuracy) of our enhanced PSSM vs. original PSSM. Enhancement experiments are conducted for low-quality proteins (Count score $\leq$ 60, Meff score $\leq$ 35) obtained from CB513, CASP11, and CASP12 datasets. Prediction experiments are conducted on CNN-based model and LSTM-based model.

Prediction model	Score range	Datasets	Original PSSM	Enhanced PSSM	Protein num
CNN-based	Count $\leq$ 60	CB513	59.106%	61.093%	36
		CASP11	64.196%	67.781%	12
		CASP12	53.300%	56.519%	3
	Meff $\leq$ 35	CB513	55.973%	56.717%	53
		CASP11	62.846%	65.732%	17
		CASP12	52.353%	54.462%	7
LSTM-based	Count $\leq$ 60	CB513	60.982%	63.041%	36
		CASP11	64.037%	64.990%	12
		CASP12	54.335%	55.865%	3
	Meff $\leq$ 35	CB513	56.929%	57.831%	53
		CASP11	63.216%	63.504%	17
		CASP12	51.493%	53.921%	7

in Table 3. The performances of extensive experiments demonstrate that our method has a significant effect on enhancing low-quality PSSM for different datasets.

5   Conclusion

We propose an innovative Bagging MSA model to enhance low-quality PSSM features of proteins, which would help promote their performance in secondary structure prediction task. We employ an unsupervised learning network to enhance the PSSM features, and two conventional deep learning prediction models as the protein secondary structure prediction networks to prove the effectiveness of our method on various datasets. Our method is the first attempt to enhance PSSM features in the field of protein research. Moreover, the generalization of our Bagging MSA makes it suitable for numerous PSSM related protein prediction tasks. PSSM features are essential for studying proteins, our method pioneer another way to address the prediction limitation for low-quality proteins.

A   Appendix

A.1   Encoding Networks

As shown in Figs. 7 and 8, we use 1d-CNN of 3 hidden layers, and 100 neurons for each layer in the local contexts feature encoding module. The window size at each layer is set to 3. And for long-distance module, two stacked bi-LSTM with 512 hidden units are used for all experiments.

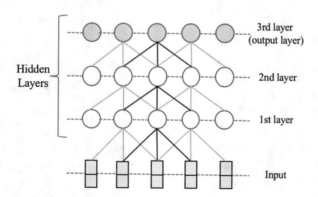

Fig. 7. Local contexts feature encoding module includes three layers of 1d-CNN and the top layer (3rd layer) is the output layer.

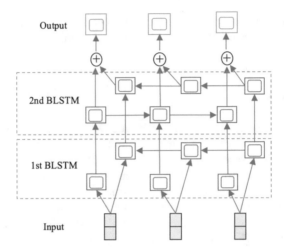

Fig. 8. Long-distance interdependencies feature encoding module includes two stacked BLSTM neural networks.

A.2   Gray-Scale Images of PSSM

As shown in Fig. 9, which is a set of gray-scale images of the original pssm (a) and enhanced pssm (b) of a protein from cb513 dataset. Where, y-axis is the length N of the protein sequence, the sample protein contains 26 residues ($N = 26$), x-axis is l, 20 plus an unknown amino acids marker ($l = 21$). Lighter colors indicate

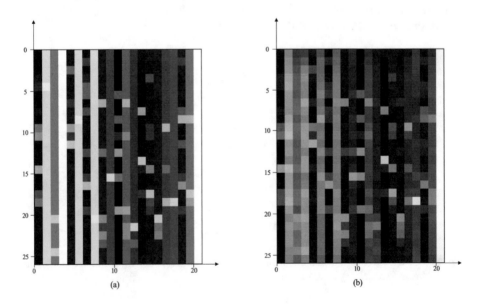

Fig. 9. Gray-scale images of the PSSMs. (a) Original PSSM of 6O4M protein; (b) Enhanced PSSM of 6O4M protein.

larger values, while darker colors indicate smaller values. See https://www.rcsb.org for the structure information of the protein (6O4M) in the example.

A.3   Jackhmmer Options for Extracting MSA

In the per-target output, report target profiles with an E-value ≤ 1.0; In the per-domain output, for target profiles that have already satisfied the per-profile reporting threshold, report individual domains with a conditional E-value of ≤ 1.0; Use a conditional E-value of ≤ 0.03 as the per-domain inclusion threshold, in targets that have already satisfied the overall per-target inclusion threshold; Obtain residue alignment probabilities from the built-in substitution matrix named BLOSUM62.

A.4   Infrastructure and Software

Our model was implemented through Pytorch package. And our models was trained in a self-hosted 16-GPU cluster platform with Intel i7 6700K @ 4.00 GHz CPU, 64 Gigabytes RAM and four Nvidia GTX 1080Ti GPUs on each workstation.

References

1. Heffernan, R., et al.: Improving prediction of secondary structure, local backbone angles, and solvent accessible surface area of proteins by iterative deep learning. Sci. Rep. **5** (2015). Article number: 11476
2. Morcos, F., et al.: Direct-coupling analysis of residue coevolution captures native contacts across many protein families. Proc. Natl. Acad. Sci. **108**(49), E1293–E1301 (2011)
3. Stormo, G.D., et al.: Use of the 'Perceptron' algorithm to distinguish translational initiation sites in E. coli. Nucleic Acids Res. **10**(9), 2997–3011 (1982)
4. Jones, D.T.: Protein secondary structure prediction based on position-specific scoring matrices. J. Mol. Biol. **292**(2), 195–202 (1999)
5. Gao, Y., et al.: RaptorX-Angle: real-value prediction of protein backbone dihedral angles through a hybrid method of clustering and deep learning. BMC Bioinform. **19**(4) (2018). Article number: 100. https://doi.org/10.1186/s12859-018-2065-x
6. Wang, L., Jiang, T.: On the complexity of multiple sequence alignment. J. Comput. Biol. **1**(4), 337–348 (1994)
7. Wang, Z., Jinbo, X.: Predicting protein contact map using evolutionary and physical constraints by integer programming. Bioinformatics **29**(13), i266–i273 (2013)
8. Teichert, F., et al.: High quality protein sequence alignment by combining structural profile prediction and profile alignment using SABERTOOTH. BMC Bioinform. **11**(1) (2010). Article number: 251
9. Remmert, M., et al.: HHblits: lightning-fast iterative protein sequence searching by HMM-HMM alignment. Nat. Methods **9**(2), 173–175 (2012)
10. Wheeler, T.J., Eddy, S.R.: nhmmer: DNA homology search with profile HMMs. Bioinformatics **29**(19), 2487–2489 (2013)
11. O'Driscoll, A., et al.: HBLAST: Parallelised sequence similarity-A Hadoop MapReducable basic local alignment search tool. J. Biomed. Inform. **54**, 58–64 (2015)

12. Oteri, F., et al.: BIS2Analyzer: a server for co-evolution analysis of conserved protein families. Nucleic Acids Res. **45**(W1), W307–W314 (2017)
13. Ju, F., et al.: Seq-SetNet: exploring sequence sets for inferring structures. arXiv preprint arXiv:1906.11196 (2019)
14. Ye, X., Wang, G., Altschul, S.F.: An assessment of substitution scores for protein profile-profile comparison. Bioinformatics **27**(24), 3356–3363 (2011)
15. Altschul, S.F., et al.: Gapped BLAST and PSI-BLAST: a new generation of protein database search programs. Nucleic Acids Res. **25**(17), 3389–3402 (1997)
16. Pauling, L., Corey, R.B., Branson, H.R.: The structure of proteins: two hydrogen-bonded helical configurations of the polypeptide chain. Proc. Nat. Acad. Sci. **37**(4), 205–211 (1951)
17. Kabsch, W., Sander, C.: Dictionary of protein secondary structure: pattern recognition of hydrogen-bonded and geometrical features. Biopolym. Orig. Res. Biomol. **22**(12), 2577–2637 (1983)
18. Suzek, B.E., et al.: UniRef: comprehensive and non-redundant UniProt reference clusters. Bioinformatics **23**(10), 1282–1288 (2007)
19. Mirdita, M., et al.: Uniclust databases of clustered and deeply annotated protein sequences and alignments. Nucleic Acids Res. **45**(D1), D170–D176 (2016)
20. Allen, D.M.: Mean square error of prediction as a criterion for selecting variables. Technometrics **13**(3), 469–475 (1971)
21. Gehring, J., et al.: Convolutional sequence to sequence learning. In: Proceedings of the 34th International Conference on Machine Learning, vol. 70. JMLR.org (2017)
22. Gehring, J., et al.: A convolutional encoder model for neural machine translation. arXiv preprint arXiv:1611.02344 (2016)
23. Wang, Z., et al.: Protein 8-class secondary structure prediction using conditional neural fields. In: 2010 IEEE International Conference on Bioinformatics and Biomedicine (BIBM). IEEE (2010)
24. Zhou, J., et al.: CNNH_PSS: protein 8-class secondary structure prediction by convolutional neural network with highway. BMC Bioinform. **19**(4) (2018). Article number: 60
25. Dos Santos, C., Gatti, M.: Deep convolutional neural networks for sentiment analysis of short texts. In: Proceedings of COLING 2014, the 25th International Conference on Computational Linguistics: Technical Papers (2014)
26. Wang, S., et al.: Protein secondary structure prediction using deep convolutional neural fields. Sci. Rep. **6** (2016). Article number: 18962
27. Heffernan, R., et al.: Capturing non-local interactions by long short-term memory bidirectional recurrent neural networks for improving prediction of protein secondary structure, backbone angles, contact numbers and solvent accessibility. Bioinformatics **33**(18), 2842–2849 (2017)
28. Hanson, J., et al.: Accurate prediction of protein contact maps by coupling residual two-dimensional bidirectional long short-term memory with convolutional neural networks. Bioinformatics **34**(23), 4039–4045 (2018)
29. Guo, Y., et al.: DeepACLSTM: deep asymmetric convolutional long short-term memory neural models for protein secondary structure prediction. BMC Bioinform. **20**(1) (2019). Article number: 341
30. Zhang, B., Li, J., Lü, Q.: Prediction of 8-state protein secondary structures by a novel deep learning architecture. BMC Bioinform. **19**(1) (2018). Article number: 293. https://doi.org/10.1186/s12859-018-2280-5
31. Andersen, C.A., Bohr, H., Brunak, S.: Protein secondary structure: category assignment and predictability. FEBS Lett. **507**(1), 6–10 (2001)

32. Penel, S., et al.: Length preferences and periodicity in β-strands. Antiparallel edge β-sheets are more likely to finish in non-hydrogen bonded rings. Protein Eng. **16**(12), 957–961 (2003)

33. Sønderby, S.K., Winther, O.: Protein secondary structure prediction with long short term memory networks. arXiv preprint arXiv:1412.7828 (2014)

34. Wang, G., Dunbrack Jr., R.L.: PISCES: a protein sequence culling server. Bioinformatics **19**(12), 1589–1591 (2003)

35. Zhou, J., Troyanskaya, O.G.: Deep supervised and convolutional generative stochastic network for protein secondary structure prediction. arXiv preprint arXiv:1403.1347 (2014)

36. Official CASP website. http://predictioncenter.org

37. Bairoch, A., et al.: The universal protein resource (UniProt). Nucleic Acids Res. **33**(suppl–1), D154–D159 (2005)

38. Eddy, S.: HMMER user's guide, vol. 2, no. 1, p. 13. Department of Genetics, Washington University School of Medicine (1992)

39. Hochreiter, S., Schmidhuber, J.: Long short-term memory. Neural Comput. **9**(8), 1735–1780 (1997)

40. Kingma, D.P., Ba, J.: Adam: a method for stochastic optimization. arXiv preprint arXiv:1412.6980 (2014)

41. Touw, W.G., et al.: A series of PDB-related databanks for everyday needs. Nucleic Acids Res. **43**(D1), D364–D368 (2014)

ASTARIX: Fast and Optimal Sequence-to-Graph Alignment

Pesho Ivanov$^{(\boxtimes)}$ ⓘ, Benjamin Bichsel ⓘ, Harun Mustafa ⓘ, André Kahles ⓘ,
Gunnar Rätsch ⓘ, and Martin Vechev ⓘ

Department of Computer Science, ETH Zurich, Zurich, Switzerland
`{firstname.lastname}@inf.ethz.ch`

Abstract. We present an algorithm for the *optimal alignment* of sequences to *genome graphs*. It works by phrasing the edit distance minimization task as finding a shortest path on an implicit alignment graph. To find a shortest path, we instantiate the A* paradigm with a novel domain-specific heuristic function that accounts for the upcoming subsequence in the query to be aligned, resulting in a provably optimal alignment algorithm called ASTARIX.

Experimental evaluation of ASTARIX shows that it is 1–2 orders of magnitude faster than state-of-the-art optimal algorithms on the task of aligning Illumina reads to reference genome graphs. Implementations and evaluations are available at https://github.com/eth-sri/astarix.

Keywords: Next-generation sequencing · Optimal alignment ·
Genome graph · Shortest path · A* algorithm

1 Introduction

The analysis and understanding of genetic variation encoded in the genome of an organism lies at the center of computational biology and medicine. Variation is usually identified through matching sequences obtained from DNA/RNA-sequencing back to a reference (genome) sequence in the process of *variant calling*, making the alignment task a core problem in sequence bioinformatics.

Historically, a single linear reference sequence has been used to represent the most common variants in a population. While providing a working abstraction for most cases, rare or sub-population specific variation is especially hard to model in this setting, creating a reference allele bias [4,35]. Consequently, in the last few years, the field has shifted first towards using sets of reference sequences, and more recently to graph data structures (so-called *genome graphs*), to represent many genomes or haplotypes simultaneously [7,9,25].

Both for sequence-to-sequence alignment and sequence-to-graph alignment, heuristics are employed to keep alignment tractable [2,9,21], especially for large populations of human-sized genomes. While such heuristics find the correct alignment for simple references, they often perform poorly in regions of very high complexity, such as in the human major histocompatibility complex (MHC) [7],

ⓒ The Author(s) 2020
R. Schwartz (Ed.): RECOMB 2020, LNBI 12074, pp. 104–119, 2020.
https://doi.org/10.1007/978-3-030-45257-5_7

in complex but rare genotypes arising from somatic-subclones in tumor sequencing data [10], or in the presence of frequent sequencing errors [29]. Importantly, these cases can be of specific clinical or biological interest, and incorrect alignment can cause severe biases for downstream analyses. For instance, the combination of high variability of MHC sequences in humans and small differences between alleles [5] leads to a risk of misclassifications due to suboptimal alignment. Guaranteeing optimal alignment against all variations represented in a graph is a major step towards alleviating those biases.

Formally, we consider the optimal *sequence-to-graph alignment* problem, the task of finding an optimal base-to-base correspondence between a query sequence and a (possibly cyclic) walk in the graph. Related alignment problems have already been formulated as graph shortest path problems [3, 16].

1.1   Related Work

Seed-and-Extend. Since optimal alignment is often intractable, many aligners use heuristics, most commonly the *seed-and-extend* paradigm [2,21,22]. In this approach, alignment initiation sites (*seeds*) are determined, which are then *extended* to form the *alignments* of the query sequence. The fundamental issue with this approach, however, is that the seeding and extension phases are mostly decoupled during alignment. Thus, an algorithm with a provably optimal extension phase may not result in optimal alignments due to the selection of a suboptimal seed in the first phase. In cases of high sequence variability, the seeding phase may even fail to find an appropriate seed from which to extend.

Accounting for Variation. First attempts to include variation into the reference data structure were made by augmenting the local alignment method to consider alternative walks during the extend step [17,30]. This approach has since been extended from the linear reference case to graph references. To represent non-reference variation of multiple references during the seeding stage, HISAT2 uses generalized compressed suffix arrays [33] to index walks in an augmented reference sequence, forming a local genome graph [19]. VG [9] uses a similar technique [32] to index variation graphs representing a population of references.

BrownieAligner, another recent work developed for local alignment of sequences to *de Bruijn* graph representations of genomic variation, features an optimal extension phase using a branch-and-bound-based early cutoff, while employing a heuristic maximal-exact-match approach for seeding [11].

Optimal Alignment. Current optimal sequence-to-graph alignment algorithms reach their worst-case $\mathcal{O}(nm)$ runtime [16]. In this light, approaches for improving the efficiency of optimal alignment have taken advantage of specialized features of modern CPUs to improve the practical runtime of the Smith-Waterman dynamic programming (DP) algorithm [34] considering all possible starting nodes. These use modern SIMD instructions (e.g. VG [9] and PASGAL [15]) or reformulations of edit distance computation to allow for bit-parallel computations in GRAPHALIGNER[1] [27]. Many of these, however, are designed only for

---

[1] We refer as BITPARALLEL to the bit-parallel DP algorithm implemented in GRAPHALIGNER tool [27].

specific types of genome graphs, such as *de Bruijn* graphs [11,23,24] and variation graphs [9]. A compromise often made when aligning sequences to cyclic graphs using algorithms reliant on directed acyclic graphs involves the computationally expensive "DAG-ification" of graph regions [9,18].

A⋆ Algorithm. We aim to guarantee optimal alignment while optimizing the average runtime to not reach its worst case complexity. While DIJKSTRA is an algorithm that explores graph nodes in the order of their distance from the start, A⋆ is a generalization of DIJKSTRA that also accounts for their distance from the target. A⋆ prioritizes the exploration of nodes that seem to be closer to the target nodes. This way, A⋆ can sometimes dramatically improve on the performance of DIJKSTRA while remaining optimal.

There has been one attempt to apply A⋆ for optimal alignment [8] which uses a heuristic function that accounts only for the length of the remaining query sequence to be aligned. However, it does not significantly outperform DIJKSTRA (in fact, it is equivalent for a zero matching cost). In contrast, the heuristic function we introduce is more informative and consistently outperforms DIJKSTRA.

1.2   Main Contributions

We introduce a novel approach, called ASTARIX, for optimal sequence-to-graph alignment based on A⋆. As with any A⋆ instantiation, the core difficulty lies in developing an accurate domain-specific heuristic which is fast to compute. We design a heuristic that accounts for the content of the upcoming query letters to be aligned, which more effectively guides the search. Our proposed heuristic has two advantages: (i) it is correctness-preserving, that is, it preserves the fact that ASTARIX finds the best alignment, yet (ii) it is practically effective in that the algorithm performs a near-optimal number of steps. Overall, this heuristic enables ASTARIX to compute the best alignment while also scaling to larger reference graph sizes when compared to existing state-of-the-art optimal aligners.

Our main contributions[2] include:

1. **ASTARIX.** An algorithm for optimal sequence-to-graph alignment based on a novel instantiation of A⋆ with an accurate domain-specific heuristic that accounts for the upcoming query letters to be aligned (Sect. 3).
2. **Algorithmic optimizations.** To ensure that ASTARIX is practical, we introduce a number of algorithmic optimizations which increase performance and decrease memory footprint (Sect. 4). We also prove that all optimizations are correctness-preserving.
3. **Thorough experimental evaluation of ASTARIX.** We demonstrate that ASTARIX is up to 2 orders of magnitude faster than other optimal aligners on various reference graphs (Sect. 5).

---

[2] The appendix with algorithms and evaluation details is included in the full version of this paper: https://www.biorxiv.org/content/10.1101/2020.01.22.915496v1.

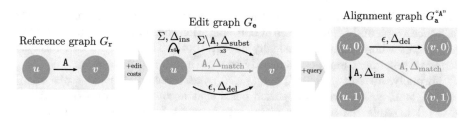

Fig. 1. Starting from the reference graph (left), we can construct the edit graph (middle) and the alignment graph G_a^q for query $q =$ "A" (right). Edges are annotated with labels and/or costs, where sets of labels represent multiple edges, one for each letter in the set (indicated by "x3" and "x4").

2 Task Description: Alignment to Reference Graphs

We now describe the task of aligning a query to a reference graph. To this end, we (i) introduce the task of optimal alignment on a *reference graph*, (ii) formalize this task in terms of an *edit graph*, and (iii) introduce an alternative formulation in terms of an *alignment graph*, which is the basis for shortest path formulations of the optimal alignment. Figure 1 summarizes these different graph types.

Reference Graph. We encode the collection of references to which we want to align in a reference graph, which captures genomic variation that a linear reference cannot express [9, 25]. We formalize a reference graph as a tuple $G_r = (V_r, E_r)$ of nodes V_r and directed, labeled edges $E_r \subseteq V_r \times V_r \times \Sigma$, where the alphabet $\Sigma = \{A, C, G, T\}$ represents the four different nucleotides. Note that in contrast to sequence graphs [28], we label edges instead of nodes.

Path, Spelling. Any path $\pi = (e_1, \ldots, e_k)$ in G_r induces a *spelling* $\sigma(\pi) \in \Sigma^*$ defined by $\sigma(e_1) \cdots \sigma(e_k)$, where $\sigma(e_i)$ is the label of edge e_i and $\Sigma^* := \bigcup_{k \in \mathbb{N}} \Sigma^k$. We note that our approach naturally handles cyclic walks and does not require cycle unrolling, a feature shared with BITPARALLEL [27] and BROWN-IEALIGNER [11] but missing from VG [9], PASGAL [15] and V-ALIGN [18].

Alignment on Reference Graph. An *alignment* of *query* $q \in \Sigma^*$ to a reference graph $G_r = (V_r, E_r)$ consists of (i) a path π in G_r and (ii) a sequence of edit operations (matches, substitutions, insertions, deletions) transforming $\sigma(\pi)$ to q.

Optimal Alignment, Edit Distance. Each edit operation is associated with a real-valued cost (Δ_{match}, Δ_{subst}, Δ_{ins}, and Δ_{del}, respectively). An optimal alignment minimizes the total cost of the edit operations converting $\sigma(\pi)$ to q. For optimal alignments, this total cost is equal to the edit distance between $\sigma(\pi)$ and q, i.e., the cheapest sequence of edit operations transforming $\sigma(\pi)$ into q.

We make the (standard) assumption that $0 \leq \Delta_{\text{match}} \leq \Delta_{\text{subst}}, \Delta_{\text{ins}}, \Delta_{\text{del}}$, which will be a prerequisite for the correctness of our approach.

Edit Graph. Instead of representing alignments as pairs of (i) paths in the reference graph and (ii) sequences of edit operations on these paths, we introduce *edit graphs* whose paths intrinsically capture both. This way, we can formally define an alignment more conveniently as a path in an edit graph.

Formally, an *edit graph* $G_e := (V_e, E_e)$ has directed, labeled edges $E_e \subseteq V_e \times V_e \times \Sigma_\epsilon \times \mathbb{R}_{\geq 0}$ with associated costs that account for edits. Here, $\Sigma_\epsilon := \Sigma \cup \{\epsilon\}$ extends the alphabet Σ by ϵ to account for deleted characters (see Fig. 1). The edit and reference graphs consist of the same vertices, i.e., $V_e = V_r$. However, E_e contains more edges than E_r to account for edits. Concretely, for each edge $(u, v, \ell) \in E_r$, E_e contains edges to account for (i) matches, by an edge $(u, v, \ell, \Delta_{\text{match}})$, (ii) substitutions, by edges $(u, v, \ell', \Delta_{\text{subst}})$ for each $\ell' \in \Sigma \backslash \ell$, (iii) deletions, by an edge $(u, v, \epsilon, \Delta_{\text{del}})$, and (iv) insertions, by edges $(u, u, \ell', \Delta_{\text{ins}})$ for each $\ell' \in \Sigma$. The spelling $\sigma(\pi) \in \Sigma^*$ of a path $\pi \in G_e$ is defined analogously to reference graphs, except that deleted letters (represented by ϵ) are ignored. The cost $\text{cost}(\pi)$ of a path $\pi \in G_e$ is the sum of all its edge costs.

Alignment on Edit Graph. An *alignment* of query q to G_r is a path π in G_e spelling q, i.e., $q = \sigma(\pi)$. An *optimal alignment* is an alignment of minimal cost.

Alignment Graph. To find an optimal alignment of q to the edit graph G_e using shortest path finding algorithms, we must ensure that only paths spelling q are considered. To this end, we introduce an alternative but equivalent formulation of alignments in terms of an *alignment graph* $G_a^q = (V_a^q, E_a^q)$.

Here, each *state* $\langle v, i \rangle \in V_a^q$ consists of a vertex $v \in V_e$ and a query position $i \in \{0, \ldots, |q|\}$ (equivalent to [28]). Traversing a state $\langle v, i \rangle \in V_a^q$ represents the alignment of the first i query characters ending at node v. In particular, query position $i = 0$ indicates that we have not yet matched any letters from the query. We note that the alignment graph explicitly depends on the query q. In particular, the example alignment graph $G_a^{\text{"A"}}$ in Fig. 1 lacks substitution edges from G_e, as their labels (C, G, T) do not match the query $q = $ "A".

We construct the alignment graph G_a^q to guarantee that any walk from a source $\langle u, 0 \rangle$ to a state $\langle v, i \rangle$ corresponds to an alignment of the first i letters of query q to G_r. As a consequence, there is a one-to-one correspondence between alignments π_e of q to G_e and paths $\pi_a^q \in G_a^q$ from sources $S := V_r \times \{0\}$ to targets $T := V_r \times \{|q|\}$, with $\text{cost}(\pi_e) = \text{cost}(\pi_a^q)$. To find the best alignment in G_e, only paths in G_a^q (walks without repeating nodes) can be considered, since repeating a node in G_a^q cannot lead to a lower cost ($\Delta_{\text{del}} \geq 0$) for the same state.

The edges $E_a^q \subseteq V_a^q \times V_a^q \times \Sigma_\epsilon \times \mathbb{R}_{\geq 0}$ are built based on the edges in E_e, except that the former (i) keep track of the position in the query i, and (ii) only contain empty edges or edges whose label matches the next query letter:

$$(u, v, \ell, w) \in E_e \implies (\langle u, i \rangle, \langle v, i+1 \rangle, \ell, w) \in E_a^q \quad \text{for } 0 \leq i < |q| \text{ with } q[i] = \ell(1)$$
$$(u, v, \epsilon, w) \in E_e \implies (\langle u, i \rangle, \langle v, i \quad \rangle, \epsilon, w) \in E_a^q \quad \text{for } 0 \leq i < |q| \quad (2)$$

Here, assuming 0-indexing, $q[i]$ is the next letter to be matched after matching i letters. Then, Eq. (1) represents matches, substitutions, and insertions (which advance the position in the query by 1), while Eq. (2) represents deletions (which do not advance the position in the query).

Algorithm 1. ASTARIX including heuristic function.

1: G_r: Reference graph   ▷ Global variables
2: d: Upcoming sequence length

3: **function** ASTARIX(q: Query)
4:      $G_a^q \leftarrow$ DEFINEALIGNMENTGRAPH(G_r, q)   ▷ Following Sect. 2
5:      $S \leftarrow \{\langle v, i \rangle \in V_a^q \mid i = 0\}$ ▷ Sources: no letter matched
6:      $T \leftarrow \{\langle v, i \rangle \in V_a^q \mid i = |q|\}$ ▷ Targets: all letters matched
7:      **return** A*(G_a^q, S, T, HEURISTIC)   ▷ A* provided in App. A.1

8: **function** HEURISTIC($\langle u, i \rangle$: State)   ▷ Heuristic: Cost of upcoming sequence
9:      $d' \leftarrow \min(d, |q| - i)$ ▷ Actual length of upcoming sequence
10:     $s \leftarrow q[i : i + d']$ ▷ Upcoming sequence (next d letters after current)
11:     **return** $h(u, s)$ ▷ Cost of aligning s to G_e starting from u

12: **function** $h(u, s)$ ▷ Cost of aligning s starting from u
13:     **return** RECURSIVEALIGN($u, s, 0.0, \infty$)   ▷ Simple branch-and-bound

Dynamic Construction. As the size of the alignment graph is $\mathcal{O}(|G_r| \cdot |q|)$, it is expensive to build it fully for every new query. Therefore, our implementation constructs the alignment graph G_a^q on-the-fly: the outgoing edges of a node are only generated on demand and are freed from memory after alignment.

3   ASTARIX: Finding Optimal Alignments Using A*

In this section, we first introduce the general A* algorithm for finding shortest paths, and the notion of an optimistic heuristic, a sufficient condition for instantiations of A* to be correct (i.e., to indeed find shortest paths). Then we instantiate A* with our domain-specific heuristic that accounts for upcoming subsequences to be aligned, and prove that this heuristic is optimistic.

3.1   Background: General A* Algorithm

Given a weighted graph $G = (V, E)$ with $E \subseteq V \times V \times \mathbb{R}_{\geq 0}$, the A* algorithm (abbreviated as A*) searches for the shortest path from sources $S \subseteq V$ to targets $T \subseteq V$. It is an extension of Dijkstra's algorithm that additionally leverages a *heuristic function* $h: V \to \mathbb{R}_{\geq 0}$ to decide which paths to explore first. If $h(u) \equiv 0$, A* is equivalent to Dijkstra's algorithm. We provide an implementation of A* and Dijkstra in App. A.1, but do not assume knowledge of either algorithm in the following. At a high level, A* maintains the set of all *explored* states, initialized with the set of sources S. Then, A* iteratively *expands* the explored state with lowest estimated cost by exploring all its neighbors, until it finds a target. Here, the cost for node u is estimated by the distance from source, called $g(u)$, plus the estimate from the heuristic $h(u)$.

Heuristic Function. The heuristic function $h(u)$ estimates the cost $h^*(u)$ of a shortest path in G from u to a target $t \in T$. Intuitively, a good heuristic correlates well with the distance from u to t.

To ensure that A* indeed finds the shortest path, h should be *optimistic*:

Definition 1 (Optimistic heuristic). *A heuristic h is optimistic if it provides a lower bound on the distance to the closest target: $\forall u.h(u) \leq h^*(u)$.*

While any optimistic h ensures that A* finds optimal alignments [6, Res. 3], the specific choice of h is critical for performance. In particular, decreasing the error $\delta(u) = h^*(u) - h(u)$ can only improve the performance of A* [6, Res. 6]. Thus, a key contribution of ours is a domain-specific heuristic h.

3.2   ASTARIX: Instantiating A*

Algorithm 1 shows an unoptimized version of ASTARIX and its heuristic function. ASTARIX expects a reference graph (Line 1) and a query (Line 3) as input, and returns an optimal alignment (Line 7) by searching for a shortest path from S to T in the alignment graph G_a^q. It is parameterized by hyper-parameters (d in Line 2, more in Sect. 4) and edit costs (implicitly provided).

The function HEURISTIC (Lines 8–11) computes a lower bound on the remaining cost of a best alignment: the minimum cost $h(u, s)$ of aligning the *upcoming sequence* s (where $|s| \leq d$) starting from node u. Importantly, s is limited to the next $d' \leq d$ letters of q, starting from query position i. Thus, computing $h(u, s)$ is substantially cheaper than aligning all remaining letters of q.

To compute $h(u, s)$ we leverage a simple branch-and-bound algorithm, provided in App. A.2. In the following, for convenience, we refer to the heuristic as h (which is parameterized by (u, s)) instead of HEURISTIC (which is parameterized by $\langle u, i \rangle$). Further, we say that h is optimistic if $h(u, s)$ is a lower bound on the cost for aligning all remaining letters (i.e., $q[i : |q|]$) starting from node u (note that s is a prefix of $q[i : |q|]$).

Theorem 1. *h is optimistic.*

Proof. h only considers the next d' letters of q instead of all remaining letters. Since all costs are non-negative, the theorem follows. $\square$

Benefit of A* Heuristic over DIJKSTRA. Figure 2 shows the benefit of using our heuristic function compared to DIJKSTRA. Here, DIJKSTRA expands states based on their distance g from the origin nodes $\langle u, 0 \rangle$ and $\langle v, 0 \rangle$. Hence, depending on tie-breaking, DIJKSTRA may expand all states with $h \leq 1$, as shown in Fig. 2. By contrast, A* chooses the next state to expand by the sum of the distance from the origin g and the heuristic h, expanding only states with $g + h \leq 1$.

Memoization. Recall that the return value of h in Line 8 only depends on u and the upcoming sequence s (which in turn depends on i and d). Thus, $h(u, s)$ can be reused for different positions across different queries in $\mathcal{O}(1)$ time, if it was computed for a previous query.

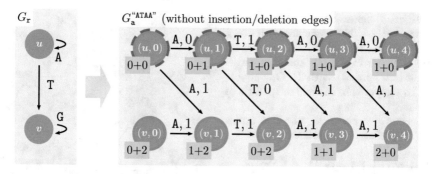

Fig. 2. The benefit of using our heuristic over DIJKSTRA. Alignment graph $G_a^{\text{"ATAA"}}$ (right) is based on reference graph G_r (left), but omits insertion and deletion edges for simplicity. The pink boxes $g + h$ indicate the distance from the sources $S = \{\langle u, 0 \rangle, \langle v, 0 \rangle\}$ (in g) and the cost of aligning the next $d = 2$ letters (in h). DIJKSTRA (resp. A*) expands states circled in blue (resp. dashed red).

4 ASTARIX Algorithm: Optimizations

We now discuss several optimizations we developed to speed up ASTARIX while preserving its optimality. These optimizations reduce preprocessing and alignment runtime as well as memory footprint (in particular for memoization).

4.1 Reducing Semi-global to Local Alignment Using a Trie

To find an optimal alignment, we generally need to consider all reference graph nodes $u \in G_r$ as possible starting nodes. Thus, optimal aligners PASGAL [15] and BITPAR-ALLEL [27] brute-force through all possible starting nodes $u \in G_r$.

To more efficiently handle arbitrary starting positions for alignments, we extend the reference graph with a trie (referred to as *suffix tree* in [8]) to effectively align from all possible starting nodes *simultaneously*.

Single Starting State. In the trie approach, abstraction nodes are added to the graph, each of which corresponds to a set of nodes in G_r that correspond to the same prefix. In the following, we formalize this approach.

Concretely, we extend G_r by a *trie of depth D*, resulting in graph $G_r^+ = (V_r^+, E_r^+)$. Our goal is that all paths in G_r that have length D and end in $v \in V_r$ correspond to paths in G_r^+ starting from a single source ϵ to $v \in V_r^+$, where ϵ represents the empty string. This correspondence ensures that it suffices to consider only paths in G_r^+ starting from the source ϵ. In particular, each alignment on G_r^+ can be translated into an alignment on G_r (we omit this translation here).

Figure 3 shows an example trie. To construct it, we first associate with every node $v \in V_r$ the set $\mathcal{S}_v$ of its D-mers (orange boxes in Fig. 3): spells of paths ending in v and of length D. Our goal is then to use paths in the trie to spell these D-mers.

Second, we construct the trie nodes from all prefixes of these D-mers:

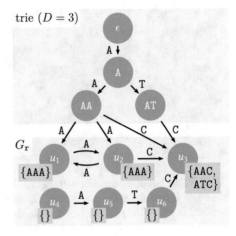

trie $(D = 3)$

Fig. 3. G_r^+ enables semi-global alignment by extending G_r with a trie.

$$V_r^+ := V_r \cup \bigcup_{v \in V_r} \left\{ s[0:i] \,\middle|\, \begin{array}{l} s \in \mathcal{S}_v, \\ 0 \le i < D \end{array} \right\}.$$

Third, we add edges within the trie, which ensure that paths from ϵ to any trie node s spell s. Formally, whenever $s \cdot \ell \in V_r^+$, we add an edge $(s, s{\cdot}\ell, \ell)$ to E_r^+, where "$\cdot$" denotes string concatenation. Finally, we add edges between the trie and the reference graph, which ensure that any D-mer of any node $v \in V_r$ can be spelled by a walk from ϵ to v. Formally, if $s \cdot \ell \in \mathcal{S}_v$, then $(s, v, \ell) \in E_r^+$.

Importantly, extending G_r to G_r^+ is compatible with the construction of the edit graph G_e, the construction of the alignment graph and all other optimizations. In particular, when searching for a shortest path in the alignment graph constructed from G_r^+, it suffices to only consider starting node $\langle \epsilon, 0 \rangle$.

Reducing Size of Trie. We can reduce the size of the trie by removing specific trie nodes. In particular, we iteratively remove each trie leaf node $s \cdot \ell \in V_r^+$ with a unique outgoing edge $(s \cdot \ell, v, \ell')$ to a reference graph node $v \in V_r$. To compensate for removing node $s \cdot \ell$, we introduce a new edge (s, u, ℓ) to a node $u \in V_r$ with an edge (u, v, ℓ') (such a node must exist according to the construction of G_r^+). For example, in Fig. 3, we (i) remove node AT including its edges $(\mathtt{A}, \mathtt{AT}, \mathtt{T})$ and $(\mathtt{AT}, u_3, \mathtt{C})$, but (ii) introduce an edge $(\mathtt{A}, u_2, \mathtt{T})$.

This optimization is lossless, as the D-mer $s \cdot \ell \cdot \ell' \in \mathcal{S}_v$ can still be spelled by the path from ϵ to s, extended by (s, u, ℓ) and (u, v, ℓ').

4.2 Greedy Match Optimization

We also employ an optimization originally developed for computing the edit distance between two strings [1,31], but which has also been used in the context of string to graph alignment [8]. We omit the correctness proof of this optimization, which is already covered in [31], and only explain the intuition behind it.

Suppose there is only one outgoing edge $e = (u, v, \ell) \in E_r$ from a node $u \in V_r$. Suppose also that while aligning a query q, we explore state $\langle u, i \rangle$ for which the next query letter $q[i]$ matches the label ℓ. In this case, we do not need to consider the edit outgoing edges, because any edit at this point can be postponed without additional cost, as $\Delta_{\text{match}} \le \min(\Delta_{\text{subst}}, \Delta_{\text{ins}}, \Delta_{\text{del}})$. Thus, we can greedily explore state $\langle v, i+1 \rangle$, aligning $q[i + 1]$ to e by using the edge $(\langle u, i \rangle, \langle v, i + 1 \rangle, \ell, \Delta_{\text{match}})$ before continuing with the A^* search. We note that this optimization is only applicable when aligning in non-branching regions of the reference graph. In particular, it is not applicable for most trie nodes (Sect. 4.1).

4.3 Speeding up Evaluation of Heuristic

In the following, we show how to reduce the runtime of evaluating the heuristic $h(u, s)$, by introducing two separate optimizations that compose naturally.

Capping Cost. We cap $h(u, s)$ at c, replacing it by $h_c(u, s) := \min(h(u, s), c)$. To achieve this, we allow RECURSIVEALIGN to ignore paths costing more than c. For large enough c, this speeds up computation without significantly decreasing the benefit of the heuristic, since nodes associated with a high heuristic value are typically not explored anyways. We investigate the effect of c in App. A.3.

Theorem 2. h_c *is optimistic.*

Proof. We have $h_c(u, s) \leq h(u, s)$ and that $h(u, s)$ is optimistic (Theorem 1). □

Capping Depth. We reduce the number of nodes that need to be considered by $h(u, s)$. To this end, we define a modified heuristic $h_d(u, s)$ that only considers nodes $R_u \subseteq V_e$ at distance at most d from u (cp. Line 2 in Algorithm 1): $R_u := \{v \in V_r \mid \exists$ path $\pi \in G_e$ from u to v with $|\pi| \leq d\}$.

If an alignment of s reaches the boundary of R_u, defined as

$$B(R_u) := \{v \in R_u \mid \exists (v, v', \ell) \in E_e \text{ with } v' \notin R_u\},$$

it is allowed to only spell a prefix of s, and the remaining unaligned letters of s are considered aligned with zero cost:

$$h_d(u, s) := \min_{\pi \in \Pi} \text{cost}(\pi), \text{ where}$$

$$\Pi := \left\{ \pi \in G_r \mid start(\pi) = u, \sigma(\pi) = s \vee \left(end(\pi) \in B(R_u) \wedge \exists i. \sigma(\pi) = s[1..i] \right) \right\}$$

Theorem 3. h_d *is optimistic.*

Proof. It suffices to show $h_d(u, s) \leq h(u, s)$ since $h(u, s)$ is optimistic. In the case where all of s is aligned, $h_d(u, s) = h(u, s)$. Otherwise, the unaligned letters of s are not penalized, so $h_d(u, s) \leq h(u, s)$. □

4.4 Partitioning Nodes into Equivalence Classes

We have shown in Sect. 3.2 how to reuse an already computed $h(u, s)$ for repeating s across different queries and query positions. In the following, we additionally aim to reuse $h(u, s)$ across different nodes u, so that $h(u, s)$ does not need to be computed for all nodes u. Intuitively, we want to assign two nodes u and v to the same equivalence class when the *graph region* considered by $h(u, s)$ is equivalent to the graph region considered by $h(v, s)$, up to renaming of nodes.

Thus, $h(u, s) = h(v, s)$ if u and v are from the same equivalence class. Therefore, we can (arbitrarily) choose a representative node $r \in V_r$ for every equivalence class, and evaluate $h(r, s)$ instead of $h(u, s)$, where r is the representative of the equivalence class of u. To look up representative nodes in $\mathcal{O}(1)$, we define a helper array *repr* with $repr[u] = r$.

Identifying Equivalence Classes. To identify the nodes belonging to the same equivalence class, we assume the optimization from Sect. 4.3, i.e., that our heuristic only considers nodes up to a distance d from u. Moreover, for performance reasons, our implementation detects only the equivalence classes of nodes u with a single outgoing path of length at least d. In this case, u and u' are in the same equivalence class if their outgoing paths spell the same sequence. In contrast, we leave nodes with forking paths in separate equivalence classes.

Note that for smaller d, the number of equivalence classes gets smaller, the reuse of the heuristic gets higher, and the memoization table has a lower memory footprint. At the same time, however, the heuristic $h_d(u, s)$ is less informative.

5   Evaluation

In this section we present a thorough experimental evaluation[3] of ASTARIX on simulated Illumina reads. Our evaluation demonstrates that:

1. ASTARIX is faster than DIJKSTRA because the heuristic reduces the number of explored states by an order of magnitude.
2. The runtime of ASTARIX scales better than state-of-the-art optimal aligners with increasing graph size, on a variety of reference graphs.

5.1   Implementation of ASTARIX and DIJKSTRA

Our ASTARIX implementation uses an adjacency list graph data structure to represent the reference and the trie in a unified way, representing each letter by a separate edge object. To represent the reverse complementary walks in G_r, the vertices are doubled, connected in the opposite direction, and labeled with complementary nucleotides (A $\leftrightarrow$ T, C $\leftrightarrow$ G). We do not limit the number of memoized heuristic function values (Sect. 3.2), but note we could do so by resetting the memoization table periodically. Our implementation of DIJKSTRA reuses the same ASTARIX codebase except the use of a heuristic function (i.e., with $h \equiv 0$).

We apply all described optimizations to ASTARIX and DIJKSTRA, except Sects. 4.3 and 4.4 which are applicable only to ASTARIX.

While the optimality of ASTARIX is not affected by its parameters, its performance is (see App. A.3 for analysis). To compare with other aligners, we use values $d = 5$, $c = 5$, $D = \lfloor \log_\Sigma |G_r| \rfloor$.

5.2   Compared Aligners: PaSGAL and BitParallel

We compare the performance of ASTARIX to that of two state-of-the-art optimal aligners: PaSGAL and BitParallel, with their default parameters. We do not compare to the exact aligner of VG as (i) its optimal alignment is intended for testing purposes only, (ii) it does not provide an interface for aligning a set of reads, and (iii) it has been consistently outperformed by PaSGAL [15].

---

[3] https://github.com/eth-sri/astarix/tree/RECOMB2020_experiments.

PASGAL is compiled with AVX2 SIMD support. The resulting alignments are not expected to match exactly between the local aligner PASGAL and the semi-global aligners (ASTARIX and BITPARALLEL) as they solve different tasks with different edit costs. Nevertheless, in analogy with the evaluations of PASGAL [15], it is still meaningful to compare performance, assuming that the dynamic programming approach of PASGAL can be adapted to semi-global alignment with similar performance.

Both BITPARALLEL and PASGAL reach their worst-case runtime complexity independent of the edit costs $\Delta = (\Delta_{\text{match}}, \Delta_{\text{subst}}, \Delta_{\text{ins}}, \Delta_{\text{del}})$. PASGAL is evaluated using its default costs $\Delta = (-1, 1, 1, 1)$ and BITPARALLEL is evaluated using the only supported costs $\Delta = (0, 1, 1, 1)$.

5.3 Setting

All evaluations were executed singled-threaded on an Intel Core i7-6700 CPU running at 3.40 GHz.

Reference Graphs and Reads. We designed three experiments utilizing three different reference graphs (in Table 1). The first is a linear graph without variation based on the *E. coli* reference genome (strain: K-12 substr. MG1655, ASM584v2 [13]). The other two are variation graphs taken from the PASGAL evaluations [15]: they are based on the Leukocyte Receptor Complex (LRC, with 1 099 856 nodes and 1 144 498 edges), and the Major Histocompatibility Complex (MHC1, with 5 138 362 nodes and 5 318 019 edges). We note that we do not evaluate on de Brujin graphs, since PASGAL does not support cyclic graphs.

For the *E. coli* dataset we used the ART tool [14] to simulate an Illumina single-end read set with 10 000 reads of length 100. For the LCR and MHC1 datasets, we sampled 20 000 single-end reads of length 100 from the already generated sets in [15] using the Mason2 [12] simulator.

For DIJKSTRA and ASTARIX, the runtime complexity depends not only on the data size, but also on the data content, including edit costs. More accurate heuristics lead to better A* performance [26], which is why we evaluate ASTARIX with costs corresponding more closely to Illumina error profiles: $\Delta = (0, 1, 5, 5)$.

Metrics. As all aligners evaluated in this work are provably optimal, we are mostly interested in their performance. To study the end-to-end performance of the optimal aligners, we use the Snakemake [20] pipeline framework to measure the execution time

Table 1. Performance of optimal aligners for different reference graphs.

Genome graph	Size	Runtime and Memory			
		ASTARIX	DIJKSTRA	PASGAL	BITPARALLEL
E. coli (linear)	~4.7 Mbp	33 s	73 s	3 272 s	4 906 s
		0.66 GB	0.66 GB	0.55 GB	0.43 GB
LCR (variant)	~1 Mbp	437 s	940 s	1 614 s	SegFault
		1.12 GB	1.09 GB	0.30 GB	
MHC1 (variant)	~5 Mbp	1 282 s	1588 s	≥7 200 s	SegFault
		4.35 GB	1.21 GB	0.87 GB	

of every aligner (including the time spent on reading and indexing the reference graph input and outputting the resulting alignments). We note that the alignment phase dominates for all tools and experiments.

To judge the potential of heuristic functions, we measure not only the runtime but also the number of states explored by ASTARIX and DIJKSTRA. This number reflects the quality of the heuristic function rather than the speed of computation of the heuristic, the implementation and the system parameters.

5.4   Comparison of Optimal Aligners

Different Reference Graphs. Table 1 shows the performance of optimal aligners across various references. On all references, ASTARIX is consistently faster than DIJK-STRA, which is consistently faster than PASGAL and BITPARALLEL. The memory usage of DIJKSTRA is within a factor of 3 compared to PASGAL and BITPARALLEL. Due to the heuristic memoization, the memory usage of ASTARIX can grow several times compared to DIJKSTRA.

Scaling with Reference Graph Size. Figure 4 compares the performance of existing optimal aligners. BITPARALLEL and PASGAL always explore all states, thus their average-case reaches the worst-case complexity of $\mathcal{O}(|G_a^q|) = \mathcal{O}(m \cdot G_r)$. Due to the trie indexing, the runtime of ASTARIX and DIJKSTRA scales in the reference size with a polynomial of power around 0.2 versus the expected linear dependency of BITPARALLEL and PASGAL.

The heuristic function of ASTARIX demonstrates a 2-fold speed-up over DIJKSTRA. This is possible due to the highly branching trie structure, which allows skipping the explicit exploration for the majority of starting nodes.

5.5   A* Speedup

To measure the speedup caused by the heuristic function, we compare the number of not only the expanded, but also of explored states (the latter number is never smaller, see Sect. 3.1 and the example in Fig. 2) between ASTARIX and DIJKSTRA on the MHC1 dataset.

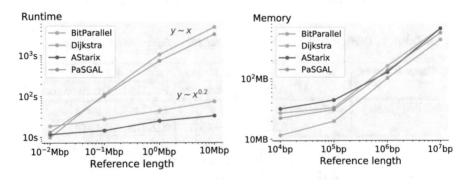

Fig. 4. Comparison of overall runtime and memory usage of optimal aligners with increasing prefixes of E. coli as references.

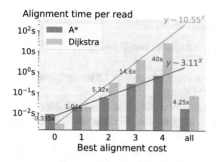 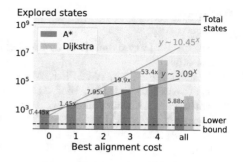

Fig. 5. Comparison of $A^\star$ and DIJKSTRA in terms of mean alignment runtime per read and mean explored states depending on the best alignment cost on MHC1.

Figure 5 demonstrates the benefit of the heuristic function in terms of both alignment time and number of explored states. Most importantly, ASTARIX scales much better with increasing number of errors in the read, compared to DIJKSTRA. More specifically, the number of states explored by DIJKSTRA, as a function of alignment cost, grows exponentially with a base of around 10, whereas the base for ASTARIX is around 3 (the empirical complexity is estimated as a best exponential fit $exploredStates \sim a \cdot score^b$).

The horizontal black line in Fig. 5 denotes the total number of states $|G_r| \cdot |q|$, which is always explored by BITPARALLEL and PASGAL. On the other hand, any aligner must explore at least $m = |q|$ states, which we show as a horizontal dashed line. This lower bound is determined by the fact that at least the states on a best alignment need to be explored.

6   Conclusion

We presented ASTARIX, an $A^\star$ algorithm to find optimal alignments, based on a domain-specific heuristic and enhanced by multiple algorithmic optimizations. Importantly, our approach allows for both cyclic and acyclic graphs including variation and de Bruijn graphs.

We demonstrated that ASTARIX scales exponentially better than DIJKSTRA with increasing (but small) number of errors in the reads. Moreover, for short reads, both ASTARIX and DIJKSTRA scale better and outperform current state-of-the-art optimal aligners with increasing genome graph size. Nevertheless, scaling optimal alignment of long reads on big graphs remains an open problem.

We expect that ASTARIX can be scaled further, to both (i) bigger graphs and (ii) longer and noisier reads. Scaling ASTARIX may require a combination of (i) the development of more clever heuristic functions (by leveraging existing work on $A^\star$ and edit distance) and (ii) algorithmic optimizations. We note that if desired, a (sub-optimal) seeding step could speed up ASTARIX by pre-filtering the starting positions, analogously to other practical aligners.

References

1. Allison, L.: Lazy dynamic-programming can be eager. Inf. Process. Lett. **43**, 207–212 (1992)
2. Altschul, S.F., Gish, W., Miller, W., Myers, E.W., Lipman, D.J.: Basic local alignment search tool. J. Mol. Biol. **215**, 403–410 (1990)
3. Antipov, D., Korobeynikov, A., McLean, J.S., Pevzner, P.A.: hybridSPAdes: an algorithm for hybrid assembly of short and long reads. Bioinformatics (Oxford, England) **32**, 1009–1015 (2016)
4. Brandt, D.Y.C., Aguiar, V.R.C., Bitarello, B.D., Nunes, K., Goudet, J., Meyer, D.: Mapping Bias Overestimates Reference Allele Frequencies at the HLA Genes in the 1000 Genomes Project Phase I Data. G3 (Bethesda, Md.) (2015)
5. Buhler, S., Sanchez-Mazas, A.: HLA DNA sequence variation among human populations: molecular signatures of demographic and selective events. PLoS One **6**, e14643 (2011)
6. Dechter, R., Pearl, J.: Generalized best-first search strategies and the optimality of A*. J. ACM (1985)
7. Dilthey, A., Cox, C., Iqbal, Z., Nelson, M.R., McVean, G.: Improved genome inference in the MHC using a population reference graph. Nat. Genet. **47**(6), 682 (2015)
8. Dox, G., Fostier, J.: Efficient algorithms for pairwise sequence alignment on graphs. Master's thesis, Ghent University (2018)
9. Garrison, E., et al.: Variation graph toolkit improves read mapping by representing genetic variation in the reference. Nat. Biotechnol. **36**, 875–879 (2018)
10. Harismendy, O., et al.: Detection of low prevalence somatic mutations in solid tumors with ultra-deep targeted sequencing. Genome Biol. **12**, R124 (2011)
11. Heydari, M., Miclotte, G., Van de Peer, Y., Fostier, J.: BrownieAligner: accurate alignment of Illumina sequencing data to de Bruijn graphs. BMC Bioinformatics (2018)
12. Holtgrewe, M.: Mason - a read simulator for second generation sequencing data. Technical report FU Berlin (2010). http://publications.imp.fu-berlin.de/962/
13. Howe, K.L., et al.: Ensembl Genomes 2020-enabling non-vertebrate genomic research. Nucleic Acids Res. **48**, D689–D695 (2020)
14. Huang, W., Li, L., Myers, J.R., Marth, G.T.: ART: a next-generation sequencing read simulator. Bioinformatics (Oxford, England) **28**, 593–594 (2012)
15. Jain, C., Misra, S., Zhang, H., Dilthey, A., Aluru, S.: Accelerating sequence alignment to graphs. In: International Parallel and Distributed Processing Symposium (IPDPS) (2019). ISSN 1530-2075
16. Jain, C., Zhang, H., Gao, Y., Aluru, S.: On the complexity of sequence to graph alignment. In: Cowen, L.J. (ed.) RECOMB 2019. LNCS, vol. 11467, pp. 85–100. Springer, Cham (2019). https://doi.org/10.1007/978-3-030-17083-7_6
17. Jean, G., Kahles, A., Sreedharan, V.T., De Bona, F., Rätsch, G.: RNA-Seq read alignments with PALMapper. Curr. Protoc. Bioinformatics **32**, 11–16 (2010)
18. Kavya, V.N.S., Tayal, K., Srinivasan, R., Sivadasan, N.: Sequence alignment on directed graphs. J. Comput. Biol. **26**, 53–67 (2019)
19. Kim, D., Paggi, J.M., Park, C., Bennett, C., Salzberg, S.L.: Graph-based genome alignment and genotyping with HISAT2 and HISAT-genotype. Nat. Biotechnol. **37**, 907–915 (2019)
20. Köster, J., Rahmann, S.: Snakemake-a scalable bioinformatics workflow engine. Bioinformatics (Oxford, England) **28**(9), 2520–2522 (2012)

21. Langmead, B., Salzberg, S.L.: Fast gapped-read alignment with Bowtie 2. Nat. Methods **9**, 357–359 (2012)
22. Li, H., Durbin, R.: Fast and accurate short read alignment with Burrows-Wheeler transform. Bioinformatics (Oxford, England) **25**, 1754–1760 (2009)
23. Limasset, A., Flot, J.F., Peterlongo, P.: Toward perfect reads: self-correction of short reads via mapping on de Bruijn graphs. Bioinformatics **36**(5), 1374–1381 (2019). btz102
24. Liu, B., Guo, H., Brudno, M., Wang, Y.: deBGA: read alignment with de Bruijn graph-based seed and extension. Bioinformatics (Oxford, England) **32**, 3224–3232 (2016)
25. Paten, B., Novak, A.M., Eizenga, J.M., Garrison, E.: Genome graphs and the evolution of genome inference. Genome Res. **27**, 665–676 (2017)
26. Pearl, J.: On the discovery and generation of certain heuristics. AI Mag. **4**, 23 (1983)
27. Rautiainen, M., Mäkinen, V., Marschall, T.: Bit-parallel sequence-to-graph alignment. Bioinformatics **35**, 3599–3607 (2019)
28. Rautiainen, M., Marschall, T.: Aligning sequences to general graphs in O(V+mE) time (2017, preprint)
29. Salmela, L., Rivals, E.: LoRDEC: accurate and efficient long read error correction. Bioinformatics (Oxford, England) **30**, 3506–3514 (2014)
30. Schneeberger, K., et al.: Simultaneous alignment of short reads against multiple genomes. Genome Biol. **10**, R98 (2009)
31. Sellers, P.H.: An algorithm for the distance between two finite sequences. J. Comb. Theory **16**, 253–258 (1974)
32. Sirén, J.: Indexing variation graphs. In: 2017 Proceedings of the Ninteenth Workshop on Algorithm Engineering and Experiments (ALENEX) (2017)
33. Sirén, J., Välimäki, N., Mäkinen, V.: Indexing graphs for path queries with applications in genome research. IEEE/ACM Trans. Comput. Biol. Bioinf. (TCBB) **11**, 375–388 (2014)
34. Smith, T.F., Waterman, M.S.: Comparison of biosequences. Adv. Appl. Math. **2**, 482–489 (1981)
35. Stevenson, K.R., Coolon, J.D., Wittkopp, P.J.: Sources of bias in measures of allele-specific expression derived from RNA-seq data aligned to a single reference genome. BMC Genom. **14**, 536 (2013)

Polynomial-Time Statistical Estimation of Species Trees Under Gene Duplication and Loss

Brandon Legried[1], Erin K. Molloy[2], Tandy Warnow[2],
and Sébastien Roch[1(✉)]

[1] University of Wisconsin-Madison, Madison, WI, USA
{blegried,roch}@math.wisc.edu
[2] University of Illinois at Urbana-Champaign, Urbana, IL, USA
{emolloy2,warnow}@illinois.edu

Abstract. Phylogenomics—the estimation of species trees from multi-locus datasets—is a common step in many biological studies. However, this estimation is challenged by the fact that genes can evolve under processes, including incomplete lineage sorting (ILS) and gene duplication and loss (GDL), that make their trees different from the species tree. In this paper, we address the challenge of estimating the species tree under GDL. We show that species trees are *identifiable* under a standard stochastic model for GDL, and that the polynomial-time algorithm ASTRAL-multi, a recent development in the ASTRAL suite of methods, is *statistically consistent* under this GDL model. We also provide a simulation study evaluating ASTRAL-multi for species tree estimation under GDL. All scripts and datasets used in this study are available on the Illinois Data Bank: https://doi.org/10.13012/B2IDB-2626814_V1.

Keywords: Species trees · Gene duplication and loss · Identifiability · Statistical consistency · Estimation · ASTRAL

1 Introduction

Phylogeny estimation is a statistically and computationally complex estimation problem, due to heterogeneity across the genome resulting from processes such as incomplete lineage sorting (ILS), gene duplication and loss (GDL), rearrangements, gene flow, horizontal gene transfer, introgression, etc. [20]. Much is known about the problem of estimating species trees in the presence of ILS, as modelled by the Multi-Species Coalescent (MSC) [17,34]. For example, because the most probable unrooted tree for every four species is the species tree on those species [1], the unrooted species tree topology is identifiable under the MSC from its gene tree distribution, and quartet-based species tree estimation methods that operate by combining gene trees (such as BUCKy-pop [18] and ASTRAL [22,24,40]) are statistically consistent estimators of the unrooted species tree topology (i.e., as the number of sampled genes increases, almost surely the tree returned by

© Springer Nature Switzerland AG 2020
R. Schwartz (Ed.): RECOMB 2020, LNBI 12074, pp. 120–135, 2020.
https://doi.org/10.1007/978-3-030-45257-5_8

these methods will be the true species tree). It is also known that concatenation (whether partitioned or unpartitioned) is not statistically consistent, and can even be positively misleading (i.e., converge to the wrong tree as the number of loci increases) [29,31]. In general, establishing whether a method is statistically consistent or not is important for understanding its performance guarantees.

Yet, correspondingly little has been established about species tree estimation in the presence of GDL. For example, although likelihood-based approaches for species tree estimation have been developed (e.g., PHYLDOG [7]), they have not been established to be statistically consistent. Key to understanding the performance of species tree estimation under GDL is whether the species tree topology itself is identifiable from the distribution it defines on the gene trees it generates. However, since gene trees can have multiple copies of each species when gene duplication occurs, this question can be formulated as: "Is the species tree identifiable from the distribution on MUL-trees?", where a MUL-tree is a tree with potentially multiple copies of each species.

In this paper, we prove that unrooted species tree topologies are identifiable from the distribution implied on MUL-trees (Sect. 3) under the simple GDL model of [2]. Furthermore, we prove that the polynomial-time method ASTRAL-multi [26], a recent variant of ASTRAL designed to enable analyses of datasets with multiple individuals per species, is statistically consistent under this model (Sect. 3). We then present an experimental study evaluating ASTRAL-multi on 16-taxon datasets simulated under the DLCoal model (a unified model of GDL and ILS) [27]; the results of this study show that when given a sufficiently large number of genes, ASTRAL-multi is competitive with other methods (e.g., Dup-Tree [4], MulRF [9], and ASTRID-multi [36], the implementation of ASTRID for multi-allele datasets) that also estimate species trees from MUL-trees (Sect. 4). We conclude with remarks about future work and implications for large-scale species tree estimation (Sect. 5).

2   Species Tree Estimation from Gene Families

Our input is a collection $\mathcal{T}$ of gene trees representing the inferred evolutionary histories of gene families. In the presence of gene duplication and loss events, such gene trees may be multi-labeled trees (MUL-trees), meaning that the same species label may be assigned to several gene copies. Our goal is to reconstruct a species tree T over the corresponding set S of species.

ASTRAL. We provide theoretical guarantees and empirically validate an approach based on ASTRAL [22] in its variant for multiple alleles [26], which we refer to as ASTRAL-multi. Following [12], the input consists of unrooted MUL-trees $\mathcal{T}$ from all gene families, where copies of a gene in a species are treated as multiple alleles within the species.

ASTRAL-multi proceeds as follows. Let S be the set of n species and let R be the set of m individuals. The input are the gene trees $\mathcal{T} = \{t_i\}_{i=1}^k$, where t_i is labeled by individuals $R_i \subseteq R$. For any (unrooted) species tree $\tilde{T}$ labeled by

S, an extended species tree $\widetilde{T}_{ext}$ labeled by R is built by adding to each leaf of $\widetilde{T}$ all individuals corresponding to that species as a polytomy. The quartet score of $\widetilde{T}$ with respect to $\mathcal{T}$ is then

$$Q_k(\widetilde{T}) = \sum_{i=1}^{k} \sum_{\mathcal{J}=\{a,b,c,d\}\subseteq R_i} \mathbf{1}(\widetilde{T}_{ext}^{\mathcal{J}}, t_i^{\mathcal{J}}), \tag{1}$$

where $\mathbf{1}(T_1, T_2)$ is the indicator that T_1 and T_2 agree and $T_1^{\mathcal{J}}$ is the restriction of T_1 to individuals $\mathcal{J}$. Run in its *exact* version (i.e., an unrooted species tree that maximizes the quartet score), ASTRAL-multi is guaranteed to find an optimal solution, but can use exponential time. The *default* mode, which runs in polynomial time, uses dynamic programming to solve a constrained version of the problem, requiring that the output tree draw its bipartitions from a set Σ of bipartitions that ASTRAL computes on the input, where Σ by construction includes all the bipartitions on S that occur in any gene tree in $\mathcal{T}$.

3   Theoretical Results

In this section, we provide theoretical guarantees for the reconstruction algorithm discussed in Sect. 2. Specifically, we establish statistical consistency under a standard model of GDL [2]. First we show that the species tree is identifiable.

3.1   Gene Duplication and Loss Model

We assume in this section that gene tree heterogeneity is due exclusively to GDL (and so no ILS) and that the true gene trees are known. That is, there is no gene tree estimation error (GTEE).

Birth-Death Process of Gene Duplication and Loss. The rooted n-species tree $T = (V, E)$ has vertices V and directed edges E with lengths (in time units) η that depend on the edge. For ease of presentation, we assume that there is a single copy of each gene at the root of T and that the rates of duplication λ and loss μ are fixed throughout T (although our proofs do not use these assumptions). Each gene tree is generated by a top-down birth-death process within the species tree. That is, on each edge, each gene copy independently duplicates at exponential rate λ and is lost at exponential rate μ; at speciation events, each gene copy bifurcates and proceeds similarly in the descendant edges. Each duplication is indicated in the gene tree by a bifurcation. The resulting gene tree is then pruned of lost copies to give the observed unrooted gene tree t_i. The gene trees $\{t_i\}_{i=1}^{k}$ are assumed independent and identically distributed. See more details in [2].

3.2   Identifiability of the Species Tree Under the GDL Model

We first show that the unrooted species tree is identifiable from the distribution of MUL-trees $\mathcal{T}$ under the GDL model over T. That is, that two distinct unrooted species trees necessarily produce different gene tree distributions.

We begin with a quick proof sketch. The idea is to show that, for each 4-tuple of species $Q = \{A, B, C, D\}$, the corresponding species quartet topology can be identified by taking an independent uniform random gene copy in each species in Q and showing that the quartet topology consistent with the species tree is most likely to result in the gene tree restricted to these copies. It should be noted that the proof is not as straightforward as it is under the multispecies coalescent [1], as we explain next. Assume the species tree restricted to Q is $((A, B), (C, D))$, let R be the most recent common ancestor of Q in T, and let a, b, c, d be random gene copies in A, B, C, D respectively.

- When all ancestral copies of a, b, c, d in R are distinct, by symmetry *all quartet topologies are equally likely*. The *ancestral copy of x in R* is the vertex of the gene tree that is ancestral to x and corresponds to a speciation event at node R of the species tree.
- When the ancestors of a and b (or c and d) in R are the same, the *species quartet topology results*.
- *However,* there are further cases. For example, if the ancestors of a and c in R coincide while being distinct from those of b and d, then the resulting quartet topology *differs* from that of the species tree.

Hence, one must carefully account for all possible cases to establish that the species quartet topology is indeed likeliest, which we do next. Our argument relies primarily on the symmetries (i.e., exchangeability) of the process.

Theorem 1 (Identifiability). *Let T be a species tree with $n \geq 4$ leaves. Then T, without its root, is identifiable from the distribution of MUL-trees $\mathcal{T}$ under the GDL model over T.*

Proof. It is known that the unrooted topology of a species tree is defined by its set of quartet trees [3]. Let $Q = \{A, B, C, D\}$ be four distinct species in T and let T^Q be the species tree restricted to Q. Assume without loss of generality that the corresponding unrooted quartet topology is $AB|CD$. Let t be a MUL-tree generated under the GDL model over T and let t^Q be its restriction to the gene copies from species in Q. Conditioning on having at least one gene copy in the species Q, independently pick a uniformly random gene copy a, b, c, d in species A, B, C, D respectively and let q be the corresponding quartet topology under t^Q. We show that the most likely outcome is $q = ab|cd$. There are two cases: T^Q is (1) balanced or (2) a caterpillar.

In case (1), let R be the most recent common ancestor of Q in T and let I be the number of gene copies exiting (forward in time) R. By the law of total probability, $\mathbf{P}'[q = ab|cd] = \mathbf{E}'[\mathbf{P}'_I[q = ab|cd]]$, where the primes indicate that we are conditioning on having at least one gene copy in each species in Q and the subscript I indicates conditioning on I. So it suffices to prove

$$\mathbf{P}'_I[q = ab|cd] > \max\left\{\mathbf{P}'_I[q = ac|bd], \mathbf{P}'_I[q = ad|bc]\right\}, \tag{2}$$

almost surely. Let $i_x \in \{1, \ldots, I\}$ be the ancestral lineage of $x \in \{a, b, c, d\}$ in R. Then

$$\mathbf{P}'_I[q = ab|cd] = \mathbf{P}'_I[i_a = i_b] + \mathbf{P}'_I[i_c = i_d] - \mathbf{P}'_I[i_a = i_b, i_c = i_d]$$
$$+ \mathbf{P}'_I[q = ab|cd \text{ and } i_a, i_b, i_c, i_d \text{ all distinct}]. \qquad (3)$$

On the other hand,

$$\mathbf{P}'_I[q = ac|bd] \leq \mathbf{P}'_I[i_b \neq i_a = i_c \neq i_d] + \mathbf{P}'_I[i_a \neq i_b = i_d \neq i_c]$$
$$+ \mathbf{P}'_I[q = ac|bd \text{ and } i_a, i_b, i_c, i_d \text{ all distinct}], \qquad (4)$$

and similarly for $\mathbf{P}'_I[q = ac|bd]$, where note that we double-counted the case $i_a = i_c \neq i_d = i_b$ to simplify the expression. By symmetry of the GDL process above R (which holds under $\mathbf{P}'_I$), the last term on the RHS of (3) and (4) are the same. The same holds for the first two terms on the RHS of (4) this time by the independence and exchangeability of the pairs (i_a, i_b) and (i_c, i_d) under $\mathbf{P}'_I$, which further implies

$$\mathbf{P}'_I[q = ab|cd] - \mathbf{P}'_I[q = ac|bd]$$
$$\geq \mathbf{P}'_I[i_a = i_b] + \mathbf{P}'_I[i_c = i_d] - \mathbf{P}'_I[i_a = i_b, i_c = i_d] - 2\mathbf{P}'_I[i_b \neq i_a = i_c \neq i_d]$$
$$= x + y - xy - 2(1-x)(1-y)\mathbf{P}'_I[i_a = i_c \mid i_a \neq i_b, i_c \neq i_d]$$
$$= x + y - xy - 2(1-x)(1-y)\mathbf{P}'_I[i_a = i_c]$$
$$= x + y - xy - 2(1-x)(1-y)\frac{1}{I} \equiv h(x, y).$$

where $x = \mathbf{P}'_I[i_a = i_b]$ and $y = \mathbf{P}'_I[i_c = i_d]$.

For fixed y, $h(x, y)$ is linear in x and $h(1, y) = 1$. So $h(\cdot, y)$ achieves its minimum at the smallest value allowed for x. The same holds for y. Intuitively, i_a and i_b are "positively correlated" so $x \geq 1/I$. We prove this formally next.

Lemma 1. *Almost surely, $x, y \geq 1/I$.*

Proof. For $j \in \{1, \ldots, I\}$, let N_j be the number of gene copies at the most recent common ancestor R' of A and B that descend from copy j in R. Upon conditioning on $(N_j)_j$, the choice of a and b is independent, with i_a and i_b being picked proportionally to the corresponding N_j's (i.e., the gene copies in R' are equally likely to have given rise to a). By the law of total probability and the fact that the quadratic mean is greater than the arithmetic mean,

$$\mathbf{P}'_I[i_a = i_b] = \mathbf{E}'_I[\mathbf{P}'_I[i_a = i_b \mid (N_j)_j]] = \mathbf{E}'_I\left[\frac{\sum_{j=1}^{I} N_j^2}{\left(\sum_{j=1}^{I} N_j\right)^2}\right] \geq \frac{1}{I},$$

and similarly for $\mathbf{P}'_I[i_c = i_d]$. □

Returning to the proof of the theorem, evaluating h at $x, y = 1/I$ gives

$$h(1/I, 1/I) = 2\frac{1}{I} - \frac{1}{I^2} - 2\frac{(I-1)^2}{I^3} = \frac{2I^2 - I}{I^3} - \frac{2I^2 - 4I + 2}{I^3} = \frac{3I - 2}{I^3} > 0.$$

That establishes (2) in case (1), which implies

$$\mathbf{P}'[q = ab|cd] > \max \left\{ \mathbf{P}'[q = ac|bd], \mathbf{P}'[q = ad|bc] \right\}, \tag{5}$$

as desired. The proof in case (2) can be found in the appendix.   □

As a direct consequence of our identifiability proof, it is straightforward to establish the statistical consistency of the following pipeline, which we refer to as ASTRAL/ONE (see also [12]): for each gene tree t_i, pick in each species a random gene copy (if possible) and run ASTRAL on the resulting set of modified gene trees $\tilde{t}_i$. The proof can be found in the appendix.

Theorem 2 (Statistical Consistency: ASTRAL/ONE). *ASTRAL/ONE is statistically consistent under the GDL model. That is, as the number of input gene trees tends toward infinity, the output of ASTRAL/ONE converges to T almost surely, when run in exact mode or in its default constrained version.*

3.3   Statistical Consistency of ASTRAL-multi Under GDL

The following consistency result is not a direct consequence of our identifiability result, although the ideas used are similar.

Theorem 3 (Statistical Consistency: ASTRAL-multi). *ASTRAL-multi, where copies of a gene in a species are treated as multiple alleles within the species, is statistically consistent under the GDL model. That is, as the number of input gene trees tends toward infinity, the output of ASTRAL-multi converges to T almost surely, when run in exact mode or in its default constrained version.*

Proof. First, we show that ASTRAL-multi is consistent when run in exact mode. The input are the gene trees $\mathcal{T} = \{t_i\}_{i=1}^k$ with t_i labelled by individuals (i.e., gene copies) $R_i \subseteq R$. Then the quartet score of $\tilde{T}$ with respect to $\mathcal{T}$ is given by (1). For any 4-tuple of gene copies $\mathcal{J} = \{a, b, c, d\}$, we define $m(\mathcal{J})$ to be the corresponding set of species. It was proved in [26] that those $\mathcal{J}$'s with fewer than 4 species contribute equally to all species tree topologies. As a result, it suffices to work with a modified quartet score

$$\tilde{Q}_k(\tilde{T}) = \sum_{i=1}^k \sum_{\substack{\mathcal{J} = \{a,b,c,d\} \subseteq R_i \\ |m(\mathcal{J})| = 4}} \mathbf{1}(\tilde{T}_{ext}^{\mathcal{J}}, t_i^{\mathcal{J}}).$$

By independence of the gene trees (and non-negativity), $\tilde{Q}_k(\tilde{T})/k$ converges almost surely to its expectation simultaneously for all unrooted species tree topologies over S.

The expectation can be simplified as

$$\mathbf{E}\left[\frac{1}{k}\widetilde{Q}_k(\widetilde{T})\right] = \mathbf{E}\left[\sum_{\substack{\mathcal{J}=\{a,b,c,d\}\subseteq R_1 \\ |m(\mathcal{J})|=4}} \mathbf{1}(\widetilde{T}_{ext}^{\mathcal{J}}, t_1^{\mathcal{J}})\right]$$

$$= \sum_{\mathcal{Q}=\{A,B,C,D\}} \mathbf{E}\left[\sum_{\mathcal{J}\subseteq R_1:m(\mathcal{J})=\mathcal{Q}} \mathbf{1}(\widetilde{T}_{ext}^{\mathcal{J}}, t_1^{\mathcal{J}})\right]. \quad (6)$$

Let $\mathcal{N}_{AB|CD}^{\mathcal{Q}}$ (respectively $\mathcal{N}_{AC|BD}^{\mathcal{Q}}, \mathcal{N}_{AD|BC}^{\mathcal{Q}}$) be the number of choices consisting of one gene copy in t_1 from each species in $\mathcal{Q}$ whose corresponding restriction $t_1^{\mathcal{Q}}$ agrees with $AB|CD$ (respectively $AC|BD$, $AD|BC$). Then each summand in (6) may be written as $\mathbf{E}[\mathcal{N}_{\widetilde{T}^{\mathcal{Q}}}^{\mathcal{Q}}]$. We establish below that this last expression is maximized at the true species tree $T^{\mathcal{Q}}$, that is,

$$\mathbf{E}[\mathcal{N}_{AB|CD}^{\mathcal{Q}}] > \max\left\{\mathbf{E}[\mathcal{N}_{AC|BD}^{\mathcal{Q}}], \mathbf{E}[\mathcal{N}_{AD|BC}^{\mathcal{Q}}]\right\}, \quad (7)$$

when (without loss of generality) $T^{\mathcal{Q}} = AB|CD$. From (6) and the law of large numbers, it will then follow that almost surely the quartet score is eventually maximized by the true species tree as $k \to +\infty$.

It remains to establish (7). Fix $\mathcal{Q} = \{A, B, C, D\}$ a set of four distinct species in T. Assume that the corresponding unrooted quartet topology in T is $AB|CD$. Let t_1 be a MUL-tree generated under the GDL model over T. Again, there are two cases: $T^{\mathcal{Q}}$ is (1) balanced or (2) a caterpillar.

In case (1), let R be the most recent common ancestor of $\mathcal{Q}$ in T and let I be the number of gene copies exiting (forward in time) R. For $j \in \{1, \ldots, I\}$, let $\mathcal{A}_j$ be the number of gene copies in A descending from j in R, and similarly define $\mathcal{B}_j$, $\mathcal{C}_j$ and $\mathcal{D}_j$. By the law of total probability, $\mathbf{E}[\mathcal{N}_{AB|CD}^{\mathcal{Q}}] = \mathbf{E}[\mathbf{E}_I[\mathcal{N}_{AB|CD}^{\mathcal{Q}}]]$. We show that, almost surely,

$$\mathbf{E}_I[\mathcal{N}_{AB|CD}^{\mathcal{Q}}] > \max\left\{\mathbf{E}_I[\mathcal{N}_{AC|BD}^{\mathcal{Q}}], \mathbf{E}_I[\mathcal{N}_{AD|BC}^{\mathcal{Q}}]\right\}, \quad (8)$$

which implies (7). By symmetry, we have $X^= \equiv \mathbf{E}_I[\mathcal{A}_j\mathcal{B}_j] = \mathbf{E}_I[\mathcal{A}_1\mathcal{B}_1]$, $Y^= \equiv \mathbf{E}_I[\mathcal{C}_j\mathcal{D}_j] = \mathbf{E}_I[\mathcal{C}_1\mathcal{D}_1]$, $X^{\neq} \equiv \mathbf{E}_I[\mathcal{A}_j\mathcal{B}_k] = \mathbf{E}_I[\mathcal{A}_1]\mathbf{E}_I[\mathcal{B}_1]$ as well as $Y^{\neq} \equiv \mathbf{E}_I[\mathcal{C}_j\mathcal{D}_k] = \mathbf{E}_I[\mathcal{C}_1]\mathbf{E}_I[\mathcal{D}_1]$ for all j, k with $j \neq k$. Hence, the expected number of pairs consisting of a single gene copy from A and B is $X = IX^= + I(I-1)X^{\neq}$. Arguing similarly to (3) and (4),

$$\mathbf{E}_I[\mathcal{N}_{AB|CD}^{\mathcal{Q}}] - \mathbf{E}_I[\mathcal{N}_{AC|BD}^{\mathcal{Q}}]$$
$$\geq (IX^=)Y + X(IY^=) - (IX^=)(IY^=) - I(I-1)X^{\neq}[2(I-1)Y^{\neq}]$$
$$= XY\left[x + y - xy - 2(1-x)(1-y)\frac{1}{I}\right],$$

where here we define $x = \frac{IX^=}{X}, y = \frac{IY^=}{Y}$. Following the argument in the proof of Theorem 1, to establish (8) it suffices to show that almost surely, $x, y \geq 1/I$. That is implied by the following positive correlation result.

Lemma 2. *Almost surely,* $X^= \geq X^{\neq}$.

Indeed, we then have: $x = \frac{IX^=}{IX^=+I(I-1)X^{\neq}} \geq \frac{IX^=}{IX^=+I(I-1)X^=} = \frac{1}{I}$.

Proof (Lemma 2). For $j \in \{1, \ldots, I\}$, let N_j be the number of gene copies at the divergence of the most recent common ancestor of A and B that are descending from j in R. Then, for $j \in \{1, \ldots, I\}$, since $\mathcal{A}_j$ and $\mathcal{B}_j$ are conditionally independent given $(N_j)_j$ under $\mathbf{E}_I$, it follows that

$$X^= = \mathbf{E}_I[\mathbf{E}_I[\mathcal{A}_j \mathcal{B}_j \mid (N_j)_j]] = \mathbf{E}_I[(N_j \alpha)(N_j \beta)] = \alpha \beta \mathbf{E}_I[N_j^2],$$

where α (respectively β) is the expected number of gene copies in A (respectively B) descending from a single gene copy in the most recent common ancestor of A and B under $\mathbf{E}_I$. Similarly, for $j \neq k \in \{1, \ldots, I\}$,

$$X^{\neq} = \mathbf{E}_I[\mathbf{E}_I[\mathcal{A}_j \mathcal{B}_k \mid (N_j)_j]] = \mathbf{E}_I[(N_j \alpha)(N_k \beta)] = \alpha \beta \mathbf{E}_I[N_j N_k] \leq \alpha \beta \mathbf{E}_I[N_j^2],$$

by Cauchy-Schwarz and $\mathbf{E}_I[N_j^2] = \mathbf{E}_I[N_k^2]$. □

We establish (8) in case (2) in the appendix. Thus, ASTRAL-multi is statistically consistent when run in exact mode, because it is guaranteed to return the optimal tree, and that is realized by the species tree. To see why the default version of ASTRAL-multi is also statistically consistent, note that the true species tree will appear as one of the input gene trees, almost surely, as the number of MUL-trees sampled tends to infinity. For instance, the probability of observing no duplications or losses is strictly positive. Furthermore, when this happens, the true species tree bipartitions are all contained in the constraint set Σ used by the default version. Hence, as the number of sampled MUL-trees increases, almost surely ASTRAL-multi will return the true species tree topology. □

4   Experiments

We performed a simulation study to evaluate ASTRAL-multi and other species tree estimation methods on 16-taxon datasets with model conditions characterized by three GDL rates, five levels of gene tree estimation error (GTEE), and four numbers of genes. Due to space constraints, we briefly describe the study here and provide details sufficient to reproduce the study on bioRxiv: https://doi.org/10.1101/821439. In addition, all scripts and datasets used in this study are available on the Illinois Data Bank: https://doi.org/10.13012/B2IDB-2626814V1.

Our simulation protocol uses parameters estimated from the 16-taxon fungal dataset studied in [12, 27]. First, we used the species tree and other parameters estimated from the fungal dataset to simulate gene trees under the DLCoal [27] model with three GDL rates (the lowest rate 1×10^{-10} reflects the GDL rate estimated from the fungal dataset, so that the two higher rates reflect more challenging model conditions). Specifically, for each GDL rate, we simulated 10 replicate datasets (each with 1000 model gene trees that deviated from the strict molecular clock) using SimPhy [21]. Although we simulated gene trees under a unified model of GDL and ILS, there was effectively no ILS in our simulated datasets (Table 1 on bioRxiv: https://doi.org/10.1101/821439). Second, for each model gene tree, we used INDELible [14] to simulate a multiple sequence alignment under the GTR+GAMMA model with parameters based on the fungal dataset. Third, we ran RAxML [32] to estimate a gene tree under the GTR+GAMMA model from each gene alignment. By varying the length of each gene alignment, four model conditions were created with 23% to 65% mean GTEE, as measured by the normalized Robinson-Foulds (RF) distance [28] between true and estimated gene trees, averaged across all gene trees. Fourth, we ran species tree estimation methods given varying numbers of gene trees as input. Finally, we evaluated species tree error as the normalized RF distance between true and estimated species trees.

In our first experiment, we explored ASTRAL-multi on both true and estimated gene trees (Fig. 1). ASTRAL-multi was very accurate on true gene trees; even with just 25 true gene trees, the average species tree error was less than 1% for the two lower GDL rates and was less than 6% for the highest GDL rate (5×10^{-10}). As expected, species tree error increased with the GDL rate, increased with the GTEE level, and decreased with the number of genes.

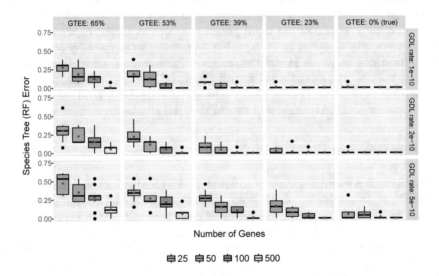

Fig. 1. ASTRAL-multi on true and estimated gene trees generated from the fungal species tree (16 taxa) under a GDL model using three different rates (subplot rows). Estimated gene trees had four different levels of gene tree estimation error (GTEE), by varying the sequence length (subplot columns). We report the average Robinson-Foulds (RF) error rate between the true and estimated species trees. There are 10 replicate datasets per model condition. Red dots indicate means, and bars indicated medians.

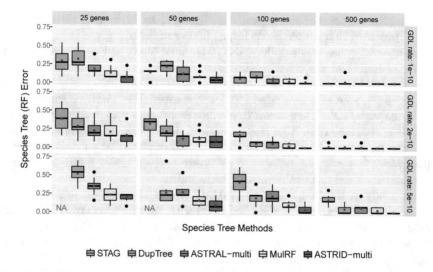

Fig. 2. Average RF tree error rates of species tree methods on estimated gene trees (mean GTEE: 53%) generated from the fungal 16-taxon species tree using three different GDL rates (subplot rows) and different numbers of genes (subplot columns). STAG failed to run on some replicate datasets for model conditions indicated by "NA", because none of the input gene trees included at least one copy of every species.

In our second experiment, we compared ASTRAL-multi to four other species tree methods (DupTree [38], MulRF [9], STAG [13], and ASTRID-multi, which is ASTRID [36] run under the multi-allele setting) that take gene trees as input. Figure 2 shows species tree error for model conditions with mean GTEE of 53%. As expected, the error increased for all methods with the GDL rate and GTEE level, and decreased with the number of genes. Differences between methods depended on the model condition. When given 500 genes, all five methods were competitive (with a slight disadvantage to STAG); a similar trend was observed when methods were given 100 genes provided that the GDL rate was one of the two lower rates. When given 50 genes, ASTRAL-multi, MulRF, and ASTRID-multi were the best methods for the two lower GDL rates. On the remaining model conditions, ASTRID-multi was the best method. Finally, STAG was unable to run on some datasets when the GDL rate was high and the number of genes was low; this result was due to STAG failing when none of the input gene trees included at least one copy of every species. Results for other GTEE levels are provided Table 2 on bioRxiv: https://doi.org/10.1101/821439, and show similar trends.

5   Discussion and Conclusion

This study establishes the identifiability of unrooted species trees under the simple model of GDL from [2] and that ASTRAL-multi is statistically consistent under this model. In our simulation study, ASTRAL-multi was accurate under challenging model conditions, characterized by high GDL rates and high GTEE, provided that a sufficiently large number of genes is given as input. When the number of genes was smaller, ASTRID-multi often had an advantage over ASTRAL-multi and the other methods.

The results of this study can be compared to the previous study by Chaudhary *et al.* [8], who also evaluated species tree estimation methods under model conditions with GDL. They found that MulRF and gene tree parsimony methods had better accuracy than NJst [19] (a method that is similar to ASTRID). Their study has an advantage over our study in that it explored larger datasets (up to 500 species); however, all genes in their study evolved under a strict molecular clock, and they did not evaluate ASTRAL-multi.

Our study is the first study to evaluate ASTRAL-multi on *estimated* gene trees, and we also explore model conditions with varying levels of GTEE. Evaluating methods under conditions with moderate to high GTEE is critical, as estimated gene trees from four recent studies [6,15,16,33] all had mean bootstrap support values below 50% (see Table 1 in [25]), suggesting high GTEE.

Our study is limited to one underlying species tree topology with 16 species. Previous studies [37] have shown that MulRF (which uses a heuristic search strategy to find solutions to its NP-hard optimization problem) is much slower than ASTRAL on large datasets, suggesting that ASTRAL-multi may dominate MulRF as the number of species increases. Hence, future studies should investigate ASTRAL-multi and other methods under a broader range of conditions, including larger numbers of species. Future research should also consider empirical performance and statistical consistency under different causes of gene tree heterogeneity.

We note with interest that the proof that ASTRAL-multi is statistically consistent is based on the fact that the most probable unrooted gene tree on four leaves (according to two ways of defining it) under the GDL model is the true species tree (equivalently, there is no anomaly zone for the GDL model for unrooted four-leaf trees). This coincides with the reason ASTRAL is statistically consistent under the MSC as well as under a model for random HGT [10,30]. Furthermore, previous studies have shown that ASTRAL has good accuracy in simulation studies where both ILS and HGT are present [11]. Hence ASTRAL, which was originally designed for species tree estimation in the presence of ILS, has good accuracy and theoretical guarantees under different sources of gene tree heterogeneity.

We also note the surprising accuracy of DupTree, MulRF, and ASTRID-multi, methods that, like ASTRAL-multi, are not based on likelihood under a GDL model. Therefore, DynaDup [5,23] is also of potential interest, as it is similar to DupTree in seeking a tree that minimizes the duploss score (though the score is modified to reflect true biological loss), but has the potential to scale to larger datasets via its use of dynamic programming to solve the optimization problem in polynomial time within a constrained search space. In addition, future research should explore these methods compared to more computationally intensive methods such as InferNetwork_ML and InferNetwork_MPL (maximum likelihood and maximum pseudo-likelihood methods in PhyloNet [35,39]) restricted so that they produce trees rather than reticulate phylogenies, or PHYLDOG [7], a likehood-based method for co-estimating gene trees and the species tree under a GDL model.

Acknowledgments. This study was supported in part by NSF grants CCF-1535977 and 1513629 (to TW) and by the Ira and Debra Cohen Graduate Fellowship in Computer Science (to EKM). SR was supported by NSF grants DMS-1614242, CCF-1740707 (TRIPODS), and DMS-1902892, as well as a Simons Fellowship and a Vilas Associates Award. BL was supported by NSF grant DMS-1614242 (to SR). All computational analyses were performed on the Illinois Campus Cluster and the Blue Waters supercomputer, computing resources that are operated and financially supported by UIUC in conjunction with the National Center for Supercomputing Applications. Blue Waters is supported by the NSF (grants OCI-0725070 and ACI-1238993) and the state of Illinois.

A   Additional Proofs

A.1   Proof of Theorem 1: case (2)

In case (2), assume that $T^{\mathcal{Q}} = (((A, B), C), D)$, let R be the most recent common ancestor of A, B, C (but not D) in $T^{\mathcal{Q}}$, and let I be the number of gene copies exiting R. As in case 1), it suffices to prove (2) almost surely. Let $i_x \in \{1, ..., I\}$ be the ancestral lineage of $x \in \{a, b, c\}$ in R. Then

$$\mathbf{P}'_I[q = ab|cd] = \mathbf{P}'_I[i_a = i_b] + \mathbf{P}'_I[q = ab|cd \text{ and } i_a, i_b, i_c \text{ all distinct}]. \quad (9)$$

On the other hand,

$$\mathbf{P}'_I[q = ac|bd] = \mathbf{P}'_I[i_b \neq i_a = i_c] + \mathbf{P}'_I[q = ac|bd \text{ and } i_a, i_b, i_c \text{ all distinct}], \quad (10)$$

with a similar result for $\mathbf{P}'_I[q = ad|bc]$. By symmetry again, the last term on the RHS of (9) and (10) are the same. This implies

$$\mathbf{P}'_I[q = ab|cd] - \mathbf{P}'_I[q = ac|bd] = \mathbf{P}'_I[i_a = i_b] - \mathbf{P}'_I[i_b \neq i_a = i_c]$$
$$= x - (1 - x)\mathbf{P}'_I[i_a = i_c|i_a \neq i_b] = x - (1 - x)\frac{1}{I} \equiv g(x),$$

where $x = \mathbf{P}'_I[i_a = i_b]$. This function g attains its minimum value at the smallest possible of x, which by Lemma 1 is $x = 1/I$. Evaluating at $x = 1/I$ gives

$$g(1/I) = \frac{1}{I} - \frac{1}{I} + \frac{1}{I^2} = \frac{1}{I^2} > 0,$$

which establishes (2) in case (2).

A.2   Proof of Theorem 2

First, we prove consistency for the exact version of ASTRAL. The input to the ASTRAL/ONE pipeline is the collection of gene trees $\mathcal{T} = \{t_i\}_{i=1}^k$, where t_i is labeled by individuals (i.e., gene copies) $R_i \subseteq R$. For each species and each gene tree t_i, we pick a uniform random gene copy, producing a new gene tree $\tilde{t}_i$. Recall that the quartet score of $\widetilde{T}$ with respect to $\widetilde{\mathcal{T}} = \{\tilde{t}_i\}_{i=1}^k$ is then

$$Q_k(\widetilde{T}) = \sum_{i=1}^k \sum_{\mathcal{J}=\{a,b,c,d\}\subseteq R_i} \mathbf{1}(\widetilde{T}_{ext}^{\mathcal{J}}, \tilde{t}_i^{\mathcal{J}}).$$

We note that the score only depends on the unrooted topology of $\widetilde{T}$. Under the GDL model, by independence of the gene trees (and non-negativity), $Q_k(\widetilde{T})/k$ converges almost surely to its expectation simultaneously for all unrooted species tree topologies over S.

For a species $A \in S$ and gene tree $\tilde{t}_i$, let A_i be the gene copy in A on $\tilde{t}_i$ if it exists and let $\mathcal{E}_i^A$ be the event that it exists. For a 4-tuple of species $\mathcal{Q} = \{A, B, C, D\}$, let $\mathcal{Q}_i = \{A_i, B_i, C_i, D_i\}$ and $\mathcal{E}_i^{\mathcal{Q}} = \mathcal{E}_i^A \cap \mathcal{E}_i^B \cap \mathcal{E}_i^C \cap \mathcal{E}_i^D$. The expectation can then be written as

$$\mathbf{E}\left[\frac{1}{k}Q_k(\widetilde{T})\right] = \sum_{\mathcal{Q}=\{A,B,C,D\}} \mathbf{E}\left[\mathbf{1}(\widetilde{T}_{ext}^{\mathcal{Q}_1}, \tilde{t}_1^{\mathcal{Q}_1}) \,\Big|\, \mathcal{E}_1^{\mathcal{Q}}\right] \mathbf{P}[\mathcal{E}_1^{\mathcal{Q}}], \tag{11}$$

as, on the event $(\mathcal{E}_1^{\mathcal{Q}})^c$, there is no contribution from $\mathcal{Q}$ in the sum over the first sample.

Based on the proof of Theorem 1, a different way to write $\mathbf{E}[\mathbf{1}(\widetilde{T}_{ext}^{\mathcal{Q}_1}, \tilde{t}_1^{\mathcal{Q}_1}) \,|\, \mathcal{E}_1^{\mathcal{Q}}]$ is in terms of the original gene tree t_1. Let a, b, c, d be random gene copies on t_1 in A, B, C, D respectively. Then if q is the topology of t_1 restricted to a, b, c, d,

$$\mathbf{E}\left[\mathbf{1}(\widetilde{T}_{ext}^{\mathcal{Q}_1}, \tilde{t}_1^{\mathcal{Q}_1}) \,\Big|\, \mathcal{E}_1^{\mathcal{Q}}\right] = \mathbf{P}'[q = \widetilde{T}^{\mathcal{Q}}].$$

From (5), we know that this expression is maximized (strictly) at the true species tree $\mathbf{P}'[q = T^{\mathcal{Q}}]$. Hence, together with (11) and the law of large numbers, almost surely the quartet score is eventually maximized by the true species tree as $k \to +\infty$. This completes the proof for the exact version.

The default version is statistically consistent for the same reason as in the proof of Theorem 3. As the number of MUL-trees sampled tends to infinity, the true species tree will appear as one of the input gene trees almost surely. So ASTRAL returns the true species tree topology almost surely as the number of sampled MUL-trees increases.

A.3   Proof of Theorem 3: case (2)

In case (2), assume that $T^{\mathcal{Q}} = (((A, B), C), D)$ and let R be the most recent common ancestor of A, B, C (but not D) in T. We want to establish (8) in this case. For $i = 1, 2, 3$, let $\mathcal{N}_{AB|CD}^{\mathcal{Q},\{i\}}$ (respectively $\mathcal{N}_{AC|BD}^{\mathcal{Q},\{i\}}$) be the number of choices consisting of one gene copy from each species in $\mathcal{Q}$ whose corresponding restriction on $t^{\mathcal{Q}}$ agrees with $AB|CD$ (respectively $AC|BD$) and where, in addition, copies of A, B, C descend from i distinct lineages in R. We make five observations:

- Contributions to $\mathcal{N}_{AB|CD}^{\mathcal{Q},\{2\}}$ necessarily come from copies in A and B descending from the same lineage in R, together with a copy in C descending from a distinct lineage and any copy in D. Similarly for $\mathcal{N}_{AC|BD}^{\mathcal{Q},\{2\}}$

- Moreover $\mathcal{N}_{AC|BD}^{\mathcal{Q},\{1\}} = 0$ almost surely, as in that case the corresponding copies from A and B coalesce (backwards in time) below R.

- Arguing as in the proof of Theorem 1, by symmetry we have the equality $\mathbf{E}_I[\mathcal{N}_{AB|CD}^{\mathcal{Q},\{3\}}] = \mathbf{E}_I[\mathcal{N}_{AC|BD}^{\mathcal{Q},\{3\}}]$.

- For $j \in \{1, \ldots, I\}$, let $\mathcal{A}_j$ be the number of gene copies in A descending from j in R, and similarly define $\mathcal{B}_j$, $\mathcal{C}_j$. Let $\mathcal{D}$ be the number of gene copies in D. Then, under the conditional probability $\mathbf{P}_I$, $\mathcal{D}$ is independent of $(\mathcal{A}_j, \mathcal{B}_j, \mathcal{C}_j)_{j=1}^I$. Moreover, under $\mathbf{P}_I$, $(\mathcal{C}_j)_{j=1}^I$ is independent of $(\mathcal{A}_j, \mathcal{B}_j)_{j=1}^I$.

- Similarly to case 1), by symmetry we have $X^= \equiv \mathbf{E}_I[\mathcal{A}_{j_1}\mathcal{B}_{j_1}] = \mathbf{E}_I[\mathcal{A}_1\mathcal{B}_1]$, $X^{\neq} \equiv \mathbf{E}_I[\mathcal{A}_{j_1}\mathcal{B}_{k_1}] = \mathbf{E}_I[\mathcal{A}_1]\mathbf{E}_I[\mathcal{B}_1]$ for all j_1, k_1 with $j_1 \neq k_1$. Define also $X = IX^= + I(I-1)X^{\neq}$, $Y \equiv \mathbf{E}_I[\mathcal{C}_1]$ and $Z \equiv \mathbf{E}_I[\mathcal{D}]$.

Putting these observations together, we obtain

$$
\begin{aligned}
\mathbf{E}_I[\mathcal{N}_{AB|CD}^{\mathcal{Q}}] &- \mathbf{E}_I[\mathcal{N}_{AC|BD}^{\mathcal{Q}}] \\
&= \mathbf{E}_I[\mathcal{N}_{AB|CD}^{\mathcal{Q},\{1\}}] + \mathbf{E}_I[\mathcal{N}_{AB|CD}^{\mathcal{Q},\{2\}}] - \mathbf{E}_I[\mathcal{N}_{AC|BD}^{\mathcal{Q},\{2\}}] \\
&= IX^=YZ + I(I-1)X^=YZ - I(I-1)X^{\neq}YZ \\
&> 0,
\end{aligned}
$$

where we used Lemma 2 on the last line.

References

1. Allman, E.S., Degnan, J.H., Rhodes, J.A.: Identifying the rooted species tree from the distribution of unrooted gene trees under the coalescent. J. Math. Biol. **62**(6), 833–862 (2011). https://doi.org/10.1007/s00285-010-0355-7

2. Arvestad, L., Lagergren, J., Sennblad, B.: The gene evolution model and computing its associated probabilities. J. ACM **56**(2), 7 (2009). https://doi.org/10.1145/1502793.1502796

3. Bandelt, H.J., Dress, A.: Reconstructing the shape of a tree from observed dissimilarity data. Adv. Appl. Math. **7**(3), 309–343 (1986). https://doi.org/10.1016/0196-8858(86)90038-2

4. Bansal, M.S., Burleigh, J.G., Eulenstein, O., Fernández-Baca, D.: Robinson-foulds supertrees. Algorithms Mol. Biol. **5**(1), 18 (2010). https://doi.org/10.1186/1748-7188-5-18

5. Bayzid, M.S., Warnow, T.: Gene tree parsimony for incomplete gene trees: addressing true biological loss. Algorithms Mol. Biol. **13**(1), 1 (2018). https://doi.org/10.1186/s13015-017-0120-1

6. Blom, M.P.K., Bragg, J.G., Potter, S., Moritz, C.: Accounting for uncertainty in gene tree estimation: summary-coalescent species tree inference in a challenging radiation of Australian lizards. Syst. Biol. **66**(3), 352–366 (2017). https://doi.org/10.1093/sysbio/syw089

7. Boussau, B., Szöllősi, G.J., Duret, L., Gouy, M., Tannier, E., Daubin, V.: Genome-scale coestimation of species and gene trees. Genome Res. **23**(2), 323–330 (2013). https://doi.org/10.1101/gr.141978.112

8. Chaudhary, R., Boussau, B., Burleigh, J.G., Fernández-Baca, D.: Assessing approaches for inferring species trees from multi-copy genes. Syst. Biol. **64**(2), 325–339 (2015). https://doi.org/10.1093/sysbio/syu128

9. Chaudhary, R., Fernández-Baca, D., Burleigh, J.G.: MulRF: a software package for phylogenetic analysis using multi-copy gene trees. Bioinformatics **31**(3), 432–433 (2014). https://doi.org/10.1093/bioinformatics/btu648

10. Daskalakis, C., Roch, S.: Species trees from gene trees despite a high rate of lateral genetic transfer: a tight bound (extended abstract). In: Proceedings of the Twenty-Seventh Annual ACM-SIAM Symposium on Discrete Algorithms, pp. 1621–1630 (2016). https://doi.org/10.1137/1.9781611974331.ch110

11. Davidson, R., Vachaspati, P., Mirarab, S., Warnow, T.: Phylogenomic species tree estimation in the presence of incomplete lineage sorting and horizontal gene transfer. BMC Genom. **16**(10), S1 (2015). https://doi.org/10.1186/1471-2164-16-S10-S1

12. Du, P., Hahn, M.W., Nakhleh, L.: Species tree inference under the multispecies coalescent on data with paralogs is accurate. bioRxiv (2019). https://doi.org/10.1101/498378

13. Emms, D., Kelly, S.: STAG: species tree inference from all genes. bioRxiv (2018). https://doi.org/10.1101/267914

14. Fletcher, W., Yang, Z.: INDELible: a flexible simulator of biological sequence evolution. Mol. Biol. Evol. **26**(8), 1879–1888 (2009). https://doi.org/10.1093/molbev/msp098

15. Hosner, P.A., Faircloth, B.C., Glenn, T.C., Braun, E.L., Kimball, R.T.: Avoiding missing data biases in phylogenomic inference: an empirical study in the landfowl (Aves: Galliformes). Mol. Biol. Evol. **33**(4), 1110–1125 (2016). https://doi.org/10.1093/molbev/msv347

16. Jarvis, E.D., Mirarab, S., et al.: Whole-genome analyses resolve early branches in the tree of life of modern birds. Science **346**(6215), 1320–1331 (2014). https://doi.org/10.1126/science.1253451

17. Kingman, J.F.C.: The coalescent. Stoch. process. Their Appl. **13**(3), 235–248 (1982). https://doi.org/10.1016/0304-4149(82)90011-4

18. Larget, B.R., Kotha, S.K., Dewey, C.N., Ané, C.: BUCKy: gene tree/species tree reconciliation with Bayesian concordance analysis. Bioinformatics **26**(22), 2910–2911 (2010). https://doi.org/10.1093/bioinformatics/btq539

19. Liu, L., Yu, L.: Estimating species trees from unrooted gene trees. Syst. Biol. **60**(5), 661–667 (2011). https://doi.org/10.1093/sysbio/syr027

20. Maddison, W.: Gene trees in species trees. Syst. Biol. **46**(3), 523–536 (1997). https://doi.org/10.1093/sysbio/46.3.523

21. Mallo, D., De Oliveira Martins, L., Posada, D.: SimPhy: phylogenomic simulation of gene, locus, and species trees. Syst. Biol. **65**(2), 334–344 (2016). https://doi.org/10.1093/sysbio/syv082

22. Mirarab, S., Reaz, R., Bayzid, M.S., Zimmermann, T., Swenson, M.S., Warnow, T.: ASTRAL: genome-scale coalescent-based species tree estimation. Bioinformatics **30**(17), i541–i548 (2014). https://doi.org/10.1093/bioinformatics/btu462

23. Mirarab, S.: DynaDup github repository: a software package for species tree estimation from rooted gene trees under gene duplication and loss. https://github.com/smirarab/DynaDup. Accessed 3 Oct 2019
24. Mirarab, S., Warnow, T.: ASTRAL-II: coalescent-based species tree estimation with many hundreds of taxa and thousands of genes. Bioinformatics **31**(12), i44–i52 (2015). https://doi.org/10.1093/bioinformatics/btv234
25. Molloy, E.K., Warnow, T.: To include or not to include: the impact of gene filtering on species tree estimation methods. Syst. Biol. **67**(2), 285–303 (2018). https://doi.org/10.1093/sysbio/syx077
26. Rabiee, M., Sayyari, E., Mirarab, S.: Multi-allele species reconstruction using ASTRAL. Mol. Phylogenet. Evol. **130**, 286–296 (2019). https://doi.org/10.1016/j.ympev.2018.10.033
27. Rasmussen, M.D., Kellis, M.: Unified modeling of gene duplication, loss, and coalescence using a locus tree. Genome Res. **22**(4), 755–765 (2012). https://doi.org/10.1101/gr.123901.111
28. Robinson, D., Foulds, L.: Comparison of phylogenetic trees. Math. Biosci. **53**(1), 131–147 (1981). https://doi.org/10.1016/0025-5564(81)90043-2
29. Roch, S., Nute, M., Warnow, T.: Long-branch attraction in species tree estimation: inconsistency of partitioned likelihood and topology-based summary methods. Syst. Biol. **68**(2), 281–297 (2018). https://doi.org/10.1093/sysbio/syy061
30. Roch, S., Snir, S.: Recovering the treelike trend of evolution despite extensive lateral genetic transfer: a probabilistic analysis. J. Comput. Biol. **20**(2), 93–112 (2013). https://doi.org/10.1089/cmb.2012.0234
31. Roch, S., Steel, M.: Likelihood-based tree reconstruction on a concatenation of aligned sequence data sets can be statistically inconsistent. Theor. Popul. Biol. **100**, 56–62 (2015). https://doi.org/10.1016/j.tpb.2014.12.005
32. Stamatakis, A.: RAxML version 8: a tool for phylogenetic analysis and post-analysis of large phylogenies. Bioinformatics **30**(9), 1312–1313 (2014). https://doi.org/10.1093/bioinformatics/btu033
33. Streicher, J.W., Schulte II, J.A., Wiens, J.J.: How should genes and taxa be sampled for phylogenomic analyses with missing data? An empirical study in iguanian lizards. Syst. Biol. **65**(1), 128–145 (2016). https://doi.org/10.1093/sysbio/syv058
34. Takahata, N.: Gene genealogy in three related populations: consistency probability between gene and population trees. Genetics **122**(4), 957–966 (1989)
35. Than, C., Ruths, D., Nakhleh, L.: PhyloNet: a software package for analyzing and reconstructing reticulate evolutionary relationships. BMC Bioinform. **9**(1), 322 (2008). https://doi.org/10.1186/1471-2105-9-322
36. Vachaspati, P., Warnow, T.: ASTRID: accurate species TRees from internode distances. BMC Genom. **16**(10), S3 (2015). https://doi.org/10.1186/1471-2164-16-S10-S3
37. Vachaspati, P., Warnow, T.: FastRFS: fast and accurate Robinson-Foulds supertrees using constrained exact optimization. Bioinformatics **33**(5), 631–639 (2016). https://doi.org/10.1093/bioinformatics/btw600
38. Wehe, A., Bansal, M.S., Burleigh, J.G., Eulenstein, O.: DupTree: a program for large-scale phylogenetic analyses using gene tree parsimony. Bioinformatics **24**(13), 1540–1541 (2008). https://doi.org/10.1093/bioinformatics/btn230
39. Wen, D., Yu, Y., Zhu, J., Nakhleh, L.: Inferring phylogenetic networks using PhyloNet. Syst. Biol. **67**(4), 735–740 (2018). https://doi.org/10.1093/sysbio/syy015
40. Zhang, C., Rabiee, M., Sayyari, E., Mirarab, S.: ASTRAL-III: polynomial time species tree reconstruction from partially resolved gene trees. BMC Bioinform. **19**(6), 153 (2018). https://doi.org/10.1186/s12859-018-2129-y

RoboCOP: Multivariate State Space Model Integrating Epigenomic Accessibility Data to Elucidate Genome-Wide Chromatin Occupancy

Sneha Mitra[1], Jianling Zhong[2], David M. MacAlpine[2,3,4], and Alexander J. Hartemink[1,2,4(✉)]

[1] Department of Computer Science, Duke University, Durham, NC 27708, USA
amink@cs.duke.edu
[2] Program in Computational Biology and Bioinformatics, Duke University, Durham, NC 27708, USA
[3] Department of Pharmacology and Cancer Biology, Duke University Medical Center, Durham, NC 27710, USA
[4] Center for Genomic and Computational Biology, Duke University, Durham, NC 27708, USA

Abstract. Chromatin is the tightly packaged structure of DNA and protein within the nucleus of a cell. The arrangement of different protein complexes along the DNA modulates and is modulated by gene expression. Measuring the binding locations and level of occupancy of different transcription factors (TFs) and nucleosomes is therefore crucial to understanding gene regulation. Antibody-based methods for assaying chromatin occupancy are capable of identifying the binding sites of specific DNA binding factors, but only one factor at a time. On the other hand, epigenomic accessibility data like ATAC-seq, DNase-seq, and MNase-seq provide insight into the chromatin landscape of all factors bound along the genome, but with minimal insight into the identities of those factors. Here, we present RoboCOP, a multivariate state space model that integrates chromatin information from epigenomic accessibility data with nucleotide sequence to compute genome-wide probabilistic scores of nucleosome and TF occupancy, for hundreds of different factors at once. RoboCOP can be applied to any epigenomic dataset that provides quantitative insight into chromatin accessibility in any organism, but here we apply it to MNase-seq data to elucidate the protein-binding landscape of nucleosomes and 150 TFs across the yeast genome. Using available protein-binding datasets from the literature, we show that our model more accurately predicts the binding of these factors genome-wide.

Keywords: Chromatin accessibility · MNase-seq · Hidden Markov model

© Springer Nature Switzerland AG 2020
R. Schwartz (Ed.): RECOMB 2020, LNBI 12074, pp. 136–151, 2020.
https://doi.org/10.1007/978-3-030-45257-5_9

1   Introduction

Chromatin is a tightly packaged structure of proteins and DNA in the nucleus of a cell. The arrangement of different proteins along the DNA determines how gene expression is regulated. Two important groups of DNA binding factors (DBFs) are transcription factors (TFs) and nucleosomes. TFs are key gene regulatory proteins that promote or suppress transcription by binding with specific sequence preferences to sites along the DNA. Nucleosomes form when 147 base pairs of DNA are wrapped around an octamer of histone proteins. They have lower sequence specificity than TFs, but exhibit preferences for a periodic arrangement of dinucleotides that facilitate DNA wrapping. Likened to beads on a string, nucleosomes are positioned fairly regularly along the DNA, occupying about 81% of the genome in the case of *Saccharomyces cerevisiae* (yeast) [14]. In taking up their respective positions, nucleosomes allow or block TFs from occupying their putative binding sites, thereby contributing to the regulation of gene expression. Revealing the chromatin landscape—how all these DBFs are positioned along the genome—is therefore crucial to developing a more mechanistic (and eventually predictive) understanding of gene regulation.

Antibody-based methods have been used extensively to assay the binding of particular DBFs at high resolution. However, such methods are limited to assaying only one factor at a time. Chromatin accessibility datasets, on the other hand, provide information about open regions of the chromatin, indirectly telling us about the regions occupied by various proteins. Many protocols can be used to generate chromatin accessibility data, including transposon insertion (ATAC-seq), enzymatic cleavage (DNase-seq), or enzymatic digestion (MNase-seq). In the latter, the endo-exonuclease MNase is used to digest unbound DNA, leaving behind undigested fragments of bound DNA. Paired-end sequencing of these fragments reveals not only their location but also their length, yielding information about the sizes of the proteins bound in different genomic regions. MNase-seq has been widely used to study nucleosome positions [3,4], but evidence of TF binding sites has also been observed in the data [10].

Several chromatin segmentation methods use epigenomic data to infer the locations of 'states' like promoters and enhancers, particularly in human and mouse genomes [1,6,11,22], but identifying the precise binding locations of myriad individual DBFs is more difficult. The high cost of repeated deep sequencing of large genomes poses a major challenge. In comparison to the complex human and mouse genomes, the problem is a bit simpler when working with the yeast genome, because it is smaller and therefore more economical to sequence deeply.

In earlier work, we proposed COMPETE to compute a probabilistic occupancy landscape of DBFs along the genome [23]. COMPETE considers DBFs binding to the genome in the form of a thermodynamic ensemble, where the DBFs are in continual competition to occupy locations along the genome and their chances of binding are affected by their concentrations, akin to a repeated game of 'musical chairs'. COMPETE output depends only on genome sequence (static) and DBF concentrations (dynamic); it is entirely theoretical, in that it makes no use of experimental chromatin data to influence its predictions of the chromatin landscape. A modified version of COMPETE was later developed to estimate DBF concentrations by maximizing the correlation between COMPETE's output and an MNase-seq signal, improving the reported binding landscape [25]. However, it still does not directly incorporate chromatin accessibility data into the model.

Here, we present RoboCOP, a new method that integrates epigenomic accessibility data and genomic sequence to produce accurate chromatin occupancy profiles of the genome. With nucleotide sequence and chromatin accessibility data as input, RoboCOP uses a multivariate hidden Markov model (HMM) to compute a probabilistic occupancy landscape of hundreds of DBFs genome-wide at single-nucleotide resolution. In this paper, we use paired-end MNase-seq data to predict TF binding sites and nucleosome positions throughout the *Saccharomyces cerevisiae* genome. We validate our TF binding site predictions using annotations reported by ChIP [15], ChIP-exo [19], and ORGANIC [13] experiments, and our nucleosome positioning predictions using high-precision annotations reported by a chemical cleavage method [2]. We find that RoboCOP provides valuable insight into the chromatin architecture of the genome, and can elucidate how it changes in response to different environmental conditions.

2   Results

2.1   MNase-seq Fragments of Different Lengths Are Informative About Different DNA Binding Factors

In Fig. 1a, we plot MNase-seq fragments around the transcription start sites (TSSs) of all yeast genes [16]. Fragments of length 127–187 (which we call

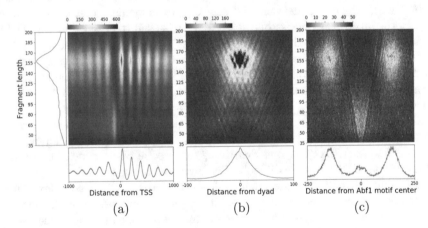

Fig. 1. (a) Heatmap of MNase-seq fragments, centered on all TSSs. Each fragment is plotted based on its length (*y*-axis) and the location of its midpoint (*x*-axis). Panels along the side and bottom show marginal densities. Heatmap reveals strong enrichment (red) of fragments corresponding to +1 nucleosomes (just downstream of TSS, lengths near 157). Upstream of TSS, in the promoter region, are many shortFrags (length ≤80). (b) Heatmap of MNase-seq fragments, centered on dyads of top 2000 well-positioned nucleosomes [2]. Fragment midpoint counts are highest at the dyad and decrease symmetrically in either direction. (c) Heatmap of MNase-seq fragments, centered on annotated Abf1 binding sites [15], showing an enrichment of shortFrags near Abf1 sites.

nucleosomal fragments, or `nucFrags` for short) occur in tandem arrays within gene bodies but are generally absent from promoters (Fig. 1a). Fragments are particularly concentrated at the +1 nucleosome position, just downstream of the TSS, because the +1 nucleosome is usually well-positioned. Furthermore, the marginal density of the midpoints of these fragments around annotated nucleosome dyads [2] peaks precisely at the dyad, with counts dropping nearly symmetrically in either direction (Fig. 1b). This makes sense because MNase digests linker regions, leaving behind undigested DNA fragments wrapped around histone octamers. So the midpoint counts of these `nucFrags` would be highest at the annotated dyads and decrease on moving away from the dyad.

In addition, it has been shown that shorter fragments in MNase-seq provide information about TF binding sites [10]. To verify that we see this signal in our data, Fig. 1a reveals that promoter regions are enriched with shorter fragments. The promoter region is often bound by specific and general TFs that aid in the transcription of genes. To ensure that the MNase-seq signal in these promoter regions is not just noise, we plot the MNase-seq midpoints around annotated TF binding sites. We choose the well-studied TF, Abf1, because it has multiple annotated binding sites across the genome. On plotting the MNase-seq midpoint counts around these annotated binding sites we notice a clear enrichment of short fragments at the binding sites (Fig. 1c). We denote these short fragments of length less than 80 as `shortFrags`. Unlike the midpoint counts of the `nucFrags` which have a symmetrically decreasing shape around the nucleosome dyads, the midpoint counts of `shortFrags` are more uniformly distributed within the binding site (Fig. 1c). The `shortFrags` signal at the Abf1 binding sites is noisier than the MNase signal associated with nucleosomes. One reason for this increased noise is that fragments protected from digestion by bound TFs may be quite small, and the smallest fragments (of length less than 27 in our case) are not even present in the dataset due to sequencing and alignment limitations.

We ignore fragments of intermediate length (81–126) in our analysis, though these could provide information about other kinds of complexes along the genome, like hexasomes [18]. Such factors would also be important for a complete understanding of the chromatin landscape, but we limit our analysis here to studying the occupancy of nucleosomes and TFs. For the subsequent sections of this paper, we only consider the midpoint counts of `nucFrags` and `shortFrags`. A representative snapshot of MNase-seq fragments is shown in Fig. 2a. We further simplify the two-dimensional plot in Fig. 2a to form two one-dimensional signals by separately aggregating the midpoint counts of `nucFrags` and `shortFrags`, as shown in Fig. 2b.

2.2 `RoboCOP` Computes Probabilistic Chromatin Occupancy Profiles

`RoboCOP` (**robo**tic **c**hromatin **o**ccupancy **p**rofiler) is a multivariate hidden Markov model (HMM) that jointly models the nucleotide sequence and the midpoint counts of `nucFrags` and `shortFrags` to learn the occupancy landscape of nucleosomes and TFs across a genome at single-nucleotide resolution. We apply `RoboCOP` on the *Saccharomyces cerevisiae* genome to predict nucleosome positions and the binding sites of 150 TFs (listed in Table S1). The HMM is

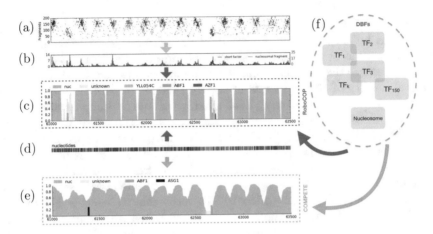

Fig. 2. (a) MNase-seq fragment midpoints in region chrI:61000–63500 of the yeast genome. Each dot at position (x, y) (red for nucFrags; blue for shortFrags) represents a fragment of length y centered on genomic coordinate x. (b) Aggregate numbers of red and blue dots. (d) Nucleotide sequence for chrI:61000–63500. (f) Set of DBFs (nucleosomes and TFs). (c) RoboCOP and (e) COMPETE outputs, with inputs depicted using green and orange arrows respectively. The score on the y-axes of (c) and (e) is the probability of that location being bound by each DBF.

structured such that each DBF corresponds to a collection of hidden states and each hidden state corresponds to a single genome coordinate. The hidden states of RoboCOP are inferred from a set of three observables at each coordinate: the nucleotide and the midpoint counts of nucFrags and shortFrags (Fig. S1). Based on these three observables, we estimate the posterior distribution over all hidden states. The resulting posterior probability of each DBF at each position in the genome provides a probabilistic profile of DBF occupancy at base-pair resolution (Fig. 2c). The inputs to RoboCOP are a set of DBFs (Fig. 2f), MNase-seq midpoint counts (Fig. 2b), and nucleotide sequence (Fig. 2d). From RoboCOP output in Fig. 2c, we observe that the nucleosome predictions line up well with the nucleosome signal in Figs. 2a,b.

RoboCOP's emission probabilities are derived from published position weight matrices [7] and MNase-seq signals around annotated DBF-occupied regions. These emission probabilities are fixed, remaining unchanged during model optimization. The transition probabilities among the DBFs, however, are unknown, so we optimize these parameters using expectation maximization (EM).

Our model contains 150 TFs and nucleosomes, but that is not all possible DBFs. Thus, factors not in the model could be responsible for the enrichment in nucFrags and shortFrags at certain locations. We observe the midpoint counts of shortFrags to be noisier, likely because of the binding of other small complexes that are not a part of the model, including general TFs. We therefore add an 'unknown factor' into the model to account for such DBFs. It is plotted in light gray in Fig. 2c. Incorporating this unknown factor reduces false positives among binding site predictions for the other TFs (Fig. S2).

2.3   RoboCOP's Use of Epigenomic Accessibility Data Improves the Resulting Chromatin Occupancy Profiles

Our group's previous work, COMPETE [23], is an HMM that computes a probabilistic occupancy landscape of the DBFs in the genome using only nucleotide sequence as input. The model output is theoretical in that it does not incorporate experimental data in learning the binding landscape of the genome. Perhaps unsurprisingly, the nucleosome positions learned by COMPETE (Fig. 2e) do not line up well with the nucleosomal signal apparent in MNase-seq data (Figs. 2a, b). The nucleosome predictions of COMPETE (Fig. 2e) are more diffuse, which is understandable because it relies entirely on sequence information, and nucleosomes have only weak and periodic sequence specificity. Because of a lack of chromatin accessibility data, COMPETE fails to identify the clear nucleosome depleted regions (all throughout the genome, as seen in Figs. 3a, b), as a result of which it fails to recognize the two Abf1 binding sites known to exist in this locus (Fig. 2e) [15]. In contrast, RoboCOP utilizes the chromatin accessibility data to accurately learn the nucleosome positions and the annotated Abf1 binding sites (Fig. 2c).

2.4   Predicted Nucleosome Positions

Nucleosomes have weak sequence specificity and can adopt alternative nearby positions along the genome. It is therefore likely that the nucleosome positions reported by one method do not exactly match those reported by another. However, since RoboCOP generates genome-wide probabilistic scores of nucleosome

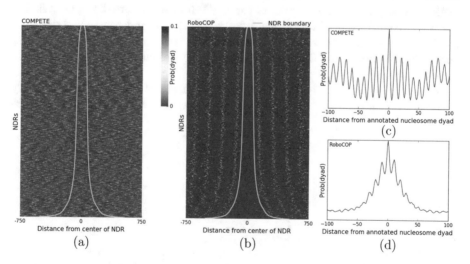

Fig. 3. Probability that positions around NDRs correspond to a nucleosome dyad, as computed by (a) COMPETE and (b) RoboCOP. Cyan lines depict experimentally determined NDR boundaries [5]. Note that Prob(dyad) computed by RoboCOP is appropriately almost always zero within NDRs, unlike COMPETE. Aggregate Prob(dyad), as computed by (c) COMPETE and (d) RoboCOP across all annotated nucleosome dyads [2]. Note that Prob(dyad) computed by RoboCOP appropriately peaks at annotated dyads.

occupancy, we can plot the probability of a nucleosome dyad, Prob(dyad), around annotated nucleosome locations [2]. We find that the RoboCOP dyad score peaks precisely at the annotated dyads (Fig. 3d), and decreases almost symmetrically in either direction. In contrast, COMPETE does not provide accurate location predictions (Fig. 3c); the oscillatory nature of the score reported by COMPETE reflects the periodic dinucleotide sequence specificity model for nucleosomes, and does not correspond well with actual nucleosome locations. When evaluated genome-wide using an F1-score (Fig. 4), the nucleosome positions called by RoboCOP are far more similar to the nucleosome annotations in [2] than are the ones called by COMPETE, which turn out to be not much better than random.

2.5 Predicted TF Binding Sites

MNase-seq is primarily used to study nucleosome positions; at present, no methods exist to predict TF binding sites from MNase-seq. It is challenging to extract TF binding sites from the noisy signal of the shortFrags generated by MNase digestion. TFs can sometimes be bound for an extremely short span of time [21] in which case the entire region could be digested by MNase, leaving behind no shortFrags signal. Nevertheless, MNase-seq data has been reported to provide evidence of TF binding [10], so we explore how well RoboCOP is able to identify TF binding sites. When we compare TF binding site predictions made by RoboCOP to predictions made by COMPETE, we see consistent but slight improvement in F1-score with RoboCOP (Fig. 4a). As a baseline, we compare these results to an approach we call FIMO-MNase, in which we simply run FIMO [8] around the peaks of midpoint counts of shortFrags. We find both RoboCOP and COMPETE are better than FIMO-MNase (Figs. 4b, c). Abf1, Reb1, and Rap1 have the most precise annotation datasets, and for these TFs in particular, both COMPETE and RoboCOP make better predictions. Overall, the highest F1-score is for Rap1 binding site predictions made by RoboCOP.

Although RoboCOP predicts the binding of a set of 150 TFs, we can only validate the binding sites of 81 of them, given available X-ChIP-chip [15], ChIP-

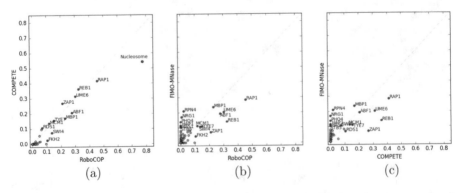

Fig. 4. Comparisons of F1-scores of TF binding site predictions by (a) RoboCOP and COMPETE, (b) RoboCOP and FIMO-MNase, and (c) COMPETE and FIMO-MNase. TFs with F1-score less than 0.1 in both methods of any given scatter plot are colored gray.

exo [19], and ORGANIC [13] datasets (Table S1). We have more precise binding sites from ChIP-exo and ORGANIC experiments for Abf1, Rap1, and Reb1. In addition, the binding sites in X-ChIP-chip data for many TFs were generated under multiple conditions [9] (Table S1) and the conditions are not specified for the reported annotations. This makes the X-ChIP-chip dataset fairly unreliable for validation purposes. In Fig. S3, we plot Venn diagrams showing the number of overlaps in the computed binding sites with the annotated binding sites for all three methods (RoboCOP, COMPETE, FIMO-MNase) and find that both COMPETE and RoboCOP have high false positives for certain motifs that are AT-rich such as Azf1 and Sfp1. Since the yeast genome is AT-rich and the shortFrags signal is noisy, any enrichment of the midpoint counts could be identified as a potential binding site. We believe prior knowledge about the occupancy of the TFs could yield higher accuracy.

2.6 RoboCOP Reveals Chromatin Dynamics Under Cadmium Stress

One of the most powerful uses of RoboCOP is that it can elucidate the dynamics of chromatin occupancy, generating profiles under changing environmental conditions. As an example, we explore the occupancy profiles of yeast cells subjected to cadmium stress for 60 minutes. We run RoboCOP separately on two MNase-seq datasets: one for a cell population before it was treated with cadmium and another 60 minutes after treatment. Cadmium is toxic to the cells and activates stress response pathways. Stress response related genes are heavily transcribed under cadmium treatment, while ribosomal genes are repressed [12]. We use RNA-seq to identify the 100 genes most up-regulated (top 100) and the 100 most down-regulated (bottom 100). As a control, we choose 100 genes with no change in transcription under cadmium treatment (mid 100) (see Table

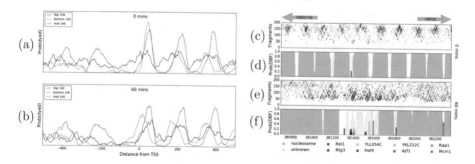

Fig. 5. Aggregate of Prob(dyad) computed by RoboCOP around the TSSs of genes most up-regulated (blue), most down-regulated (green), and unchanged in transcription (red), (a) before and (b) 60 min after treating cells with cadmium. After treatment, we see the +1 nucleosome closing in on the promoters of repressed genes (green) but opening up the promoters of highly transcribed genes (blue). MNase-seq fragment plot and RoboCOP-predicted occupancy profile of HSP26 promoter at chrII:380700–382350, (c, d) before and (e, f) after treatment with cadmium. HSP26 is highly expressed under cadmium stress, and its promoter exhibits much TF binding after treatment, most prominently by Rap1, known to bind the HSP26 promoter under stress response.

S2 for the three gene lists). Plotting the RoboCOP-predicted Prob(dyad) around the TSSs of the three gene groups, we notice that the nucleosomes in the top 100 genes are generally less well-positioned in comparison to the other groups of genes (Fig. 5a, b). Because of the uncertainty in the nucleosome positions of the top 100 genes, Prob(dyad) does not have any sharp peaks (blue curve in Fig. 5a, b). On the other hand, Prob(dyad) has sharp peaks indicating well-positioned nucleosomes for bottom 100 and mid 100 genes (green and red curves in Fig. 5a, b). This suggests RoboCOP-predicted Prob(dyad) can be used to classify 'fuzzy' or less well-positioned nucleosomes in the genome. Additionally, we see that the +1 nucleosomes of the top 100 genes move downstream after treatment with cadmium, thereby opening up the promoter region. In contrast, the +1 nucleosomes of the bottom 100 genes move upstream and close in on the promoter region to repress transcription.

HSP26, a key stress response gene, is among the top 100 most up-regulated genes. We can use RoboCOP to study how the chromatin landscape changes in the HSP26 promoter under cadmium stress. In Figs. 5c–f, we notice the HSP26 promoter opening up under stress, with shifts in nucleosomes leading to more TF binding in the promoter. From the shortFrags midpoint counts, RoboCOP identifies multiple potential TF binding sites, most prominently for Rap1. Rap1 is known to relocalize to the promoter region of HSP26 during general stress response [17]; antibody-based methods could be used to validate whether Rap1 binds in the HSP26 promoter under cadmium treatment in particular.

In comparison, COMPETE fails to capture the dynamics of chromatin occupancy because it does not incorporate chromatin accessibility information into its model. We ran COMPETE with the RoboCOP-trained DBF weights for the two time points of cadmium treatment and found that COMPETE generates binding landscapes for the two time points that are nearly identical (Fig. S4). This is a key difference between RoboCOP and COMPETE: being able to incorporate experimental chromatin accessibility data allows RoboCOP to provide a more accurate binding profile for cell populations undergoing dramatic chromatin changes.

The above analysis highlights the utility of RoboCOP. Because RoboCOP models DBFs competing to bind the genome, it produces a probabilistic prediction of the occupancy level of each DBF at single-nucleotide resolution. As the chromatin architecture changes under different environmental conditions, RoboCOP is able to elucidate the dynamics of chromatin occupancy. The cadmium treatment experiment shows that the predictions made by RoboCOP can be used both to study overall changes for groups of genes (Fig. 5a), and to focus on specific genomic loci to understand their chromatin dynamics (Fig. 5b).

3   Methods

3.1   RoboCOP Model Structure and Transition Probabilities

RoboCOP is a multivariate hidden Markov model (HMM) for computing a genome-wide chromatin occupancy profile using nucleotide sequence and epigenomic accessibility data (here MNase-seq) as observables. The HMM structure has been adapted from [23]. Let the number of TFs be K. Let $\pi_1, \ldots, \pi_K$ denote the models for the TFs, and let π_{K+1} denote the model for nucleosomes. To simplify notation, we consider an unbound DNA nucleotide to be occupied by a special 'empty' DBF [25]; suggestively, let π_0 denote this model. In summary, we have a total of $K+2$ DBFs in the model. We use a central non-emitting ('silent') state to simplify state transitions among the DBFs in the model. The HMM may transition from this central non-emitting state to any one of the DBF models (including for unbound DNA); at the end of each DBF, the HMM always transitions back to the central silent state (Fig. S5). This approach assumes DBFs bind independently of their neighbors, and each DBF therefore has just a single transition probability associated with it. The transition probabilities from the central state to the various DBFs are denoted $\{\alpha_0, \ldots, \alpha_{K+1}\}$.

Each hidden state represents a single genome coordinate. An unbound DNA nucleotide is length one, so its model π_0 has just a single hidden state. The other DBFs (nucleosomes and TFs) have binding sites of greater length and are thus modeled using collections of multiple hidden states. For TF k with a binding site of length L_k, the HMM either transitions through L_k hidden states of its binding motif or L_k hidden states of the reverse complement of its binding motif. An additional non-emitting state is added as the first hidden state of the TF model π_k, allowing the HMM to transition through the forward or reverse complement of the motif with equal probability (Fig. S6a). The complete TF model π_k therefore has a total of $2L_k + 1$ hidden states. Once the HMM enters the hidden states for either the forward or reverse motif, it transitions through the sequence of hidden states with probability 1 between consecutive hidden states. On reaching the final hidden state of either motif, the HMM transitions back to the central silent state with probability 1. Likewise, once the HMM enters the nucleosome model π_{K+1}, it transitions through a sequence of hidden states corresponding to 147 nucleotides, after which it transitions back to the central silent state (Fig. S6b). The nucleosome model differs from the TF models in that the latter are modeled with simple PWM motifs, while the former is implemented using a dinucleotide sequence specificity model.

Suppose the sequence of hidden states for the entire genome of length G is denoted as $z_1, \ldots, z_G$. Then the transition probabilities satisfy the following:

- $P(z_{g+1} = \pi_{k,l+1} | z_g = \pi_{k,l}) = 1$ whenever $l < L_k$. Within a DBF, the HMM only transitions to that DBF's next state and not any other state, until it reaches the end of the DBF.
- $P(z_{g+1} = \pi_{k_1,1} | z_g = \pi_{k_2,L_{k_2}}) = P(z_{g+1} = \pi_{k_1,1}) = \alpha_{k_1}$. The transition probability to the first state of a DBF is a constant, independent of which DBF the HMM visited previously.
- $P(z_{g+1} | z_g) = 0$ for all other cases.

The HMM always starts in the central non-emitting state with probability 1; this guarantees that it cannot start in the middle of a DBF.

3.2 RoboCOP Emission Probabilities

The HMM employed by RoboCOP is multivariate, meaning that each hidden state is responsible for emitting multiple observables per position in the genome. In our case, these observables are modeled as independent, conditioned on the hidden state, but adding dependence would be straightforward. For a genome of length G, the sequences of observables being explained by the model are: (i) nucleotide sequence $\{s_1, \ldots, s_G\}$, (ii) midpoint counts of MNase-seq nucFrags $\{l_1, \ldots, l_G\}$, and (iii) midpoint counts of MNase-seq shortFrags $\{m_1, \ldots, m_G\}$. For any position g in the genome, the hidden state z_g is thus responsible for emitting a nucleotide s_g, a number l_g of midpoints of nucFrags, and a number m_g of midpoints of shortFrags (Fig. S1). Since these three observations are independent of one another given the hidden state z_g, the hidden states have an emission model for each of the three observables, and the joint probability of the multivariate emission is the product of the emission probabilities of the three observables.

For the TF models $\pi_1, \ldots, \pi_K$, emission probabilities for nucleotide sequences are represented using PWMs. For each of our 150 TFs, we use the PWM of its primary motif reported in [7] (except for Rap1, where we use the more detailed motifs in [19]). For the nucleosome model π_{K+1}, the emission probability for a nucleotide sequence of length 147 can be represented using a position-specific dinucleotide model [20]. To represent this dinucleotide model, the number of hidden states in π_{K+1} is roughly 4×147. We use the same dinucleotide model that was used earlier in COMPETE [23].

As described earlier, the two-dimensional MNase-seq data is used to compute two one-dimensional signals. The midpoint counts of nucFrags are primarily used for learning nucleosome positions and the midpoint counts of shortFrags are used for learning the TF binding sites. In both cases, a negative binomial (NB) distribution is used to model the emission probabilities. We use two sets of NB distributions to model the midpoint counts of nucFrags. One distribution, $NB(\mu_{nuc}, \phi_{nuc})$, explains the counts of nucFrags at the nucleosome positions and another distribution, $NB(\mu_{l_b}, \phi_{l_b})$, explains the counts of nucFrags elsewhere in the genome. Since the midpoint counts of nucFrags within a nucleosome are not uniform (Fig. 1b), we model each of the 147 positions separately. To obtain $\boldsymbol{\mu}_{nuc}$ and ϕ_{nuc}, we collect the midpoint counts of nucFrags in a window of size 147 centered on the annotated nucleosome dyads of the top 2000 well-positioned nucleosomes [2] and estimate 147 NB distributions using maximum likelihood estimate (MLE). The 147 estimated values of μ are denoted as $\boldsymbol{\mu}_{nuc}$. The mean of the 147 estimated values of ϕ is denoted as ϕ_{nuc} (shared across all 147 positions). Quantile-quantile plots show the resulting NB distributions to be a good fit (Fig. S7). To compute $NB(\mu_{l_b}, \phi_{l_b})$, we estimate an NB distribution for the midpoint counts of nucFrags at the linker regions of the same set of 2000 nucleosomes using MLE. The linker length is chosen to be 15 bases long on either side of the nucleosome [5].

Similarly, we model the midpoint counts of shortFrags using two distributions where one of them, $NB(\mu_{TF}, \phi_{TF})$, explains the counts of shortFrags at the TF binding sites, while the other, $NB(\mu_{m_b}, \phi_{m_b})$, explains the counts else-where. To estimate the parameters of $NB(\mu_{TF}, \phi_{TF})$, we collect the midpoint counts of `shortFrags` at the annotated Abf1 and Reb1 binding sites from [15] and fit an NB distribution using MLE. A quantile-quantile plot shows the NB distribution to be a good fit (Fig. S8). We chose Abf1 and Reb1 for fitting the distribution because these TFs have many binding sites in the genome and the binding sites are often less noisy. For parameterizing $NB(\mu_{m_b}, \phi_{m_b})$, we compute the midpoint counts of `shortFrags` at the same linker regions used earlier and estimate the NB distribution using MLE.

3.3 `RoboCOP` Parameter Updates

The transition probabilities between hidden states within a DBF can only be 0 or 1 (except for the two transition probabilities from each TF model's first, non-emitting state to either its forward or reverse motif, which are 0.5). Consequently, only the transition probabilities $\{\alpha_0, \ldots, \alpha_{K+1}\}$ from the central silent state to the first state of each DBF are unknown. We initialize these probabilities as described below, and then iteratively update them using Baum-Welch until convergence to a local optimum of the likelihood.

To initialize the probabilities, we assign weight 1 to the 'empty' DBF (repre-senting an unbound DNA nucleotide) and 35 to the nucleosome. To each TF, we assign a weight which is that TF's dissociation constant K_D (or alternatively, a multiple thereof: $8K_D$, $16K_D$, $32K_D$, or $64K_D$). Finally, we transform these weights into a proper probability distribution to yield the initial probabilities.

To update α_k, the transition probability from the central silent state to the first hidden state $\pi_{k,1}$ of DBF k, we compute:

$$\alpha_k = \frac{\sum_{g=1}^{G} P(\pi_{k,1}|\boldsymbol{\theta}^*, \boldsymbol{s}, \boldsymbol{l}, \boldsymbol{m})}{\sum_{k'=0}^{K+1} \sum_{g=1}^{G} P(\pi_{k',1}|\boldsymbol{\theta}^*, \boldsymbol{s}, \boldsymbol{l}, \boldsymbol{m})}$$

Here, $\boldsymbol{\theta}^*$ represents all the model parameters. We find the likelihood converges within 10 iterations (Fig. S9) and the optimized transition probabilities for each DBF almost always converge to the same final values regardless of how we ini-tialize the weights (Fig. S10). We find convergence is faster for most DBFs when we initialize TF weights to K_D rather than multiples thereof (Fig. S10).

We do not use any prior information about the transition probabilities of the DBFs. We find that a few TFs such as Azf1 and Smp1 can have a large number of binding sites in the genome that are potential false positives. To curb the number of binding site predictions for such TFs, we apply a threshold on the TF transition probabilities. The threshold δ is chosen to be two standard deviations more than the mean of the initial transition probabilities of the TFs (Fig. S11). Therefore, after the Baum-Welch step in every iteration, an additional modified Baum-Welch step is computed as follows:

$$\alpha_k = \begin{cases} (1-n\delta)\dfrac{\sum\limits_{g=1}^{G} P(\pi_{k,1}|\boldsymbol{\theta}^*,\boldsymbol{s},\boldsymbol{l},\boldsymbol{m})}{\sum_{k'=0,\alpha_{k'}<\delta}^{K+1} \sum\limits_{g=1}^{G} P(\pi_{k',1}|\boldsymbol{\theta}^*,\boldsymbol{s},\boldsymbol{l},\boldsymbol{m})} & , \text{ if } \alpha_k < \delta \\[3ex] \delta & , \text{ otherwise} \end{cases}$$

Here n is the number of TFs that have transition probability more than δ. So, all the TFs whose transition probability would be more than δ are set instead to δ, and the remaining TFs have a regular Baum-Welch update of their transition probabilities. We find that this approach reduces the number of false positives (Fig. S12). Using an informed prior might be an alternative mechanism for yielding a more accurate binding profile for such TFs.

To ensure fair comparisons between RoboCOP and COMPETE, we ran COMPETE using the same parameters estimated by RoboCOP. Therefore, the output profiles of the two methods highlight the differences in the results that occur because of the inclusion of chromatin accessibility data.

3.4   Implementation Details for Posterior Decoding

RoboCOP employs posterior decoding to infer probabilistic occupancy profiles of protein-DNA binding. The motivation behind posterior decoding is that it represents the thermodynamic ensemble of potential binding configurations; the resulting probability distribution sheds light on the many different ways proteins may be bound to the genome across a cell population (applying Viterbi decoding would not provide a probabilistic landscape, but only a single, most likely chromatin configuration).

As a multivariate HMM, RoboCOP has a time complexity of $O(GN^2)$ and a space complexity of $O(GN)$ (for a genome of length G and where N denotes the total number of hidden states). The high complexity makes it difficult to decode the entire genome at once. To reduce the computational complexity of RoboCOP, we perform posterior decoding separately on blocks of the genome of length 5000, with an overlap of 1000 bases, and stitch results together. This ensures that the model has a sufficiently long sequence to learn an accurate chromatin landscape, but not so long that we run out of space. In addition, we use only the longest chromosome (chrIV) to train DBF transition probabilities with Baum-Welch, and then undertake posterior decoding genome-wide.

3.5   TF and Nucleosome Predictions and Validation

We use posterior probabilities of TF occupancy from RoboCOP and COMPETE output to identify binding sites, calling all sites whose probability is at least 0.1. In the case of Rap1 which has multiple PWMs, the maximum probability among the PWMs is chosen at every position. The same comparison is applied when choosing between the forward and reverse complement of the motif. For validation, a site is considered a true positive (TP) if it overlaps with an annotated binding site for that TF, and a false positive (FP) otherwise. If an annotated TF binding site does not overlap any of our predictions, it is a false negative (FN). We use the sacCer2 (June 2008) genome version in our analyses.

We called nucleosomes from RoboCOP and COMPETE output using a greedy algorithm, as described previously [24]. Briefly, nucleosome dyads with decreasing probability were iteratively selected. A window of size 101 around the selected dyad was removed from future rounds of dyad selection (this window size was chosen to allow mild overlap between adjacent nucleosome locations). The nucleosome annotations in [2] contain 67548 nucleosomes. We selected the same number of top scoring nucleosomes from the output of RoboCOP and COMPETE. A nucleosome position was considered to be a true positive (TP) if the distance between the predicted and annotated dyad was less than 50 bases.

3.6   FIMO-MNase

MNase-seq midpoint counts of shortFrags (length less than 80) are smoothed using a window of size 21. Peaks are detected if they have a value greater than 2 with consecutive peaks being at least 25 bases apart. Peaks for midpoint counts of nucFrags are detected if they have a value greater than 1 and are at least 100 bases apart. To prevent nucleosomal peaks occluding peaks of shortFrags, peaks of midpoint counts of shortFrags within 60 bases of peaks of midpoint counts of nucFrags are removed. After these steps, we detect 4137 peaks of shortFrags genome-wide. FIMO [8] is run using PWMs from [7] on 50-bp windows centered on the peak sites with a p-value cutoff of 10^{-4}.

3.7   Data Access

MNase-seq and RNA-seq of yeast cells before and after cadmium treatment is available at https://doi.org/10.7924/r4hx1b43s. Code and supplementary material may be downloaded from https://github.com/HarteminkLab/RoboCOP.

4   Discussion

RoboCOP is a novel method that utilizes a multivariate HMM to generate a probabilistic occupancy profile of the genome by integrating chromatin accessibility data with nucleotide sequence. We choose to apply the model to the yeast genome because of the availability of high quality MNase-seq data and the small size of the genome, which simplifies computation. Chromatin accessibility data from MNase-seq, DNase-seq, and ATAC-seq are generally noisy, so it is a challenging task to infer precise genome-wide DBF occupancy from the data, particularly for TFs. While alternative approaches using peak identification or footprint identification followed by TF-labeling with FIMO [8] can offer some insight into protein-DNA binding, we observe that RoboCOP performs notably better, presumably because it considers all DBFs together in a joint model that incorporates the thermodynamic competition among DBFs (including nucleosomes).

RoboCOP improves upon COMPETE in a number of ways: it slightly improves TF binding site predictions, it markedly improves nucleosome positioning predictions, and it uses experimental data to learn DBF transition probabilities in a principled way. When these same transition probabilities are provided to COMPETE, its TF binding site predictions are similar to RoboCOP's because of the generally high sequence specificity of TFs, but its nucleosome position predictions are much worse because of the weak sequence specificity of nucleosomes. In future work, it might be possible to improve the learned transition probabilities further through the use of prior information.

Finally, we note that RoboCOP can be used to study the chromatin architecture of the genome under varying conditions, an important task to which COMPETE is unsuited. Because RoboCOP uses data to model a collection of DBFs competing to bind to the genome, we can observe dynamic levels of occupancy for different DBFs under different environmental conditions. Since gene expression also varies in response to changing environmental conditions, we believe RoboCOP will help elucidate how the dynamics of chromatin occupancy and the dynamics of gene expression interrelate.

Acknowledgments. The authors would like to thank Heather MacAlpine and Vinay Tripuraneni for generating the MNase-seq data, and Greg Crawford, Raluca Gordân, Ed Iversen, Trung Tran, Yulong Li, and Albert Xue for helpful comments and feedback during the development of RoboCOP.

References

1. Benner, P., Vingron, M.: ModHMM: A modular supra-Bayesian genome segmentation method. In: Cowen, L.J. (ed.) RECOMB 2019. LNCS, vol. 11467, pp. 35–50. Springer, Cham (2019). https://doi.org/10.1007/978-3-030-17083-7_3
2. Brogaard, K., Xi, L., Wang, J.P., Widom, J.: A map of nucleosome positions in yeast at base-pair resolution. Nature **486**(7404), 496–501 (2012)
3. Chen, K., et al.: DANPOS: Dynamic analysis of nucleosome position and occupancy by sequencing. Genome Res. **23**(2), 341–351 (2013)
4. Chen, W., Liu, Y., Zhu, S., Green, C.D., Wei, G., Han, J.D.J.: Improved nucleosome-positioning algorithm iNPS for accurate nucleosome positioning from sequencing data. Nat. Commun. **5**(1), 4909 (2014)
5. Chereji, R.V., Ramachandran, S., Bryson, T.D., Henikoff, S.: Precise genome-wide mapping of single nucleosomes and linkers in vivo. Genome Biol. **19**(1), 19 (2018)
6. Ernst, J., Kellis, M.: ChromHMM: Automating chromatin-state discovery and characterization. Nat. Methods **9**(3), 215–216 (2012)
7. Gordân, R., Murphy, K.F., McCord, R.P., Zhu, C., Vedenko, A., Bulyk, M.L.: Curated collection of yeast transcription factor DNA binding specificity data reveals novel structural and gene regulatory insights. Genome Biol. **12**(12), R125 (2011)
8. Grant, C.E., Bailey, T.L., Noble, W.S.: FIMO: Scanning for occurrences of a given motif. Bioinformatics **27**(7), 1017–1018 (2011)
9. Harbison, C.T., et al.: Transcriptional regulatory code of a eukaryotic genome. Nature **431**(7004), 99–104 (2004)

10. Henikoff, J.G., Belsky, J.A., Krassovsky, K., MacAlpine, D.M., Henikoff, S.: Epigenome characterization at single base-pair resolution. Proc. Natl. Acad. Sci. U. S. A. **108**(45), 18318–18323 (2011)

11. Hoffman, M.M., Buske, O.J., Wang, J., Weng, Z., Bilmes, J.A., Noble, W.S.: Unsupervised pattern discovery in human chromatin structure through genomic segmentation. Nat. Methods **9**(5), 473–476 (2012)

12. Hosiner, D., Gerber, S., Lichtenberg-Fraté, H., Glaser, W., Schüller, C., Klipp, E.: Impact of acute metal stress in Saccharomyces cerevisiae. PLoS ONE **9**(1), e83330 (2014)

13. Kasinathan, S., Orsi, G.A., Zentner, G.E., Ahmad, K., Henikoff, S.: High-resolution mapping of transcription factor binding sites on native chromatin. Nat. Methods **11**(2), 203–209 (2014)

14. Lee, W., et al.: A high-resolution atlas of nucleosome occupancy in yeast. Nat. Genet. **39**(10), 1235–1244 (2007)

15. MacIsaac, K.D., Wang, T., Gordon, D.B., Gifford, D.K., Stormo, G.D., Fraenkel, E.: An improved map of conserved regulatory sites for Saccharomyces cerevisiae. BMC Bioinform. **7**(1), 113 (2006)

16. Park, D., Morris, A.R., Battenhouse, A., Iyer, V.R.: Simultaneous mapping of transcript ends at single-nucleotide resolution and identification of widespread promoter-associated non-coding RNA governed by TATA elements. Nucl. Acids Res. **42**(6), 3736–3749 (2014)

17. JM, Platt, et al.: Rap1 relocalization contributes to the chromatin-mediated gene expression profile and pace of cell senescence. Genes Dev. **27**(12), 1406–1420 (2013)

18. Rhee, H.S., Bataille, A.R., Zhang, L., Pugh, B.F.: Subnucleosomal structures and nucleosome asymmetry across a genome. Cell **159**(6), 1377–1388 (2014)

19. Rhee, H.S., Pugh, B.F.: Comprehensive genome-wide protein-DNA interactions detected at single-nucleotide resolution. Cell **147**(6), 1408–1419 (2011)

20. Segal, E., et al.: A genomic code for nucleosome positioning. Nature **442**(7104), 772–778 (2006)

21. Sung, M.H., Guertin, M.J., Baek, S., Hager, G.L.: DNase footprint signatures are dictated by factor dynamics and DNA sequence. Mol. Cell **56**(2), 275–285 (2014)

22. Tarbell, E.D., Liu, T.: HMMRATAC: A Hidden Markov ModeleR for ATAC-seq. Nucl. Acid Res. **47**, e91 (2019)

23. Wasson, T., Hartemink, A.J.: An ensemble model of competitive multi-factor binding of the genome. Genome Res. **19**(11), 2101–2112 (2009)

24. Zhong, J., Luo, K., Winter, P.S., Crawford, G.E., Iversen, E.S., Hartemink, A.J.: Mapping nucleosome positions using DNase-seq. Genome Res. **26**(3), 351–364 (2016)

25. Zhong, J., Wasson, T., Hartemink, A.J.: Learning protein-DNA interaction landscapes by integrating experimental data through computational models. Bioinformatics **30**(20), 2868–2874 (2014)

Representation of k-mer Sets Using Spectrum-Preserving String Sets

Amatur Rahman[1]($\boxtimes$) and Paul Medvedev[1,2,3]($\boxtimes$)

[1] Department of Computer Science and Engineering,
The Pennsylvania State University, State College, USA
{aur1111,pzm11}@psu.edu
[2] Department of Biochemistry and Molecular Biology,
The Pennsylvania State University, State College, USA
[3] Center for Computational Biology and Bioinformatics,
The Pennsylvania State University, State College, USA

Abstract. Given the popularity and elegance of k-mer based tools, finding a space-efficient way to represent a set of k-mers is important for improving the scalability of bioinformatics analyses. One popular approach is to convert the set of k-mers into the more compact set of unitigs. We generalize this approach and formulate it as the problem of finding a smallest spectrum-preserving string set (SPSS) representation. We show that this problem is equivalent to finding a smallest path cover in a compacted de Bruijn graph. Using this reduction, we prove a lower bound on the size of the optimal SPSS and propose a greedy method called UST that results in a smaller representation than unitigs and is nearly optimal with respect to our lower bound. We demonstrate the usefulness of the SPSS formulation with two applications of UST. The first one is a compression algorithm, UST-Compress, which we show can store a set of k-mers using an order-of-magnitude less disk space than other lossless compression tools. The second one is an exact static k-mer membership index, UST-FM, which we show improves index size by 10–44% compared to other state-of-the-art low memory indices. Our tool is publicly available at: https://github.com/medvedevgroup/UST/.

1 Introduction

Algorithms based on k-mers are now amongst the top performing tools for many bioinformatics analyses. Instead of working directly with reads or alignments, these tools work with the set of k-mer substrings present in the data, often relying on specialized data structures for representing sets of k-mers (for a survey, see [1]). Since modern sequencing datasets are huge, the space used by such data structures is a bottleneck when attempting to scale up to large databases. For example, as part of our group's work on building indices for RNA-seq data, we

---

A full version of this paper is available as a preprint https://doi.org/10.1101/2020.01.07.896928.

© Springer Nature Switzerland AG 2020
R. Schwartz (Ed.): RECOMB 2020, LNBI 12074, pp. 152–168, 2020.
https://doi.org/10.1007/978-3-030-45257-5_10

are storing gzipped k-mer set files from about 2,500 experiments [2]. Though this is only a fraction of experiments in the SRA, it already consumes 6 TB of space. For these and other applications, the development of space-efficient representations of k-mer sets can improve scalability and enable novel biological discoveries.

Conway and Bromage [3] showed that at least $\log \binom{4^k}{n}$ bits are needed to losslessly store a set of n k-mers, in the worst case. However, a set of k-mers generated from a sequencing experiment typically exhibits the spectrum-like property [1] and contains a lot of redundant information. Therefore, in practice, most data structures can substantially improve on that bound [4].

A common way to reduce the redundancy in a k-mer set K is to convert it into a set of *maximal unitigs*. A unitig is a non-branching path in the de Bruijn graph, a graph whose nodes are the k-mers of K and edges are the overlaps between k-mers. A unitig u can be written as a string $spell(u)$ of length $|u|+k-1$, such that the k-mers of u are exactly the k-mer substrings of $spell(u)$. For example, the unitig (AAC, ACG, CGT) is spelled as $AACGT$. This gives a way to represent $|u|$ k-mers using $|u|+k-1$ characters, instead of $k|u|$ characters used by a naive approach. When unitigs are long, as they are in real data, the space savings are significant. The idea can be extended to store the whole set K, because the set of maximal unitigs U forms a decomposition of K, and, therefore, has the nice property that $x \in K$ iff x is a substring of $spell(u)$, for some $u \in U$.

The maximal unitigs U can be computed efficiently [5–7] and combined with an auxiliary index to obtain a membership data structure (i.e. one that can efficiently determine if a k-mer belongs to K or not). In particular, Unitigs-FM [4] and deGSM [7] uses the FM-index as the auxiliary index, Pufferfish [8] and BLight [9] uses a minimum perfect hash function, and Bifrost [10] uses a minimizer hash table. Alternatively, U can be compressed to obtain a compressed disk representation of K, albeit without efficient support for membership queries prior to decompression.

While unitigs conveniently fit the needs of those applications, we observe in this paper that they are not necessarily the best that can be done. Concretely, we claim that what makes U useful in these scenarios is that they are a type of *spectrum-preserving string set (SPSS) representation* of K, which we define to be a set of strings X such that a k-mer is in K iff it is a substring of a string in X. (This is in contrast to the way unitigs are used in assembly, where it is crucial that they are not chimeric [11].) The *weight* of X is the number of characters it contains. In this paper, we explore the idea of low weight representations and their applicability. In particular, are there representations with a smaller weight than U that can be efficiently computed? What is the lowest weight that is achievable by a representation? Can such representations seamlessly replace unitig representations in downstream applications, and can they improve space performance?

In this paper, we show that the problem of finding a minimum weight SPSS representation is equivalent to finding a smallest path cover in a compacted de Bruijn graph (Sect. 3). We use the reduction to give a lower bound on the weight which could be achieved by any SPSS representation (Sect. 4), and we give an efficient greedy algorithm UST to find a representation that improves on U (Sect. 5) and is empirically near-optimal. We demonstrate the usefulness of our representation using two applications (Sect. 6). One, we combine it with an FM-index into a membership data structure called UST-FM, and, two, we combine it with a general compression algorithm to give a compression algorithm called UST-Compress. Both applications result in a substantial space decrease over state-of-the-art (Sect. 7), demonstrating the usefulness of SPSS representations. Our software is freely available at https://github.com/medvedevgroup/UST/.

1.1 Related Work

The idea of using a SPSS for a membership index was previously independently described in a PhD thesis [12] and questions similar to the ones in our paper are simultaneously and independently studied in [13]. The idea of greedily gluing unitigs (as UST does) has previously appeared in read compression [14], where contigs greedily constructed from the reads and the reads were stored as alignments to these contigs. The idea also appeared in the context of sequence assembly, where a greedy traversal of an assembly graph was used as an intermediate step during assembly [15, 16].

The compression of k-mer sets has not been extensively studied, except in the context of how k-mer counters store their output [17–20]. DSK [18] uses an HDF5-based encoding, KMC3 [17] combines a dense storage of prefixes with a sparse storage of suffixes, and Squeakr [20] uses a counting quotient filter [21]. The compression of read data, on the other hand, stored in either unaligned or aligned formats, has received a lot of attention [22–24]. In the scenario where the k-mer set to be compressed was originally generated from FASTA files by a k-mer counter, an alternate to k-mer compression is to compress the original FASTA file and use a k-mer counter as part of the decompression to extract the k-mers on the fly. This approach is unsatisfactory because (1) as we show in this paper, it takes substantially more space than direct k-mer compression, (2) k-mer counting on the fly adds significant time and memory to the decompression process, and (3) there are applications where the k-mer set cannot be reproduced by simply counting k-mers in a FASTA file, e.g. when it is a product of a multi-sample error correction algorithm [25]. Furthermore, there are applications where the k-mer set is not related to sequence read data at all, e.g. a universal hitting set [26], a chromosome-specific reference dictionary [27], or a winnowed min-hash sketch (for example as in [28], or see [29, 30] for a survey).

Membership data structures for k-mer sets were surveyed in a recent paper [1]. In addition to the unitig-based approaches already mentioned, other exact representations include succinct de Bruijn graphs (referred to as BOSS [31]) and their variations [32, 33], dynamic de Bruijn graphs [34, 35], and Bloom filter tries [36]. Some data structures are non-static, i.e. they provide the ability to insert and/or delete k-mers. However, such operations are not needed in many read-only applications, where the cost of supporting them can be avoided. Membership data structures can be extended to associate additional information with each k-mer, for instance an abundance count (e.g. deBGR [37]) or a color class (for a short overview, see [1]).

2 Definitions

Strings: In this paper, we assume all strings are over the alphabet $\Sigma = \{A, C, G, T\}$. The *length* of string x is denoted by $|x|$. A string of length k is called a k-mer. For a set of strings S, $weight(S) = \sum_{x \in S} |x|$ denotes the total count of characters. We write $x[i..j]$ to denote the substring of x from the ith to the jth character, inclusive. We define $suf_k(x)$ (respectively, $pre_k(x)$) to be the last (respectively, first) k characters of x. For x and y with $suf_{k-1}(x) = pre_{k-1}(y)$, we define *gluing* x and y as $x \odot y = x \cdot y[k..|y|]$. For $s \in \{0, 1\}$, we define $orient(x, s)$ to be x if $s = 0$ and to be the reverse complement of x if $s = 1$. A string x is *canonical* if x is the lexicographically smaller of x and its reverse complement. To *canonize* x is to replace it by its canonical version (i.e. $\min_i(orient(x, i))$). We say that x_0 and x_1 have a (s_0, s_1)-*oriented-overlap* if $suf_{k-1}(orient(x_0, 1 - s_0) = pre_{k-1}(orient(x_1, s_1))$. Intuitively, such an overlap exists between two strings if we can orient them in such a way that they are glueable. We define the k-*spectrum* $sp^k(x)$ as the multi-set of all canonized k-mer substrings of x. The k-spectrum for a set of strings S is defined as $sp^k(S) = \bigcup_{x \in S} sp^k(x)$.

Bidirected Graphs: A *bidirected graph* G is a pair (V, E) where the set V are called vertices and E is a set of edges. An edge e is a 4-tuple (u_0, s_0, u_1, s_1), where $u_i \in V$ and $s_i \in \{0, 1\}$, for $i \in \{0, 1\}$. Intuitively, every vertex has two sides, and an edge connects to a side of a vertex. Note that there can be multiple edges between two vertices, but only one edge once the sides are fixed. An edge is a *loop* if $u_0 = u_1$. Given a non-loop edge e that is incident to a vertex u, we denote $side(u, e)$ as the side of u to which it is incident. We say that a vertex u is *isolated* if it has no edge incident to it and is a *dead-end* if it has exactly one side to which no edges are incident. Define n_{dead} and n_{iso} as the number of dead-end and isolated vertices, respectively. A sequence $w = (u_0, e_1, u_1, \ldots, e_n, u_n)$ is a *walk* if for all $1 \leq i \leq n$, e_i is incident to u_{i-1} and to u_i, and for all $1 \leq i \leq n-1$, $side(u_i, e_i) = 1 - side(u_i, e_{i+1})$. Vertices $u_1, \ldots, u_{n-1}$ are called *internal* and u_0 and u_n are called *endpoints*. A walk can also be a single vertex, in which case it is considered to have no internal vertex and one endpoint. A *path cover* W of G is a set of walks such that every vertex is in exactly one walk in W and no walk visits a vertex more than once.

Bidirected DNA Graphs: A *bidirected DNA graph* is a bidirected graph G where every vertex u has a string label $lab(u)$, and for every edge $e = (u_0, s_0, u_1, s_1)$, there is a (s_0, s_1)-oriented-overlap between $lab(u_0)$ and $lab(u_1)$. G is said to be *overlap-closed* if there is an edge for every such overlap. Let $w = (u_0, e_1, u_1, \ldots, e_n, u_n)$ be a walk. Define $x_0 = orient(lab(u_0), 1 - side(u_0, e_1))$ and, for $1 \leq i \leq n$, $x_i = orient(lab(u_i), side(u_i, e_{i-1}))$. The *spelling* of a walk is defined as $spell(w) = x_0 \odot \cdots \odot x_n$. (The fact that the x_i's are glueable in this way can be derived from definitions.) If W is a set of walks, then define $spell(W) = \bigcup_{w \in W} spell(w)$.

De Bruijn Graphs: Let K be a set of canonical k-mers. The node-centric *bidirected de Bruijn graph*, denoted by $dBG(K)$, is the overlap-closed bidirected DNA graph where the vertices and their labels correspond to K. Figure 1A shows an example. In this paper, we will assume that $dBG(K)$ is not just a single cycle; such a case is easy to handle in practice but is a space-consuming corner-case in all the analyses. A walk in $dBG(K)$ is a *unitig* if all its vertices have in- and out-degrees of 1, except that the first vertex can have any in-degree and the last vertex can have any out-degree. A single vertex is also a unitig. A unitig is *maximal* if it is not a sub-walk of another unitig. It was shown in [5] that if $dBG(K)$ is not a cycle, then a unitig cannot visit a vertex more than once,

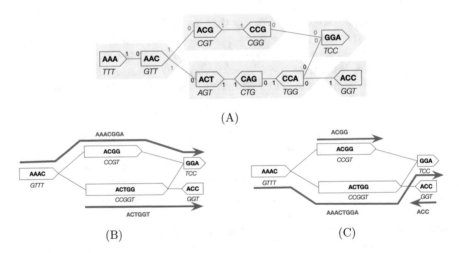

(A)

(B)   (C)

Fig. 1. *(A)* An example of a de Bruijn graph for a set K with 9 3-mers. The 0 side of a vertex is drawn flat and the 1 side pointy. The text in each vertex is its label, i.e. what is spelled by a walk going in the direction of the pointy end. The string below the vertex is the reverse complement of its label, which is what is spelled by a walk going in the opposite direction. The maximal unitigs are shown by filled in gray arrows. *(B)* The compacted de Bruijn graph for the same set K. Each vertex corresponds to a maximal unitig in the top graph. Each vertex's label corresponds to the spelling of the corresponding unitig and is shown inside the vertex; the reverse complement of the label is written below in italics. One possible path cover is five walks, each corresponding to a single vertex; the spelling of this cover is $\{AAAC, ACGG, ACTGG, GGA, ACC\}$, which is the unitig SPSS representation of K. A better path cover of size 2 that could potentially be found by our UST algorithm is shown. It corresponds to SPSS representation $\{AAACGGA, ACTGGT\}$. It is easy to verify that this path cover has minimum size, and, by Theorem 1, the corresponding representation has minimum weight (13). *(C)* Another path cover that could potentially be found by UST. It has size 3 and is suboptimal.

and the set of maximal unitigs forms a unique decomposition of the vertices in $dBG(K)$ into non-overlapping walks. The bidirected *compacted de Bruijn graph* of K, denoted by $cdBG(K)$, is the overlap-closed bidirected DNA graph where the vertices are the maximal unitigs of $dBG(K)$, and the labels of the vertices are the spellings of the unitigs. Figure 1B shows an example.

3   Equivalence of SPSS Representations and Path Covers

For this section, we fix K to be a canonical set of k-mers. A set of strings X is said to be a *spectrum-preserving string set (SPSS) representation* of K iff their k-spectrums are equal and each string in X is of length $\geq k$. For brevity, we say X *represents* K. Note that because in our definitions K is a set (i.e. no duplicates) and the k-spectrum is a multi-set, this effectively restricts X to not contain duplicate k-mers. See Fig. 1BC for examples. In this paper, we consider the problem of finding a minimum weight SPSS representation of K. In this section, we will show that it is equivalent to the problem of finding a smallest path cover of $cdBG(K)$, in the following sense:

Theorem 1. *Let X^{opt} be a minimum weight SPSS representation of K. Let W^{opt} be the smallest path cover on $cdBG(K)$. Then, $weight(X^{opt}) = |K| + |W^{opt}|(k-1)$.*

First, we show that the weight of a SPSS representation is a linear increasing function of its size (i.e. the number of strings it contains) and, hence, finding a SPSS representation of minimum weight is equivalent to finding one of min size.

Lemma 1. *Let X be SPSS representing K. Then, $weight(X) = |K| + |X|(k-1)$.*

Proof. Every string x of length $\geq k$ contains $|x| - k + 1$ k-mers. X has $|K|$ k-mers, since X and K have the same k-spectrum. Combining these, $|K| = \sum_{x \in X}(|x| - k + 1) = weight(X) - |X|(k-1)$.

The intuition behind Theorem 1 is that there is a natural size-preserving bijection between path covers of $dBG(K)$ and SPSS representations of K. Since it is more efficient to work with compacted de Bruijn graphs, we would like this to hold for $cdBG(K)$ as well. However, the path covers with an endpoint at an internal vertex of a unitig in $dBG(K)$ do not project onto $cdBG(K)$. Nevertheless, this is not an issue since such covers are necessarily non-optimal.

Lemma 2. *Let W be a path cover of $cdBG(K)$. Then $spell(W)$ represents K.*

Proof. By construction, all strings in $spell(W)$ are at least k-long, so we only need to show that the spectrum of $spell(W)$ is K. Let W^0 be the path cover with every vertex as its own walk. We can view W as being constructed from W^0 by repeatedly taking a pair of walks that share endpoints and joining them together. We prove the Lemma by induction. For the base case, $spell(W^0)$ are the unitigs of $dBG(K)$, which, by definition, have the same spectrum as K [5]. Now let W^i

be the path cover after i walk-joins. Then W^{i+1} is the result of joining some two walks w and w' into w''. Observe that joining walks preserves the k-spectrum of their spellings, i.e $sp^k(spell(w)) \cup sp^k(spell(w')) = sp^k(spell(w''))$. Combining with the inductive hypothesis for W^i, $sp^k(spell(W^i)) = sp^k(spell(W^{i+1}))$.

Lemma 3. *Let X be a smallest SPSS representation of K. Then there exists a path cover W of $cdBG(K)$ with $|W| = |X|$.*

Proof. Let $X = \{x_1, \ldots, x_m\}$. Every string x_i is spelled by a walk w'_i in $dBG(K)$, visiting the sequence of its canonized constituent k-mers. Since X is spectrum preserving with respect to K, it contains every k-mer in K exactly once; therefore, $\{w'_1, \ldots, w'_m\}$ is a path cover of $dbG(K)$.

Since X has the smallest number of strings, the endpoints of w'_i cannot be on internal vertices of unitigs, otherwise there would exist another string x_j that could be glued with x_i to form a smaller SPSS representing K. Therefore, there exists a corresponding walk w_i in $cdBG(K)$ such that $spell(w_i) = spell(w'_i) = x_i$. Hence, the set of walks $W = \{w_1, \ldots, w_m\}$ is a path cover of $cdBG(K)$.

Now we can prove Theorem 1.

Proof. By Lemma 1, X^{opt} has minimum size and, hence, by Lemma 3, there exists a path cover W with $|W| = |X^{\mathrm{opt}}|$. By the optimality of W^{opt}, $|W^{\mathrm{opt}}| \leq |W| \leq |X^{\mathrm{opt}}|$. Next, by Lemma 2, $spell(W^{\mathrm{opt}})$ represents K and, by definition, $|spell(W^{\mathrm{opt}})| = |W^{\mathrm{opt}}|$. Since X^{opt} has minimum size, $|X^{\mathrm{opt}}| \leq |spell(W^{\mathrm{opt}})| = |W^{\mathrm{opt}}|$. This proves $|X^{\mathrm{opt}}| = |W^{\mathrm{opt}}|$. Lemma 1 then implies the Theorem.

4   Lower Bound on the Weight of a SPSS Representation

In this section, we will prove a lower bound on the size of a path cover of a bidirected graph, which, by Theorem 1, gives a lower bound on the weight of any SPSS representation. Finding the minimum size of a path cover in general directed graphs is NP-hard, since a directed graph has a Hamiltonian path if and only if it has a path cover of size 1. However, we do not know the complexity of the problem when restricted to compacted de Bruijn graphs of k-mer sets. The minimum size of a path cover is known to be bounded from above by the maximum size of an independent set (at least for directed graphs [38]); however, finding a maximum independent set is itself NP-hard. We therefore take a different approach.

For this section, let $G = (V, E)$ be a bidirected graph without loops and let W be a path cover. A *vertex-side* is a pair (u, su), where $u \in V$ and $su \in \{0, 1\}$. For a non-isolated vertex u, we say (u, su) is a *dead-side* if there are no edges incident to $(u, 1 - su)$. Note that the number of dead-sides is by definition the number of dead-end vertices. Consider a walk $(v_0, e_1, \ldots, e_n, v_n)$ with $n \geq 1$. Denote its *endpoint-sides* as $(v_0, side(v_0, e_1))$ and $(v_n, side(v_n, e_n))$. If a walk contains just one vertex (v_0), then denote its *endpoint-sides* as $(v_0, 0)$ and $(v_0, 1)$.

We observe that every walk in a path cover must have two unique endpoint-sides. Our strategy is to give a lower bound on the number of endpoint-sides, thereby giving a lower bound on the size of a path cover. We know, for instance, that dead-sides must be endpoint-sides and we also know that the sides of an isolated vertex must be endpoint-sides. For other cases, we cannot predict exactly the endpoint-sides, but we can create disjoint sets of vertex-sides (which we call special neighborhoods) such that, for each set, we can guarantee that all but one of its vertex-sides are endpoint-sides. Formally, for a vertex-side (u, su), its *special neighborhood* $B_{u,su}$ is the set of vertex-sides (v, sv) such that there exists and edge between (u, su) and $(v, 1 - sv)$ and it is the only edge incident on $(v, 1 - sv)$. A vertex-side which belongs to a special neighborhood is called a *special-side*. Figure 2 shows an example. Our key lemma is that all but one member of a special neighborhood must be an endpoint-side:

Lemma 4. *For a vertex-side (u, su), there must be at least $|B_{u,su}| - 1$ endpoint-sides of W in $B_{u,su}$.*

Proof. Assume without loss of generality that $|B_{u,su}| > 1$, since the lemma is vacuous otherwise. Let $(v, sv) \in B_{u,su}$ be a vertex-side that is not an endpoint side in W, and let w_v be the walk containing v. Since, in particular, (v, sv) is not an endpoint-side of w_v, then w_v must contain an edge incident to $(v, 1 - sv)$. By definition of special neighborhood, the only such edge is incident to (u, su). By definition of a path cover, there can only be one walk in W that contains an edge incident to (u, su) and it can contain only one such edge. Hence, there can only be one $(v, sv) \in B_{u,su}$ that is not an endpoint-side. ∎

Next, we show that the special neighborhoods are disjoint, and we can therefore define $n_{sp} = \sum_{u \in V, su \in \{0,1\}} \max(0, |B_{u,su}| - 1)$ as a lower bound on the number of special-sides that are endpoint-sides:

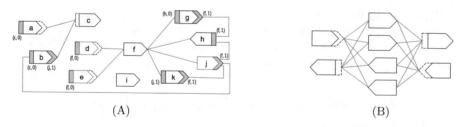

(A)   (B)

Fig. 2. (A) An example compacted de Bruijn graph (labels not shown), with a distinct ID for each vertex shown inside the vertex. The dashed hollow sides of the vertices are dead sides and the solid gray sides are special-sides. Each special-side is additionally labeled with the vertex-side to whose special neighborhoods it belongs. For example, the special neighborhood of vertex-side (c, 0) contains two vertex-sides, namely the blunt gray sides of vertices a and b, corresponding to $|B_{c,0}| = 2$. In this example, $n_{dead} = 4$, $n_{sp} = 6$, and $n_{iso} = 1$. By Theorem 2, the minimum size of a path cover is 6, and one can indeed find a path cover of this size in the graph. (B) In this example, $n_{dead} = 4$, $n_{sp} = n_{iso} = 0$, resulting in a lower bound of 2 on the size of a path cover. However, a quick inspection tells us that the the optimal size of a path cover is 4. This shows that our lower bound is not theoretically tight.

Lemma 5. *There are at least n_{sp} special-sides that are endpoint-sides of W.*

Proof. We claim that $B_{u,su} \cap B_{v,sv} = \emptyset$ for all $(u, su) \neq (v, su)$. Let $(w, sw) \in B_{u,su} \cap B_{v,sv}$. By definition of $B_{u,su}$, the only edge touching $(w, 1 - sw)$ is incident to (u, su). Similarly, by definition of B_v, the only edge touching $(w, 1 - sw)$ is incident to (v, sv). Hence, $(u, su) = (v, sv)$. The Lemma follows by applying Lemma 4 to each $B_{u,su}$ and summing the result.

Finally, we are ready to prove our lower bound on the size of path cover.

Theorem 2. $|W| \geq \lceil (n_{dead} + n_{sp})/2 \rceil + n_{iso}$.

Proof. Define a walk as *isolated* if it has only one vertex and that vertex is isolated. There are exactly n_{iso} isolated walks in W. Next, dead-sides are trivially endpoint-sides of a non-isolated walk in W. By Lemma 5, so are at least n_{sp} of the special-sides. Since the set of dead-sides and the set of special-sides are, by their definition, disjoint, the number of distinct endpoint-sides of non-isolated walks is at least $n_{dead} + n_{sp}$. Since every walk in a path cover must have exactly two distinct endpoint-sides, there must be at least $\lceil (n_{dead} + n_{sp})/2 \rceil$ non-isolated walks.

By applying Theorem 1 to Theorem 2 and observing that loops do not effect path covers, we get a lower bound on the minimum weight of any SPSS representation:

Corollary 1. *Let K be a set of canonical k-mers and let X^{opt} be its minimum weight SPSS representation. Then, $weight(X^{opt}) \geq |K| + (k - 1)(\lceil (n_{dead} + n_{sp})/2 \rceil + n_{iso})$, where n_{dead}, n_{iso}, and n_{sp} are defined with respect to the graph obtained by removing loops from cdBG(K).*

We note that the lower bound is not tight, as in the example of Fig. 2B; it can likely be improved by accounting for higher-order relationships in G. However, the empirical gap between our lower bound and algorithm is so small (Sect. 7) that we did not pursue this direction.

5   The UST Algorithm for Computing a SPSS Representation

In this section, we describe our algorithm called UST (UNITIG-STITCH) for computing a SPSS representation of a set of k-mers K. We first use the Bcalm2 tool [5] to construct $cdBG(K)$, then find a path cover W of $cdBG(K)$, and then output $spell(W)$, which by Lemma 2 is a SPSS representation of K.

UST constructs a path cover W by greedily exploring the vertices, with each vertex explored exactly once. We maintain the invariant that W is a path cover over all the vertices explored up to that point, and that the currently explored vertex is an endpoint of a walk in W. To start, we pick an arbitrary vertex u, add a walk consisting of only u to W, and start an *exploration* from u.

An exploration from u works as follows. First, we mark u as explored. Let w_u be the walk in W that contains u as an endpoint, and let su be the endpoint-side of u in w_u. We then search for an edge $e = (u, 1 - su, v, sv)$, for some v and sv. If we find such an edge and v has not been explored, then we extend w_u with e and start a new exploration from v. If v has been explored and is an endpoint vertex of a walk w_v in W, then we merge w_u and w_v together if the orientations allow (i.e. if $1 - sv$ is the side at which w_v is incident to v) and start a new exploration from an arbitrary unexplored vertex. In all other cases (i.e. if e is not found, if the orientations do not allow merging w_v with w_u, or if v in internal vertex in w_v), we start a new exploration from an arbitrary unexplored vertex. The algorithm terminates once all the vertices have been explored. It follows directly via the loop invariant that the algorithm finds a path cover, though we omit an explicit proof.

In our implementation, we do not store the walks W explicitly but rather just store a walk ID at every vertex along with some associated information. This makes the algorithm run-time and memory linear in the number of vertices and the number of edges, except for the possibility of needing to merge walks (i.e. merging of w_u and w_v). But we implement these operations using a union-find data structure, making the total time near-linear.

We note that UST's path cover depends on the arbitrary choices of which vertex to explore. Figure 1C gives an example where this leads to sub-optimal results. However, our results indicate that UST cannot be significantly improved in practice, at least for the datasets we consider (Sect. 7).

6   Applications

We apply UST to solve two problems. First, we use it to construct a *compression* algorithm UST-Compress. UST-Compress supports only compression and decompression and not membership and is intended to reduce disk space. We take K as input (in the binary output format of either DSK [18] or Jellyfish [19]), run UST on K, and then compress the resulting SPSS using a generic nucleotide compressor MFC [39]. UST-Compress can also be run in a mode that takes as input a count associated with each k-mer. In this mode, it outputs a list of counts in the order of their respective k-mers in the output SPSS representation (this is a trivial modification to UST). This list is then compressed using the generic LZMA compression algorithm. Note that we use MFC and LZMA due to their superior compression ratios, but other compressors could be substituted. To decompress, we simply run the MFC or LZMA decompressing algorithm.

Second, we use UST to construct an *exact static membership data structure* UST-FM. Given K, we first run UST on K, and then construct an FM-index [40] (as implemented in [41]) on top of the resulting SPSS representation. The FM-index then supports membership queries. In comparison to hash-based approaches, the FM-index does not support insertion or deletion; on the other hand, it allows membership queries of strings shorter than k.

7 Empirical Results

We use different types of publicly available sequencing data because each type may result in a de Bruijn graph with different properties and may inherently be more or less compressible. Our datasets include human, bacterial, and fish samples; they also include genomic, metagenomic, and RNA-seq data Each dataset was k-mer counted using DSK [18], using $k = 31$ with singleton k-mers removed. While these are not the optimal values for each of the respective applications, it allows us to have a uniform comparison across datasets. In addition, we k-mer count one of the datasets with $k = 61$, removing singletons, in order to study the effect of k-mer size. All our experiments were run on a server with a Intel(R) Xeon(R) CPU E5-2683 v4 @ 2.10 GHz with 64 cores and 512 GB of memory. All tested algorithms were verified for correctness in all datasets. Reproducibility details are available at https://github.com/medvedevgroup/UST/tree/master/experiments.

7.1 Evaluation of the UST Representation

We compare our UST representation against the unitig representation as well as against the SPSS lower bound of Corollary 1 (the full paper has a deeper breakdown). UST reduces the number of nucleotides (i.e. weight) compared to the unitigs by 10–32%, depending on the dataset. The number of nucleotides obtained is always within 3% of the SPSS lower bound; in fact, when considering the gap between the unitig representation and the lower bound, UST closes 92–99% of that gap. These results indicate that our greedy algorithm is a nearly optimal SPSS representation, on these datasets. They also indicate that the lower bound of Corollary 1, while not theoretically tight, is nearly tight on the type of real data captured by our experiments.

Table 1. Comparison of different string set representations and the SPSS lower bound. The second column shows $|K|$. For a representation X, the number of strings is $|X|$ and the number of nucleotides per distinct k-mer is $weight(X)/|K|$. Unitigs were computed using BCALM2.

Dataset	# distinct k-mers	SPSS lower bound		UST		unitigs	
		# strings	nt/k-mer	# strings	nt/k-mer	# strings	nt/k-mer
Zebrafish RNA-seq	124,740,993	3,979,856	1.96	4,174,867	2.00	7,775,719	2.87
Human RNA-seq	101,017,526	3,924,803	2.17	4,132,115	2.23	7,665,682	3.28
Human chromosome 14	99,941,572	2,235,267	1.67	2,386,324	1.72	4,871,245	2.46
Whole human genome	391,766,120	13,964,825	2.07	14,423,449	2.10	19,581,835	2.50
Human gut metagenome	103,814,001	1,517,107	1.34	1,522,139	1.34	2,187,669	1.49
Human RNA-seq ($k = 61$)	75,013,109	2,651,729	3.12	2,713,825	3.17	4,371,173	4.50

7.2 Evaluation of UST-Compress

We measure the compressed space-usage (Table 2), compression time and memory (Table 3), and decompression time and memory. We compare against the following lossless compression strategies: (1) the binary output of the k-mer counters DSK [18], KMC [17], and Squeakr-exact [20]; (2) the original FASTA sequences, with headers removed; (3) the maximal unitigs; and (4) the BOSS representation [31] (as implemented in COSMO [42]). In all cases, the stored data is additionally compressed using MFC (for nucleotide sequences, i.e. 2 and 3) or LZMA (for binary data, i.e. 1 and 4). The second strategy (which we already discussed in Sect. 1.1) is not a k-mer compression strategy per say, but it is how many users store their data in practice. The fourth strategy uses BOSS, the empirically most space efficient exact membership data structure according to a recent comparison [35]. We include this comparison to measure the advantage that can be gained by not needing to support membership queries. Note that strategies 1 and 2 retain count information, unlike strategies 3 and 4. Squeakr-exact also has an option to store only the k-mers, without counts.

Table 2. Space usage of UST-Compress and others. We show the average number of bits per distinct k-mer in the dataset. All files are compressed with MFC or LZMA, in addition to the tool shown in the column name. Squeakr-exact's implementation is limited to $k < 32$ [20] and so it could not be run for $k = 61$.

Dataset	With counts					Without counts			
	Squeakr	KMC	DSK	FASTA	UST-Compress	Squeakr	BOSS	Unitigs	UST-Compress
Zebrafish RNA-seq	91	41	47	33	5.4	45	5.9	5.0	3.6
Human RNA-seq	94	41	48	41	6.3	41	6.9	5.8	4.1
Human chromosome 14	98	43	48	49	5.8	41	5.5	4.3	3.1
Whole human genome	85	41	43	17	4.7	40	7.0	4.7	4.1
Human gut metagenome	90	46	51	23	4.2	44	5.3	3.0	2.7
Human RNA-seq (k = 61)	–	82	77	41	6.4	–	9.0	5.5	4.3

First, we observe that compared to the compressed native output of k-mer counters, UST-Compress reduces the space by roughly an order-of-magnitude; this however comes at an expense of compression time. When the value of k is increased, this improvement becomes even higher; as k nearly doubles, the UST-Compress output size remains the same, however, the compressed binary files output by k-mer counters approximately double in size. Our results indicate that when disk space is a more limited resource than compute time, SPSS-based compression can be very beneficial. Second, we observe a 4–8x space improvement compared to just compressing the reads FASTA file. In this case, however, the extra time needed for UST compression is balanced by the extra time needed to recount the k-mers from the FASTA file. Therefore, if all that is used downstream are the k-mers and possibly their counts, then SPSS-based compression is again very beneficial. Third, UST-Compress uses between 39 and 48% less space than BOSS, with comparable construction time and memory. Fourth, compared to the other SPSS-based compression (based on maximal unitigs), UST-Compress uses 10 to 29% less space, but has 10 to 24% slower compression times (with the exception of the $k = 61$ dataset, where it compresses 6% faster). The ratio of space savings after compression closely parallels the ratio of the weights of the two SPSS representations (Table 1). Fifth, we note that the best compression ratios achieved are significantly better than the worst case Conway Bromage lower bound of >35 bits per k-mer for the $k = 31$ datasets and 95 bits per k-mer for the $k = 61$ dataset. Finally, we note that the differences in the peak construction memory, and the total decompression run time and memory (<2 min and <1 GB for UST-Compress, respectively, table not shown) were negligible.

We also compressed a subset of samples from a de-noised index of 450,000 microbial DNA data used recently in large scale indexing projects of BIGSI [43] and COBS [44]. Each sample consists of error-corrected 31-mers (without abundance information) from a corresponding sequencing experiment, natively stored as bzipped McCortex binary file (see [43,44] for details). We downloaded 19,000 of these files from [45]. We ran UST-Compress, which reduced the disk space from 507 GB to 14.7 GB, a 35x reduction. The compression took a total of 82 hours and a peak memory of 3 GB (using one core).

Table 3. Time and peak memory usage of UST-Compress (without counts) and others during compression. For BOSS and unitigs, the times are separated according to the two steps of compression: running the core algorithm (Cosmo and bcalm2) followed by the generic compressor (respectively, LZMA and MFC). For UST-Compress, the first step is exactly the same as for unitigs (Bcalm2), so the column is not repeated.

Dataset	Time (minutes)									Peak memory (GB)		
	BOSS			unitigs			UST-Compress			BOSS	unitigs	UST-Compress
	Cosmo	LZMA	Total	bcalm2	MFC	Total	UST	MFC	Total			
Zebrafish RNA-seq	6.3	0.7	7.0	3.0	1.5	4.4	1.5	0.9	5.3	4.0	3.1	3.1
Human RNA-seq	4.0	0.8	4.8	4.7	1.3	5.9	1.6	0.8	7.1	3.6	3.4	3.4
Human chromosome 14	4.9	0.5	5.4	2.1	1.0	3.1	1.1	0.7	3.9	4.2	3.4	3.4
Whole human genome	17.3	3.0	20.3	10.4	2.2	12.5	4.1	1.9	16.3	4.0	4.3	4.3
Human gut metagenome	6.6	0.7	7.3	3.2	0.9	4.0	0.5	0.8	4.5	3.3	3.9	3.9
Human RNA-seq (k = 61)	4.4	0.6	5.0	3.6	3.9	7.5	1.1	2.4	7.1	4.3	2.3	2.3

7.3   Evaluation of UST-FM

We measure the memory taken by the data structure (Table 4), the query times (Table 5), and the time and memory taken during construction (table in the full paper) We compare UST-FM against two other space-efficient exact static membership data structures for k-mer sets. The first builds the FM index on top of the maximal unitigs (we refer to this as unitig-FM, but it referred to originally as dbgfm in [4]). The second is BOSS, which, as mentioned previously, was shown [35] to have superior space usage. We did not compare against the Bloom filter trie [36], which is fast but uses an order of magnitude more memory than BOSS [35]. Other data structures, such as Pufferfish [8], blight [9], and Bifrost [10], implement more sophisticated operations and hence use significantly more memory than BOSS. Moreover, these make use of a unitig SPSS representation and hence could potentially themselves incorporate the UST approach.

First, the UST-FM index is 25–44% smaller and the queries are 4 to 11 times faster compared to BOSS; however, it takes 2 to 5 times longer to build. This time is dominated by FM-index construction [41], rather than by UST. Second, the UST-FM index is 10–32% smaller than the unitigs-FM index, with negligibly faster query time. Finally, the memory use during construction was similar for all approaches.

Table 4. UST-FM data structure size, shown in the average number of bits per distinct k-mer in the dataset. This was measured by taking the peak memory usage during membership queries.

Dataset	BOSS	unitigs-FM	UST-FM
Zebrafish RNA-seq	7.5	7.9	5.5
Human RNA-seq	9.0	9.2	6.3
Human chromosome 14	8.7	6.9	4.8
Whole human genome	7.7	6.8	5.7
Human gut metagenome	8.8	5.4	4.9
Human RNA-seq (k = 61)	13.4	13.6	10.0

Table 5. UST-FM query time (in seconds) for two sets of 10,000 k-mers each, using the Human RNA-seq indices. The first set contains k-mers drawn from the dataset, so that UST-FM returns a hit. The second set takes randomly generated k-mers which were verified to not be present in the dataset. We measured the query times (per k-mer) after the index was already loaded into memory.

		BOSS	unitigs-FM	UST-FM
$k = 31$	$x \in K$	3.80	0.51	0.49
	$x \notin K$	1.48	0.38	0.37
$k = 61$	$x \in K$	15.25	1.61	1.58
	$x \notin K$	5.10	0.35	0.37

8   Conclusion

In this paper, we define the notion of a spectrum-preserving string set representation of a set of k-mers, give a lower bound on what could be achieved by such a representation, and give an algorithm to compute a representation that comes close to the lower bound. We demonstrate the applicability of the SPSS definition by using our algorithm to substantially improve space efficiency of the state-of-the-art in two applications.

A natural question is why we limit ourselves to SPSS representations. One can imagine alternative strategies, such as allowing a k-mer to appear more than once in the string set, or allowing other types of characters. In fact, for any concrete application, one might argue that a SPSS representation is too restrictive and can be improved. However, we chose to focus on SPSS representations because they are the common denominator in the applications of unitig-based representations we have observed [4, 8–10]. In this way, they retain broad applicability, as opposed to more specialized representations.

One limitation of UST is the time and memory needed to run Bcalm2 as a first step. Bcalm2 works by repeatedly gluing k-mers into longer strings, taking care to never glue across a unitig boundary. However, this care is wasted in our case, since UST then greedily glues across unitig boundaries anyway. Therefore, a potentially significant speedup and memory reduction of UST would be to implement it as a modification of Bcalm2, as opposed to running on top of it. This can keep the high-level algorithm the same but change the implementation to work directly on the k-mer set by incorporating algorithmic aspects of Bcalm2.

Acknowledgements. We are grateful to Rayan Chikhi for feedback and help with modifying Bcalm2. PM and AR were supported by NSF awards 1453527 and 1439057. AR is supported by NIH Computation, Bioinformatics, and Statistics training program.

References

1. Chikhi, R., Holub, J., Medvedev, P.: Data structures to represent sets of k-long DNA sequences. arXiv:1903.12312 [cs, q-bio], March 2019
2. Harris, R.S., Medvedev, P.: Improved representation of sequence bloom trees. bioRxiv (2018)
3. Conway, T.C., Bromage, A.J.: Succinct data structures for assembling large genomes. Bioinformatics **27**(4), 479–486 (2011)
4. Chikhi, R., Limasset, A., Jackman, S., Simpson, J.T., Medvedev, P.: On the representation of de Bruijn graphs. In: Sharan, R. (ed.) RECOMB 2014. LNCS, vol. 8394, pp. 35–55. Springer, Cham (2014). https://doi.org/10.1007/978-3-319-05269-4_4
5. Chikhi, R., Limasset, A., Medvedev, P.: Compacting de Bruijn graphs from sequencing data quickly and in low memory. Bioinformatics **32**(12), i201–i208 (2016)
6. Pan, T., Nihalani, R., Aluru, S.: Fast de Bruijn graph compaction in distributed memory environments. IEEE/ACM Trans. Comput. Biol. Bioinf. **17**, 136–148 (2018)

7. Guo, H., Fu, Y., Gao, Y., Li, J., Wang, Y., Liu, B.: deGSM: memory scalable construction of large scale de Bruijn graph. IEEE/ACM Trans. Comput. Biol. Bioinf. (2019)
8. Almodaresi, F., Sarkar, H., Srivastava, A., Patro, R.: A space and time-efficient index for the compacted colored de Bruijn graph. Bioinformatics **34**(13), i169–i177 (2018)
9. Marchet, C., Kerbiriou, M., Limasset, A.: Indexing de Bruijn graphs with minimizers. bioRxiv (2019)
10. Holley, G., Melsted, P.: Bifrost-highly parallel construction and indexing of colored and compacted de Bruijn graphs, p. 695338. bioRxiv (2019)
11. Medvedev, P.: Modeling biological problems in computer science: a case study in genome assembly. Brief. Bioinform. **20**(4), 1376–1383 (2018)
12. Břinda, K.: Novel computational techniques for mapping and classifying next-generation sequencing data. Ph.D. dissertation, Université Paris-Est, November 2016. https://doi.org/10.5281/zenodo.1045317
13. Břinda, K., Baym, M., Kucherov, G.: Simplitigs as an efficient and scalable representation of de Bruijn graphs. bioRxiv (2020)
14. Jones, D.C., Ruzzo, W.L., Peng, X., Katze, M.G.: Compression of next-generation sequencing reads aided by highly efficient de novo assembly. Nucleic Acids Res. **40**(22), e171–e171 (2012)
15. Haas, B.J., et al.: De novo transcript sequence reconstruction from RNA-seq using the Trinity platform for reference generation and analysis. Nat. Protoc. **8**(8), 1494 (2013)
16. Kolmogorov, M., Yuan, J., Lin, Y., Pevzner, P.A.: Assembly of long, error-prone reads using repeat graphs. Nat. Biotechnol. **37**(5), 540 (2019)
17. Kokot, M., Długosz, M., Deorowicz, S.: KMC 3: counting and manipulating k-mer statistics. Bioinformatics **33**(17), 2759–2761 (2017)
18. Rizk, G., Lavenier, D., Chikhi, R.: DSK: k-mer counting with very low memory usage. Bioinformatics **29**(5), 652–653 (2013)
19. Marçais, G., Kingsford, C.: A fast, lock-free approach for efficient parallel counting of occurrences of k-mers. Bioinformatics **27**(6), 764–770 (2011)
20. Pandey, P., Bender, M.A., Johnson, R., Patro, R.: Squeakr: an exact and approximate k-mer counting system. Bioinformatics **34**(4), 568–575 (2017)
21. Pandey, P., Bender, M.A., Johnson, R., Patro, R.: A general-purpose counting filter: making every bit count. In: Proceedings of the 2017 ACM International Conference on Management of Data, pp. 775–787. ACM (2017)
22. Hosseini, M., Pratas, D., Pinho, A.: A survey on data compression methods for biological sequences. Information **7**(4), 56 (2016)
23. Hernaez, M., Pavlichin, D., Weissman, T., Ochoa, I.: Genomic data compression. Ann. Rev. Biomed. Data Sci. **2**, 19–37 (2019)
24. Numanagić, I., et al.: Comparison of high-throughput sequencing data compression tools. Nat. Methods **13**(12), 1005 (2016)
25. Yang, X., Chockalingam, S.P., Aluru, S.: A survey of error-correction methods for next-generation sequencing. Brief. Bioinform. **14**(1), 56–66 (2012)
26. Orenstein, Y., Pellow, D., Marçais, G., Shamir, R., Kingsford, C.: Designing small universal k-mer hitting sets for improved analysis of high-throughput sequencing. PLoS Comput. Biol. **13**(10), e1005777 (2017)
27. Rangavittal, S., Stopa, N., Tomaszkiewicz, M., Sahlin, K., Makova, K.D., Medvedev, P.: DiscoverY: a classifier for identifying Y chromosome sequences in male assemblies. BMC Genomics **20**(1), 641 (2019)

28. Sahlin, K., Medvedev, P.: *De Novo* clustering of long-read transcriptome data using a greedy, quality-value based algorithm. In: Cowen, L.J. (ed.) RECOMB 2019. LNCS, vol. 11467, pp. 227–242. Springer, Cham (2019). https://doi.org/10.1007/978-3-030-17083-7_14

29. Marçais, G., Solomon, B., Patro, R., Kingsford, C.: Sketching and sublinear data structures in genomics. Ann. Rev. Biomed. Data Sci. **2**, 93–118 (2019)

30. Rowe, W.P.: When the levee breaks: a practical guide to sketching algorithms for processing the flood of genomic data. Genome Biol. **20**(1), 199 (2019)

31. Bowe, A., Onodera, T., Sadakane, K., Shibuya, T.: Succinct de Bruijn graphs. In: Raphael, B., Tang, J. (eds.) WABI 2012. LNCS, vol. 7534, pp. 225–235. Springer, Heidelberg (2012). https://doi.org/10.1007/978-3-642-33122-0_18

32. Boucher, C., Bowe, A., Gagie, T., Puglisi, S.J., Sadakane, K.: Variable-order de Bruijn graphs. In: Data Compression Conference, pp. 383–392. IEEE (2015)

33. Belazzougui, D., Gagie, T., Mäkinen, V., Previtali, M., Puglisi, S.J.: Bidirectional variable-order de Bruijn graphs. In: Kranakis, E., Navarro, G., Chávez, E. (eds.) LATIN 2016. LNCS, vol. 9644, pp. 164–178. Springer, Heidelberg (2016). https://doi.org/10.1007/978-3-662-49529-2_13

34. Belazzougui, D., Gagie, T., Mäkinen, V., Previtali, M.: Fully dynamic de Bruijn graphs. In: Inenaga, S., Sadakane, K., Sakai, T. (eds.) SPIRE 2016. LNCS, vol. 9954, pp. 145–152. Springer, Cham (2016). https://doi.org/10.1007/978-3-319-46049-9_14

35. Crawford, V.G., Kuhnle, A., Boucher, C., Chikhi, R., Gagie, T.: Practical dynamic de Bruijn graphs. Bioinformatics **34**(24), 4189–4195 (2018)

36. Holley, G., Wittler, R., Stoye, J.: Bloom Filter Trie: an alignment-free and reference-free data structure for pan-genome storage. Algorithms Mol. Biol. **11**(1), 3 (2016). https://doi.org/10.1186/s13015-016-0066-8

37. Pandey, P., Bender, M.A., Johnson, R., Patro, R.: deBGR: an efficient and near-exact representation of the weighted de Bruijn graph. Bioinformatics **33**(14), i133–i141 (2017)

38. Diestel, R.: Graph Theory, vol. 101 (2005)

39. Pinho, A.J., Pratas, D.: MFCompress: a compression tool for FASTA and multi-FASTA data. Bioinformatics **30**(1), 117–118 (2013)

40. Ferragina, P., Manzini, G.: Opportunistic data structures with applications. In: Proceedings 41st Annual Symposium on Foundations of Computer Science, pp. 390–398. IEEE (2000)

41. https://github.com/jts/dbgfm

42. https://github.com/cosmo-team/cosmo/tree/VARI

43. Bradley, P., den Bakker, H.C., Rocha, E.P., McVean, G., Iqbal, Z.: Ultrafast search of all deposited bacterial and viral genomic data. Nat. Biotechnol. **37**(2), 152 (2019)

44. Bingmann, T., Bradley, P., Gauger, F., Iqbal, Z.: COBS: a compact bit-sliced signature index. arXiv preprint arXiv:1905.09624 (2019)

45. http://ftp.ebi.ac.uk/pub/software/bigsi/nat_biotech_2018/ctx/

NetMix: A Network-Structured Mixture Model for Reduced-Bias Estimation of Altered Subnetworks

Matthew A. Reyna[1,2], Uthsav Chitra[1], Rebecca Elyanow[1,3],
and Benjamin J. Raphael[1(✉)]

[1] Department of Computer Science, Princeton University, Princeton, NJ 08544, USA
braphael@princeton.edu
[2] Department of Biomedical Informatics, Emory University, Atlanta, GA 30306, USA
[3] Department of Computer Science, Brown University, Providence, RI 02912, USA

Abstract. A classic problem in computational biology is the identification of *altered subnetworks*: subnetworks of an interaction network that contain genes/proteins that are differentially expressed, highly mutated, or otherwise aberrant compared to other genes/proteins. Numerous methods have been developed to solve this problem under various assumptions, but the statistical properties of these methods are often unknown. For example, some widely-used methods are reported to output very large subnetworks that are difficult to interpret biologically. In this work, we formulate the identification of altered subnetworks as the problem of estimating the parameters of a class of probability distributions which we call the Altered Subset Distribution (ASD). We derive a connection between a popular method, jActiveModules, and the maximum likelihood estimator (MLE) of the ASD. We show that the MLE is *statistically biased*, explaining the large subnetworks output by jActiveModules. We introduce NetMix, an algorithm that uses Gaussian mixture models to obtain less biased estimates of the parameters of the ASD. We demonstrate that NetMix outperforms existing methods in identifying altered subnetworks on both simulated and real data, including the identification of differentially expressed genes from both microarray and RNA-seq experiments and the identification of cancer driver genes in somatic mutation data.

Keywords: Interaction networks · Network anomaly · Maximum likelihood estimation · Gene expression · Cancer

M. A. Reyna and U. Chitra—These authors contributed equally.

Electronic supplementary material The online version of this chapter (https://doi.org/10.1007/978-3-030-45257-5_11) contains supplementary material, which is available to authorized users.

R. Schwartz (Ed.): RECOMB 2020, LNBI 12074, pp. 169–185, 2020.
https://doi.org/10.1007/978-3-030-45257-5_11

1 Introduction

A standard paradigm in computational biology is to use interaction networks as prior knowledge in the analysis of high-throughput 'omics data, with applications in protein function prediction [17,22,56,64,69], gene expression [15,24,25,40,80], germline variants [11,35,37,46,47], somatic variants in cancer [34,48,55,57,74,76], and other data [9,12,18,27,31,51,67,78]. One classic approach is to identify *active*, or *altered*, subnetworks of an interaction network that contain outlier measurements. The altered subnetwork problem takes as input: (1) an interaction network whose nodes are biological entities (e.g., genes or proteins) and whose edges represent biological interactions (e.g., physical or genetic interactions, co-expression, etc.); and (2) a measurement or score for each node. The goal is to find high-scoring subnetworks that correspond to functionally related or correlated alterations. This problem was introduced in [40] for gene expression analysis, where gene scores were derived from *p*-values of differential expression. [40] developed the jActiveModules algorithm to solve this problem and identify altered subnetworks of differentially expressed genes. Subsequently, [24] introduced heinz as "the first approach that really tackles and solves the original problem raised by [40] to optimality." jActiveModules and heinz have become widely-used tools with diverse applications; a few recent examples include mass-spectrometry proteomics [43,49], damaging *de novo* mutations in schizophrenia and other neurological disorders [16,28], and single-cell RNA-seq [29,44,75].

In the past two decades, many algorithms have been developed to identify altered subnetworks in biological data (reviewed in [18,23,54,55]). Each publication describing a new algorithm demonstrates the performance of their algorithm on specific biological datasets, and many of these publications also benchmark their algorithm against existing algorithms on real and/or simulated data. However, few of these publications prove theoretical guarantees for their algorithm's performance on a well-defined generative model of the data. Thus, the true performance of these algorithms is often unknown. Indeed, recent benchmarking studies (e.g., [8,32]) of several widely used network algorithms – including jActiveModules and heinz – show considerable disagreement between subnetworks identified by different methods on the same biological datasets. Moreover, these benchmarking studies (and many others) do not compare network algorithms against single-gene tests that do not use the network; thus, the tacit assumption that interaction networks always improve gene prioritization is often not tested.

Separately, many publications in the statistics and machine learning literature investigate the problem of *detecting* whether or not a network contains an anomalous subnetwork, or a *network anomaly*, e.g., [1,3–6,70–73]. These papers describe specific generative models of network anomalies and use a rigorous hypothesis-testing framework to prove asymptotic results regarding the conditions under which it is possible to detect a network anomaly. Importantly, these papers also provide theoretical guarantees about conditions under which a network contributes to anomaly detection. However, the network anomaly literature does not specifically address the altered subnetwork problem studied in computational biology, with three key differences. First, the *detection* problem of deciding whether or not an altered subnetwork exists is not the same as the *estimation* problem of identifying the nodes in an altered subnetwork. Second, biological networks have a finite size, and it is unclear what guarantees the asymptotic results provide for finite-size networks. Finally, the topological constraints on network anomalies are different from those considered in computational biology.

In this paper, we aim to bridge the gap between the theoretical guarantees in the network anomaly literature and the practical problem of identifying altered subnetworks in biological data. We provide a rigorous formulation of the *Altered Subnetwork Problem*, the problem that jActiveModules [40], heinz [24], and other methods aim to solve. Our formulation of the Altered Subnetwork Problem is inspired by the generative model used in the network anomaly literature, but requires that the altered subnetwork is a connected subnetwork, a constraint motivated by the topology of signaling pathways [10,42] and by the seminal works of [40] and [24].

We show that the Altered Subnetwork Problem is equivalent to estimating the parameters of a distribution which we define as the *Altered Subset Distribution (ASD)*. We prove that the jActiveModules problem [40] is equivalent to finding a maximum likelihood estimator (MLE) of the parameters of the ASD for connected subgraphs. At the same time, we demonstrate that if (1) the size of the altered subnetwork is moderately small and (2) the scores of nodes inside and outside of the altered subnetwork are not well-separated, then the MLE is a *biased* estimator of the size of the altered subnetwork. This statistical bias provides a rigorous explanation for the large subnetworks produced by jActive-Modules [40]. We also show that the size of the altered subnetworks identified by heinz [24] is biased for most choices of its user-defined parameter.

We introduce a new algorithm, NetMix, that combines a Gaussian mixture model and a combinatorial optimization algorithm to identify altered subnetworks. We show that NetMix is a reduced-bias estimator of the size of the altered subnetwork. We demonstrate that NetMix outperforms other methods for identifying altered subnetworks on simulated data, gene expression data, and somatic mutation data.

2   Altered Subnetworks, Altered Subsets, and Maximum Likelihood Estimation

2.1   Altered Subnetwork Problem

Let $G = (V, E)$ be a biological interaction network with a measurement, or score, X_v for each vertex $v \in V$. We assume that there is a connected subnetwork A in G, the *altered subnetwork*, whose scores are derived from a different distribution than the scores of the vertices not in A (Fig. 1). The goal of the Altered Subnetwork Problem is to find A. The problem is defined formally as follows.

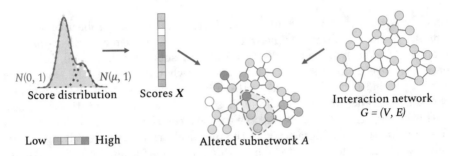

Fig. 1. Altered Subnetwork Problem. Measurements, or scores, $\mathbf{X}$ from a high-throughput experiment are drawn from one of two distributions: genes/proteins in an altered subnetwork A of an interaction network $G = (V, E)$ have scores drawn from an altered distribution $N(\mu, 1)$ with $\mu > 0$, while genes/proteins not in A have scores drawn from a background distribution $N(0, 1)$. The difficulty in identifying A depends on the separation μ between the distributions and the size $|A|$ of the altered subnetwork.

Altered Subnetwork Problem (ASP). Let $G = (V, E)$ be a graph with vertex scores $\mathbf{X} = (X_v)_{v \in V}$, and let $A \subseteq V$ be a connected subgraph of G. Suppose that

$$X_v \overset{\text{i.i.d.}}{\sim} \begin{cases} D_A, & \text{if } v \in A, \\ D_B, & \text{if } v \in V \setminus A, \end{cases} \tag{1}$$

where D_A is the *altered* distribution and D_B is the *background* distribution. Given G and $\mathbf{X}$, find A.

The seminal algorithm for solving the ASP is jActiveModules [40]. jActive-Modules takes as input a p-value p_v for each vertex v; e.g., a p-value of differential gene expression. Under the null hypothesis, the p-values p_v across genes are distributed according to the uniform distribution $U(0, 1)$. jActiveModules transforms the p-values into scores $X_v = \Phi^{-1}(1 - p_v)$, where Φ is the CDF of a standard normal distribution. Thus, jActiveModules solves the ASP with background distribution $D_B = N(0, 1)$. jActiveModules aims to find a connected subgraph $\hat{A}$ that maximizes[1] $\Gamma(S) = \frac{1}{\sqrt{|S|}} \sum_{v \in S} X_v$, i.e.,

$$\hat{A} = \underset{\text{connected } S \subseteq V}{\operatorname{argmax}} \Gamma(S) = \underset{\text{connected } S \subseteq V}{\operatorname{argmax}} \frac{1}{\sqrt{|S|}} \sum_{v \in S} X_v. \tag{2}$$

The presentation of jActiveModules in [40] does not specify the altered distribution D_A. However, in Sect. 2.2, we argue that the choice of the objective function in (2) implicitly assumes that $D_A = N(\mu, 1)$ for some parameter $\mu > 0$. Thus, we define the normally distributed ASP as follows.

---

[1] jActiveModules actually maximizes a normalized scan statistic $\Gamma_{\text{norm}}(S)$. We show in the full paper [65] that maximizing $\Gamma_{\text{norm}}(S)$ is equivalent to maximizing the unnormalized scan statistic $\Gamma(S)$ when the data is generated from normal distributions.

Normally Distributed Altered Subnetwork Problem. Let $G = (V, E)$ be a graph with vertex scores $\mathbf{X} = (X_v)_{v \in V}$, and let $A \subseteq V$ be a connected subgraph of G. Suppose that for some $\mu > 0$,

$$X_v \overset{\text{i.i.d.}}{\sim} \begin{cases} N(\mu, 1), & \text{if } v \in A, \\ N(0, 1), & \text{if } v \in V \setminus A. \end{cases} \tag{3}$$

Given G and $\mathbf{X}$, find A.

The Normally Distributed ASP has a sound statistical interpretation: if the p-values p_v of the genes are derived from an asymptotically normal test statistic, as is often the case, then the transformed p-values $X_v = \Phi^{-1}(1 - p_v)$ are distributed as $N(0, 1)$ for genes satisfying the null hypothesis and $N(\mu, 1)$ for genes satisfying the alternative hypothesis [38]. Normal distributions have been used to model transformed p-values from differential gene expression experiments [52, 60, 79].

More generally, the Normally Distributed Altered Subnetwork Problem is related to a larger class of *network anomaly* problems, which have been studied extensively in the machine learning and statistics literature [1, 3–6, 70–73]. To better understand the relationships between these problems and the algorithms developed to solve them, we will describe a generalization of the Altered Subnetwork Problem. We start by defining the following distribution, which generalizes the connected subnetworks in the Normally Distributed Altered Subnetwork Problem to any family of altered subsets.

Normally Distributed Altered Subset Distribution (ASD). Let $n > 0$ be a positive integer, let $\mathcal{S}$ be a family of subsets of $\{1, \ldots, n\}$, and let $A \in \mathcal{S}$. $\mathbf{X} = (X_1, \ldots, X_n)$ is distributed according to the *Normally Distributed Altered Subset Distribution* $\mathrm{ASD}_{\mathcal{S}}(A, \mu)$ provided

$$X_i \overset{\text{i.i.d.}}{\sim} \begin{cases} N(\mu, 1), & \text{if } i \in A, \\ N(0, 1), & \text{if } i \notin A. \end{cases} \tag{4}$$

Here, $\mu > 0$ is the *mean* of the ASD and A is the *altered* subset of the ASD.

More generally, the Altered Subset Distribution can be defined for any background distribution D_B and altered distribution D_A. We will restrict ourselves to normal distributions in accordance with the Normally Distributed Altered Subnetwork Problem, and we will subsequently assume normal distributions in both the Altered Subset Distribution and the Altered Subnetwork Problem.

The distribution in the Altered Subnetwork Problem is the $\mathrm{ASD}_{\mathcal{S}}(A, \mu)$, where the family $\mathcal{S}$ of subsets are connected subgraphs of the network G. In this terminology, the Altered Subnetwork Problem is the problem of estimating the parameters A and μ of the Altered Subset Distribution given data $\mathbf{X} \sim \mathrm{ASD}_{\mathcal{S}}(A, \mu)$ and knowledge of the parameter space $\mathcal{S}$ of altered subnetworks A. Thus, we generalize the Altered Subnetwork Problem to the ASD Estimation Problem, defined as follows.

ASD Estimation Problem. Let $\mathbf{X} = (X_1, \ldots, X_n) \sim \mathrm{ASD}_{\mathcal{S}}(A, \mu)$. Given $\mathbf{X}$ and $\mathcal{S}$, find A and μ.

The ASD Estimation Problem is a general problem of estimating the parameters of a *structured* alternative distribution. Different choices of S for the ASD Estimation Problem yield a number of interesting problems, some of which have been previously studied.

- $S = \mathcal{P}_n$, the power set of all subsets of $\{1, \ldots, n\}$. We call the distribution $\text{ASD}_{\mathcal{P}_n}(A, \mu)$ the *unstructured* ASD.
- $S = \mathcal{C}_G$, the set of all connected subgraphs of a graph $G = (V, E)$. We call $\text{ASD}_{\mathcal{C}_G}(A, \mu)$ the *connected* ASD. The connected ASD Estimation Problem is equivalent to the Altered Subnetwork Problem described above.
- $S = \mathcal{D}_G(\rho)$, the set of all subgraphs of a graph $G = (V, E)$ with edge density $\geq \rho$. [7,30,77] identify altered subnetworks with high edge density, and [2] identifies altered subnetworks with edge density $\rho = 1$, i.e., cliques.
- $S = \mathcal{N}_G = \{\mathcal{N}(v) : v \in V\}$, the set of all first-order network neighborhoods of a graph $G = (V, E)$. [14,36] use first-order network neighborhoods to prioritize cancer genes.
- $S \subset \mathcal{P}_n$, a family of subsets. Typically, $|S| \ll |\mathcal{P}_n|$ and S is not defined in terms of a graph. A classic example is gene set analysis; see [39] for a review.

2.2   Bias in Maximum Likelihood Estimation of the ASD

One reasonable approach for solving the ASD Estimation Problem is to compute a maximum likelihood estimator (MLE) for the parameters of the ASD. We derive the MLE below and show that it has undesirable statistical properties. All proofs are in the supplement [65].

Theorem 1. *Let* $\mathbf{X} \sim \text{ASD}_S(A, \mu)$. *The maximum likelihood estimators (MLEs)* $\hat{A}_{ASD}$ *and* $\hat{\mu}_{ASD}$ *of* A *and* μ, *respectively, are*

$$\hat{A}_{ASD} = \underset{S \in S}{\text{argmax}}\, \Gamma(S) = \underset{S \in S}{\text{argmax}}\, \frac{1}{\sqrt{|S|}} \sum_{v \in S} X_v \text{ and } \hat{\mu}_{ASD} = \frac{1}{|\hat{A}_{ASD}|} \sum_{v \in \hat{A}_{ASD}} X_v.$$

(5)

The maximization of Γ over S in (5) is a version of the *scan statistic*, a commonly used statistic to study point processes on lines and rectangles under various distributions [26,45]. Comparing (5) and (2), we see that jActiveModules [40] computes the scan statistic over the family $S = \mathcal{C}_G$ of connected subgraphs of the graph G. Thus, although jActiveModules [40] neither specifies the anomalous distribution D_A nor provides a statistical justification for their subnetwork scoring function, Theorem 1 above shows that jActiveModules implicitly assumes that D_A is a normal distribution, and that jActiveModules aims to solve the Altered Subnetwork Problem by finding the MLE $\hat{A}_{ASD}$.

Despite this insight that jActiveModules computes the MLE, it has been observed that jActiveModules often identifies large subnetworks. [58] notes that the subnetworks identified by jActiveModules are large and "hard to interpret biologically". They attribute the tendency of jActiveModules to identify large

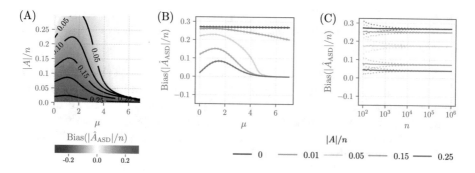

Fig. 2. Scores $\mathbf{X} \sim \text{ASD}_{\mathcal{P}_n}(A, \mu)$ are distributed according to the unstructured ASD. (A) Bias($|\hat{A}_{\text{ASD}}|/n$) in the maximum likelihood estimate of the altered subset size $|A|/n$ as a function of $|A|/n$ and the mean μ for $n = 10^4$. (B) Bias($|\hat{A}_{\text{ASD}}|/n$) for $n = 10^4$ and several values of $|A|/n$. Dotted lines indicate first and third quartiles in the estimate of the bias. (C) Bias($|\hat{A}_{\text{ASD}}|/n$) as a function of n for $\mu = 3$ and for several values of $|A|/n$.

subnetworks to the fact that a graph typically has more large subnetworks than small ones. While this observation about the relative numbers of subnetworks of different sizes is correct, we argue that this tendency of jActiveModules to identify large subnetworks is due to a more fundamental reason: the MLE $\hat{A}_{\text{ASD}}$ is a *biased* estimator of A.

First, we recall the definitions of bias and consistency for an estimator $\hat{\theta}$.

Definition 1. *Let $\hat{\theta} = \hat{\theta}(\mathbf{X})$ be an estimator of a parameter θ given observed data $\mathbf{X} = (X_1, \ldots, X_n)$. (a) The* bias *in the estimator $\hat{\theta}$ of θ is $Bias_\theta(\hat{\theta}) = E[\hat{\theta}] - \theta$. We say that $\hat{\theta}$ is a* biased *estimator of θ if $Bias_\theta(\hat{\theta}) \neq 0$ and is* unbiased *otherwise. (b) We say that $\hat{\theta}$ is a* consistent *estimator of θ if $\hat{\theta} \xrightarrow{P} \theta$, where $\xrightarrow{P}$ denotes convergence in probability, and is* inconsistent *otherwise.*

When it is clear from context, we omit the subscript θ and write Bias($\hat{\theta}$) for the bias of estimator $\hat{\theta}$.

Let $\mathbf{X} \sim \text{ASD}_{\mathcal{P}_n}(A, \mu)$ be distributed according to the unstructured ASD. We observe that the estimators $|\hat{A}_{\text{ASD}}|/n$ and $\hat{\mu}_{\text{ASD}}$ are both biased and inconsistent when both $|A|/n$ and μ are moderately small (Fig. 2). We summarize these observations in the following conjecture.

Conjecture. Let $\mathbf{X} = (X_1, \ldots, X_n) \sim \text{ASD}_{\mathcal{P}_n}(A, \mu)$. Then there exist $\mu_0 > 0$ and $\beta > 0$ such that, if $\mu < \mu_0$ and $|A|/n < \beta$, then $|\hat{A}_{\text{ASD}}|/n$ and $\hat{\mu}_{\text{ASD}}$ are biased and inconsistent estimators of $|A|/n$ and μ, respectively.

Although we do not have a proof of the above conjecture, we prove the following results that partially explain the bias and inconsistency of the estimators $|\hat{A}_{\text{ASD}}/n|$ and μ_{ASD}. For the bias, we prove the following.

Theorem 2. *Let* $\mathbf{X} = (X_1, \ldots, X_n) \sim \mathrm{ASD}_{\mathcal{P}_n}(A, \mu)$, *where* $A = \emptyset$. *Then* $|\hat{A}_{ASD}| = cn$ *for sufficiently large* n *and with high probability, where* $0 < c < 0.35$ *is independent of* n.

Empirically, we observe $c \approx 0.27$, i.e., $\hat{A}_{\mathrm{ASD}}$ contains more than a quarter of the scores (Fig. 2). This closely aligns with the observation in [58] that jActive-Modules reports subnetworks that contain approximately 29% of all nodes in the graph. Based on Theorem 2, one may suspect that $|\hat{A}_{\mathrm{ASD}}| \approx cn$ when μ or $|A|/n$ is sufficiently small, providing some intuition for why $|\hat{A}_{\mathrm{ASD}}|/n$ is biased. For inconsistency, we prove that the bias is independent of n.

Theorem 3. *Let* $\mathbf{X} = (X_1, \ldots, X_n) \sim \mathrm{ASD}_{\mathcal{P}_n}(A, \mu)$ *with* $|A| = \theta(n)$. *For sufficiently large* n, $\mathrm{Bias}(|\hat{A}_{\mathrm{ASD}}|/n)$ *and* $\mathrm{Bias}(\hat{\mu}_{ASD})$ *are independent of* n.

3   The NetMix Algorithm

Following the observation that the maximum likelihood estimators of the distribution $\mathrm{ASD}_{\mathcal{P}_n}(A, \mu)$ are biased, we aim to find a less biased estimator by explicitly modeling the distribution of the scores $\mathbf{X}$. In this section, we derive a new algorithm, NetMix, that solves the Altered Subnetwork Problem by fitting a Gaussian mixture model (GMM) to $\mathbf{X}$.

3.1   Gaussian Mixture Model

We start by recalling the definition of a GMM.

Gaussian Mixture Model. Let $\mu > 0$ and $\alpha \in (0, 1)$. X is distributed according to the *Gaussian mixture model* $\mathrm{GMM}(\alpha, \mu)$ with parameters α and μ provided

$$X \sim \alpha N(\mu, 1) + (1 - \alpha)N(0, 1). \tag{6}$$

Given data $\mathbf{X} = (X_1, \ldots, X_n)$, we define $\hat{\mu}_{\mathrm{GMM}}$ and $\hat{\alpha}_{\mathrm{GMM}}$ to be the MLEs for μ and α, respectively, obtained by fitting a GMM to $\mathbf{X}$. In practice, $\hat{\mu}_{\mathrm{GMM}}$ and $\hat{\alpha}_{\mathrm{GMM}}$ are obtained by the EM algorithm, which is known to converge to the MLEs as the number of samples goes to infinity [20, 81]. Furthermore, if $X_i \overset{\mathrm{i.i.d.}}{\sim} \mathrm{GMM}(\mu, \alpha)$ are distributed according to the GMM with $\alpha \neq 0$, then $\hat{\mu}_{\mathrm{GMM}}$ and $\hat{\alpha}_{\mathrm{GMM}}$ are consistent (and therefore asymptotically unbiased) estimators of μ and α, respectively [13].

Analogously, by fitting a GMM to data $\mathbf{X} \sim \mathrm{ASD}_{\mathcal{P}_n}(A, \mu)$ from the unstructured ASD, we observe that $\hat{\alpha}_{\mathrm{GMM}}$ is a less biased estimator of $|A|/n$ than $|\hat{A}_{\mathrm{ASD}}|/n$ (Fig. 3A, B). We also observe that $\hat{\alpha}_{\mathrm{GMM}}$ is a consistent estimator of $|A|/n$ (Fig. 3C). We summarize our findings in the following conjecture.

Conjecture. Let $\mathbf{X} \sim \mathrm{ASD}_{\mathcal{P}_n}(A, \mu)$ with $|A| > 0$, and let $\hat{A}_{\mathrm{ASD}}$ be the MLE of A as defined in (5). Let $\hat{\alpha}_{\mathrm{GMM}}$ and $\hat{\mu}_{\mathrm{GMM}}$ be the MLEs of α and μ obtained by fitting a GMM to $\mathbf{X}$. Then $\mathrm{Bias}_{|A|/n}(\hat{\alpha}_{\mathrm{GMM}}) < \mathrm{Bias}_{|A|/n}(|\hat{A}_{\mathrm{ASD}}|/n)$. Moreover, $\hat{\alpha}_{\mathrm{GMM}}$ and $\hat{\mu}_{\mathrm{GMM}}$ are consistent estimators of $|A|/n$ and μ, respectively.

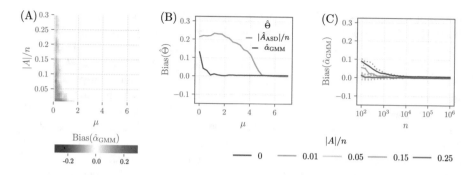

Fig. 3. Scores $\mathbf{X} \sim \mathrm{ASD}_{\mathcal{P}_n}(A, \mu)$ are distributed according to the unstructured ASD, and parameters $\hat{\alpha}_{\mathrm{GMM}}$ and $\hat{\mu}_{\mathrm{GMM}}$ are obtained by the EM algorithm. (A) $\mathrm{Bias}(\hat{\alpha}_{\mathrm{GMM}})$ as a function of the mean μ and altered subnetwork size $|A|/n$ for $n = 10^4$. Compare with Fig. 2A. (B) $\mathrm{Bias}(\hat{\alpha}_{\mathrm{GMM}})$ and $\mathrm{Bias}(|\hat{A}_{\mathrm{ASD}}|/n)$ as functions of the mean μ for $|A|/n = 0.05$ and $n = 10^4$. (C) $\mathrm{Bias}(\hat{\alpha}_{\mathrm{GMM}})$ as a function of n for mean $\mu = 3$ and several values of $|A|/n$. Compare with Fig. 2C.

3.2 NetMix Algorithm

We derive an algorithm, NetMix, that uses the maximum likelihood estimators (MLEs) $\hat{\mu}_{\mathrm{GMM}}$ and $\hat{\alpha}_{\mathrm{GMM}}$ from the GMM to solve the Altered Subnetwork Problem. Note that the GMM is *not* identical to ASD, the distribution that generated the data. Despite this difference in distributions, the above conjecture provides justification that the GMM will yield less biased estimators of A and μ than the MLEs of the ASD distribution.

Given a graph $G = (V, E)$ and scores $\mathbf{X} = (X_v)_{v \in V}$, NetMix first computes the *responsibility* $r_v = \Pr(v \in A \mid X_v)$, or the probability that $v \in A$, for each vertex $v \in V$. The responsibilities r_v are computed from the GMM MLEs $\hat{\mu}_{\mathrm{GMM}}$ and $\hat{\alpha}_{\mathrm{GMM}}$ (which are estimated by the EM algorithm [21]) by:

$$\hat{r}_v = \frac{\hat{\alpha}_{\mathrm{GMM}}\phi(X_v - \hat{\mu}_{\mathrm{GMM}})}{\hat{\alpha}_{\mathrm{GMM}}\phi(X_v - \hat{\mu}_{\mathrm{GMM}}) + (1 - \hat{\alpha}_{\mathrm{GMM}})\phi(X_v)}, \tag{7}$$

where ϕ is the PDF of the standard normal distribution.

Next, NetMix aims to find a connected subgraph C of size $|C| \approx n\alpha$ that maximizes $\sum_{v \in C} r_v$. In order to find such a subgraph, NetMix assigns a weight $w(v) = \hat{r}_v - \tau$ to each vertex v, where τ is chosen so that approximately $n\hat{\alpha}_{\mathrm{GMM}}$ nodes have non-negative weights. NetMix then computes the maximum weight connected subgraph (MWCS) $\hat{A}_{\mathrm{NetMix}}$ in G by adapting the integer linear program in [24]. The use of τ is motivated by the observation that, if $\hat{\alpha}_{\mathrm{GMM}} \approx \alpha$, then we expect $|\hat{A}_{\mathrm{NetMix}}| \approx n\hat{\alpha}_{\mathrm{GMM}} \approx n\alpha \approx |A|$. The formal description of the NetMix algorithm is in the full version of the paper [65], and the implementation is available online at https://github.com/raphael-group/netmix.

NetMix bears some similarities to heinz [24], another algorithm to identify altered subnetworks. However, there are two important differences. First, heinz does not solve the Altered Subnetwork Problem defined in the previous section. Instead, heinz models the vertex scores (assumed to be p-values) with a beta-uniform mixture (BUM) distribution. The motivation for the BUM is based on an empirical goodness-of-fit in [63]; however, later work by the same author [62] observes that the BUM tends to underestimate the number of p-values drawn from the altered distribution. Second, heinz requires that the user specify a False Discovery Rate (FDR) and shifts the p-values according to this FDR. We show below that nearly all choices of the FDR lead to a biased estimate of $|A|$. Moreover, the manually selected FDR allows users to selectively tune the value of this parameter to influence which genes are in the inferred altered subnetwork, analogous to "p-hacking" [33, 41, 59]. Indeed, published analyses using heinz [16, 32, 44] use a wide range of FDR values. See the full version [65] for more details on the differences between heinz and NetMix. Despite these limitations, the ILP given in heinz to solve the MWCS problem is useful for implementing NetMix and for computing the scan statistic (2) used in jActiveModules (see below).

4   Results

We compared NetMix to jActiveModules [40] and heinz [24] on simulated instances of the Altered Subnetwork Problem and on real datasets, including differential gene expression experiments from the Expression Atlas [61] and somatic mutations in cancer. The details of our experiments on real datasets are in the full version of the paper [65].

Since jActiveModules is accessible only through Cytoscape [68] and not a command-line interface, we implemented jActiveModules*, which computes the scan statistic (5) by adapting the ILP in heinz[2]. jActiveModules* outputs the global optimum of the scan statistic, while jActiveModules relies on heuristics (simulated annealing and greedy search) and may output a local optimum.

4.1   Simulated Data

We compared NetMix, jActiveModules*, and heinz on simulated instances of the Altered Subnetwork Problem using the HINT+HI interaction network [48], a combination of binary and co-complex interactions in HINT [19] with high-throughput derived interactions from the HI network [66] as the graph G. For each instance, we randomly selected a connected subgraph $A \subseteq V$ with size $|A| = 0.05n$ using the random walk method of [50], and drew a sample $\mathbf{X} \sim \mathrm{ASD}_{\mathcal{C}_G}(A, \mu)$. We ran each method on $\mathbf{X}$ and G to obtain an estimate $\hat{A}$ of the altered subnetwork A. We ran heinz with three different choices of the FDR parameter (FDR $= 0.001, 0.1, 0.5$) to reflect the variety of FDRs used in practice.

---

[2] The scan statistic (2) is the maximization of a non-linear objective function, but for fixed subnetwork size $|S|$ the objective function is linear. We computed the scan statistic by modifying the ILP in heinz [24] to find a subnetwork of a fixed size, and running this ILP over all possible subnetwork sizes.

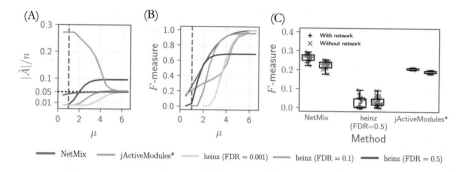

Fig. 4. Comparison of altered subnetwork identification methods on simulated instances of the Altered Subnetwork Problem using the HINT+HI interaction network with $n = 15074$ nodes, and where the altered subnetwork A has size $|A| = 0.05n$. [Dashed vertical line ($\mu = 1$) represents the smallest μ such that one can detect whether G contains an altered subnetwork (Formally, μ is the smallest mean such that the hypotheses $H_0 : X \sim \mathrm{ASD}_{\mathcal{C}_G}(\emptyset, 0)$ and $H_1 : X \sim \mathrm{ASD}_{\mathcal{C}_G}(A, \mu)$ are asymptotically distinguishable. See [73] for details.). (A) Size $|\hat{A}|/n$ of identified altered subnetwork $\hat{A}$ as a function of mean μ. (B) F-measure for $\hat{A}$ as a function of μ. (C) F-measure for $\hat{A}$ at $\mu = 1$, comparing performance with the network (left series for each method) and without the network (right series for each method).

We found that NetMix output subnetworks whose size $|\hat{A}_{\mathrm{NetMix}}|$ was very close to the true size $|A|$ across all values of μ in the simulations (Fig. 4A). In contrast, jActiveModules* output subnetworks that were much larger than the implanted subnetwork for $\mu < 5$. This behavior is consistent with our conjectures above about the large bias in the maximum likelihood estimator $\hat{A}_{\mathrm{ASD}}$ for the unstructured ASD. Note that $\mu > 5$ corresponds to a large separation between the background and alternative distributions, and the network is not needed to separate these two distributions.

We also quantified the overlap between the true altered subnetwork A and the subnetwork $\hat{A}$ output by each method using the F-measure, finding that NetMix outperforms other methods across the full range of μ (Fig. 4B). heinz requires the user to select an FDR value, and we find that the size of the output subnetwork and the F-measure vary considerably for different FDR values (Fig. 4A, 4B). When μ was small, a high FDR value (FDR = 0.5) yielded the best performance in terms of F-measure. However, when μ was large, a low FDR value (FDR = 0.001) gave better performance. While there are FDR values where the performance of heinz is similar to NetMix, the user *does not know what FDR value to select* for any given input, as the values of μ and the size $|A|$ of the altered subnetwork are unknown.

The bias in $|\hat{A}|/n$ observed using jActiveModules* with the interaction network (Fig. 4A) was similar to the bias for the unstructured ASD (Fig. 2A). Thus, we also evaluated how much benefit the network provided for each method. For small μ, we found that NetMix had a small but noticeable gain in performance when using the network; in contrast, other methods had nearly the same performance with or without the network (Fig. 4C). These results emphasize the importance of evaluating network methods on simulated data *and* demonstrating that a network method outperforms a single-gene test; neither of these were done in the jActiveModules [40] and heinz [24] papers, nor are they common in many other papers on biological network analysis.

5   Discussion

In this paper, we revisit the classic problem of identifying altered subnetworks in high-throughput biological data. We formalize the Altered Subnetwork Problem as the estimation of the parameters of the Altered Subset Distribution (ASD). We show that the seminal algorithm for this problem, jActiveModules [40], is equivalent to a maximum likelihood estimator (MLE) of the ASD. At the same time, we show that the MLE is a biased estimator of the altered subnetwork, with especially large positive bias for small altered subnetworks. This bias explains previous reports that jActiveModules tends to output large subnetworks [58].

We leverage these observations to design NetMix, a new algorithm for the Altered Subnetwork Problem. We show that NetMix outperforms existing methods on simulated and real data. NetMix fits a Gaussian mixture model (GMM) to observed node scores and then finds a maximum weighted connected subgraph using node weights derived from the GMM. heinz [24], another widely used method for altered subnetwork identification, also derives node weights from a mixture model (a beta-uniform mixture of p-values) and finds a maximum weighted connected subgraph. However, heinz does not solve the Altered Subnetwork Problem in a strict sense; rather, heinz requires users to choose a parameter (an FDR estimate for the mixture fit) that implicitly constrains the size of the identified subnetwork. This user-defined parameter encourages p-hacking [33,41,59], and we find that nearly all values of this parameter lead to biased estimates of the size of the altered subnetwork.

We note a number of directions for future work. The first is to generalize our theoretical contributions to the identification of *multiple* altered subnetworks, a situation which is common in biological applications where multiple biological processes may be perturbed [53]. While it is straightforward to run NetMix iteratively to identify multiple subnetworks – as jActiveModules does – a rigorous assessment of the identification of multiple altered subnetworks would be of interest. Second, our results on simulated data (Sect. 4.1) show that altered subnetwork methods have only marginal gains over simpler methods that rank vertices without information from network interactions. We hypothesize that this is because connectivity is not a strong constraint for biological networks; indeed the biological interaction networks that we use have both small diameter and small average shortest path between nodes (see the supplement for specific statistics). Specifically, we suspect that most subsets of nodes are "close" to a connected subnetwork in such biological networks, and thus the MLE of connected altered subnetworks has similar bias as the MLE of the unstructured altered subset distribution. In contrast, for other network topologies like the line graph, connectivity is a much stronger topological constraint (see the supplement for a brief comparison of different topologies). It would be useful to investigate this hypothesis and characterize the conditions when networks provide benefit for finding altered subnetworks. In particular, other topological constraints such as dense subgraphs [7,30], cliques [2], and subgraphs resulting from heat diffusion and network propagation processes [18,48,76,77] have been used used to model altered subnetworks in biological data. Generalizing the theoretical results in this paper to these other topological constraints may be helpful for understanding the parameter regimes where these topological constraints provide signal for identification of altered subnetorks. Finally, we note that biological networks often have substantial ascertainment bias, with more interactions annotated for well-studied genes [36,66], and these well-studied genes in turn may also be more likely to have outlier measurements/scores. Thus, any network method should carefully quantify the regime where it outperforms straightforward approaches – e.g., methods based on ranking nodes by gene scores or node degree – both on well-calibrated simulations and on real data.

Acknowledgments. We thank Mohammed El-Kebir for assistance with implementing jActiveModules* by modifying the ILP in heinz. We thank David Tse for directing us to the network anomaly literature. M.A.R. was supported in part by the National Cancer Institute of the NIH (Cancer Target Discovery and Development Network grant U01CA217875). B.J.R. was supported by US National Institutes of Health (NIH) grants R01HG007069 and U24CA211000.

References

1. Addario-Berry, L., Broutin, N., Devroye, L., Lugosi, G., et al.: On combinatorial testing problems. Ann. Stat. **38**(5), 3063–3092 (2010)
2. Amgalan, B., Lee, H.: WMAXC: a weighted maximum clique method for identifying condition-specific sub-network. PLoS ONE **9**(8), e104993 (2014)
3. Arias-Castro, E., Candès, E.J., Durand, A.: Detection of an anomalous cluster in a network. Ann. Stat. **39**(1), 278–304 (2011)
4. Arias-Castro, E., Candès, E.J., Helgason, H., Zeitouni, O.: Searching for a trail of evidence in a maze. Ann. Stat. **36**(4), 1726–1757 (2008)
5. Arias-Castro, E., et al.: Adaptive multiscale detection of filamentary structures in a background of uniform random points. Ann. Stat. **34**(1), 326–349 (2006)
6. Arias-Castro, E., et al.: Distribution-free detection of structured anomalies: permutation and rank-based scans. J. Am. Stat. Assoc. **113**(522), 789–801 (2018)
7. Ayati, M., et al.: MOBAS: identification of disease-associated protein subnetworks using modularity-based scoring. EURASIP J. Bioinform. Syst. Biol. **2015**, 7 (2015)
8. Batra, R., Alcaraz, N., Gitzhofer, K., et al.: On the performance of de novo pathway enrichment. NPJ Syst. Biol. Appl. **3**(1), 6 (2017)
9. Berger, B., et al.: Computational solutions for omics data. Nat. Rev. Genet. **14**(5), 333 (2013)
10. Bhalla, U.S., Iyengar, R.: Emergent properties of networks of biological signaling pathways. Science **283**(5400), 381–387 (1999)

11. Califano, A., et al.: Leveraging models of cell regulation and GWAS data in integrative network-based association studies. Nat. Genet. **44**(8), 841–847 (2012)
12. Chasman, D., Siahpirani, A.F., Roy, S.: Network-based approaches for analysis of complex biological systems. Curr. Opin. Biotechnol. **39**, 157–166 (2016)
13. Chen, J.: Consistency of the MLE under mixture models. Statist. Sci. **32**(1), 47–63 (2017)
14. Cho, A., et al.: MUFFINN: cancer gene discovery via network analysis of somatic mutation data. Genome Biol. **17**(1), 129 (2016)
15. Cho, D.Y., Kim, Y.A., Przytycka, T.M.: Network biology approach to complex diseases. PLoS Comput. Biol. **8**(12), 1–11 (2012)
16. Choi, J., Shooshtari, P., Samocha, K.E., Daly, M.J., Cotsapas, C.: Network analysis of genome-wide selective constraint reveals a gene network active in early fetal brain intolerant of mutation. PLoS Genet. **12**(6), e1006121 (2016)
17. Chua, H.N., et al.: Exploiting indirect neighbours and topological weight to predict protein function from protein-protein interactions. Bioinformatics **22**(13), 1623–1630 (2006)
18. Cowen, L., Ideker, T., Raphael, B.J., Sharan, R.: Network propagation: a universal amplifier of genetic associations. Nat. Rev. Genet. **18**(9), 551–562 (2017)
19. Das, J., Yu, H.: HINT: high-quality protein interactomes and their applications in understanding human disease. BMC Syst. Biol. **6**(1), 92 (2012)
20. Daskalakis, C., et al.: Ten steps of EM suffice for mixtures of two Gaussians. In: Proceedings of the 2017 Conference on Learning Theory, pp. 704–710 (2017)
21. Dempster, A.P., et al.: Maximum likelihood from incomplete data via the EM algorithm. J. Roy. Stat. Soc. Ser. B (Methodol.) **39**(1), 1–38 (1977)
22. Deng, M., et al.: Prediction of protein function using protein-protein interaction data. J. Comput. Biol. **10**(6), 947–960 (2003)
23. Dimitrakopoulos, C.M., Beerenwinkel, N.: Computational approaches for the identification of cancer genes and pathways. Wiley Interdisc. Rev. Syst. Biol. Med. **9**(1), e1364 (2017)
24. Dittrich, M.T., et al.: Identifying functional modules in protein-protein interaction networks: an integrated exact approach. Bioinformatics **24**(13), i223–i231 (2008)
25. de la Fuente, A.: From 'differential expression' to 'differential networking' - identification of dysfunctional regulatory networks in diseases. Trends Genet. **26**(7), 326–333 (2010)
26. Glaz, J., Naus, J., Wallenstein, S.: Scan Statistics. Springer, New York (2001). https://doi.org/10.1007/978-1-4757-3460-7
27. Gligorijević, V., Pržulj, N.: Methods for biological data integration: perspectives and challenges. J. R. Soc. Interface **12**(112), 20150571 (2015)
28. Gulsuner, S., et al.: Spatial and temporal mapping of de novo mutations in schizophrenia to a fetal prefrontal cortical network. Cell **154**(3), 518–529 (2013)
29. Guo, M., et al.: SLICE: determining cell differentiation and lineage based on single cell entropy. Nucleic Acid Res. **45**(7), e54 (2016)
30. Guo, Z., et al.: Edge-based scoring and searching method for identifying condition-responsive protein-protein interaction sub-network. Bioinformatics **23**(16), 2121–2128 (2007)
31. Halldórsson, B.V., Sharan, R.: Network-based interpretation of genomic variation data. J. Mol. Biol. **425**(21), 3964–3969 (2013)
32. He, H., Lin, D., Zhang, J., Wang, Y., Deng, H.W.: Comparison of statistical methods for subnetwork detection in the integration of gene expression and protein interaction network. BMC Bioinformatics **18**(1), 149 (2017)

33. Head, M.L., Holman, L., Lanfear, R., Kahn, A.T., Jennions, M.D.: The extent and consequences of P-Hacking in science. PLoS Biol. **13**(3), e1002106 (2015)

34. Hofree, M., Shen, J.P., Carter, H., Gross, A., Ideker, T.: Network-based stratification of tumor mutations. Nat. Methods **10**(11), 1108–1115 (2013)

35. Hormozdiari, F., et al.: The discovery of integrated gene networks for autism and related disorders. Genome Res. **25**(1), 142–154 (2015)

36. Horn, H., Lawrence, M.S., et al.: NetSig: network-based discovery from cancer genomes. Nat. Methods **15**(1), 61–66 (2017)

37. Huang, J.K., Carlin, D.E., et al.: Systematic evaluation of molecular networks for discovery of disease genes. Cell Syst. **6**(4), 484–495 (2018)

38. Hung, H.M.J., O'Neill, R.T., Bauer, P., Kohne, K.: The behavior of the P-value when the alternative hypothesis is true. Biometrics **53**(1), 11–22 (1997)

39. Hung, J.H., et al.: Gene set enrichment analysis: performance evaluation and usage guidelines. Brief. Bioinform. **13**(3), 281–291 (2011)

40. Ideker, T., et al.: Discovering regulatory and signalling circuits in molecular interaction networks. Bioinformatics **18**(suppl 1), S233–S240 (2002)

41. Ioannidis, J.P.: Why most published research findings are false. PLoS Med. **2**(8), e124 (2005)

42. Kelley, B.P., Yuan, B., Lewitter, F., Sharan, R., Stockwell, B.R., Ideker, T.: PathBLAST: a tool for alignment of protein interaction networks. Nucleic Acid Res. **32**(suppl 2), W83–W88 (2004)

43. Kim, M., Hwang, D.: Network-based protein biomarker discovery platforms. Genomics Inform. **14**(1), 2 (2016)

44. Klimm, F., et al.: Functional module detection through integration of single-cell RNA sequencing data with protein-protein interaction networks. bioRxiv (2019)

45. Kulldorff, M.: A spatial scan statistic. Commun. Stat. Theor. Methods **26**(6), 1481–1496 (1997)

46. Lee, I., et al.: Prioritizing candidate disease genes by network-based boosting of genome-wide association data. Genome Res. **21**(7), 1109–1121 (2011)

47. Leiserson, M.D., Eldridge, J.V., Ramachandran, S., Raphael, B.J.: Network analysis of GWAS data. Curr. Opin. Genet. Dev. **23**(6), 602–610 (2013)

48. Leiserson, M.D., et al.: Pan-cancer network analysis identifies combinations of rare somatic mutations across pathways and protein complexes. Nat. Genet. **47**(2), 106–114 (2015)

49. Liu, J.J., Sharma, K., Zangrandi, L., et al.: In vivo brain GPCR signaling elucidated by phosphoproteomics. Science **360**(6395) (2018)

50. Lu, X., Bressan, S.: Sampling connected induced subgraphs uniformly at random. In: Ailamaki, A., Bowers, S. (eds.) SSDBM 2012. LNCS, vol. 7338, pp. 195–212. Springer, Heidelberg (2012). https://doi.org/10.1007/978-3-642-31235-9_13

51. Luo, Y., Zhao, X., et al.: A network integration approach for drug-target interaction prediction and computational drug repositioning from heterogeneous information. Nat. Commun. **8**(1), 573 (2017)

52. McLachlan, G., et al.: A simple implementation of a normal mixture approach to differential gene expression in multiclass microarrays. Bioinformatics **22**(13), 1608–1615 (2006)

53. Menche, J., et al.: Disease networks. Uncovering disease-disease relationships through the incomplete interactome. Science **347**(6224), 1257601–1257601 (2015)

54. Mitra, K., et al.: Integrative approaches for finding modular structure in biological networks. Nat. Rev. Genet. **14**, 719 (2013)

55. Mutation Consequences and Pathway Analysis Working Group of the International Cancer Genome Consortium, et al.: Pathway and network analysis of cancer genomes. Nat. Methods **12**, 615 (2015)
56. Nabieva, E., et al.: Whole-proteome prediction of protein function via graph-theoretic analysis of interaction maps. Bioinformatics **21**, i302–i310 (2005)
57. Nibbe, R.K., Koyutürk, M., Chance, M.R.: An integrative-omics approach to identify functional sub-networks in human colorectal cancer. PLoS Comput. Biol. **6**(1), e1000639 (2010)
58. Nikolayeva, I., Pla, O.G., Schwikowski, B.: Network module identification-a widespread theoretical bias and best practices. Methods **132**, 19–25 (2018)
59. Nuzzo, R.: How scientists fool themselves-and how they can stop. Nat. News **526**(7572), 182 (2015)
60. Pan, W., et al.: A mixture model approach to detecting differentially expressed genes with microarray data. Funct. Integr. Genomics **3**(3), 117–124 (2003). https://doi.org/10.1007/s10142-003-0085-7
61. Petryszak, R., et al.: Expression atlas update: an integrated database of gene and protein expression in humans, animals and plants. Nucleic Acids Res. **44**(D1), D746–D752 (2015)
62. Pounds, S., Cheng, C.: Improving false discovery rate estimation. Bioinformatics **20**(11), 1737–1745 (2004)
63. Pounds, S., Morris, S.W.: Estimating the occurrence of false positives and false negatives in microarray studies by approximating and partitioning the empirical distribution of p-values. Bioinformatics **19**(10), 1236–1242 (2003)
64. Radivojac, P., Clark, W.T., et al.: A large-scale evaluation of computational protein function prediction. Nat. Methods **10**(3), 221 (2013)
65. Reyna, M.A., Chitra, U., et al.: Netmix: a network-structured mixture model for reduced-bias estimation of altered subnetworks. bioRxiv (2020). https://www.biorxiv.org/content/early/2020/01/19/2020.01.18.911438
66. Rolland, T., et al.: A proteome-scale map of the human interactome network. Cell **159**(5), 1212–1226 (2014)
67. Roy, S., Ernst, J.O.: Identification of functional elements and regulatory circuits by drosophila modencode. Science **330**(6012), 1787–1797 (2010)
68. Shannon, P., et al.: Cytoscape: a software environment for integrated models of biomolecular interaction networks. Genome Res. **13**(11), 2498–2504 (2003)
69. Sharan, R., Ulitsky, I., Shamir, R.: Network-based prediction of protein function. Mol. Syst. Biol. **3**(1), 88 (2007)
70. Sharpnack, J., Singh, A.: Near-optimal and computationally efficient detectors for weak and sparse graph-structured patterns. In: IEEE GlobalSIP (2013)
71. Sharpnack, J., Singh, A., Rinaldo, A.: Changepoint detection over graphs with the spectral scan statistic. In: Artificial Intelligence and Statistics, pp. 545–553 (2013)
72. Sharpnack, J., et al.: Detecting anomalous activity on networks with the graph Fourier scan statistic. IEEE Trans. Signal Process. **64**(2), 364–379 (2016)
73. Sharpnack, J.L., et al.: Near-optimal anomaly detection in graphs using Lovasz extended scan statistic. In: Advance Neural Information Processing Systems (2013)
74. Shrestha, R., Hodzic, E., et al.: Hit'ndrive: patient-specific multidriver gene prioritization for precision oncology. Genome Res. **27**(9), 1573–1588 (2017)
75. Soul, J., et al.: PhenomeExpress: a refined network analysis of expression datasets by inclusion of known disease phenotypes. Sci. Rep. **5**, 8117 (2015)
76. Vandin, F., Upfal, E., Raphael, B.J.: Algorithms for detecting significantly mutated pathways in cancer. J. Comput. Biol. **18**(3), 507–522 (2011)

77. Vanunu, O., et al.: Associating genes and protein complexes with disease via network propagation. PLoS Comput. Biol. **6**(1), e1000641 (2010)
78. Wang, X., et al.: HTSanalyzeR: an R/Bioconductor package for integrated network analysis of high-throughput screens. Bioinformatics **27**(6), 879–880 (2011)
79. Wang, Y.H., Bower, N.I., et al.: Gene expression patterns during intramuscular fat development in cattle. J. Anim. Sci. **87**(1), 119–130 (2009)
80. Xia, J., et al.: Networkanalyst for statistical, visual and network-based meta-analysis of gene expression data. Nat. Protoc. **10**, 823 (2015)
81. Xu, J., Hsu, D., Maleki, A.: Global analysis of expectation maximization for mixtures of two gaussians. In: Advances in Neural Information Processing (2016)

Stochastic Sampling of Structural Contexts Improves the Scalability and Accuracy of RNA 3D Module Identification

Roman Sarrazin-Gendron[1] , Hua-Ting Yao[1,2] , Vladimir Reinharz[3,4] ,
Carlos G. Oliver[1], Yann Ponty[2] , and Jérôme Waldispühl[1(✉)]

[1] School of Computer Science, McGill University, Montreal, Canada
`jeromew@cs.mcgill.ca`
[2] LIX, CNRS UMR 7161, Ecole Polytechnique, Palaiseau, France
`yann.ponty@lix.polytechnique.fr`
[3] Center for Soft and Living Matter, Institute for Basic Science, Ulsan, South Korea
[4] Département d'informatique, Université du Québec à Montréal, Montreal, Canada

Abstract. RNA structures possess multiple levels of structural organization. Secondary structures are made of canonical (i.e. Watson-Crick and Wobble) helices, connected by loops whose local conformations are critical determinants of global 3D architectures. Such local 3D structures consist of conserved sets of non-canonical base pairs, called RNA modules. Their prediction from sequence data is thus a milestone toward 3D structure modelling. Unfortunately, the computational efficiency and scope of the current 3D module identification methods are too limited yet to benefit from all the knowledge accumulated in modules databases. Here, we introduce BayesPairing 2, a new sequence search algorithm leveraging secondary structure tree decomposition which allows to reduce the computational complexity and improve predictions on new sequences. We benchmarked our methods on 75 modules and 6380 RNA sequences, and report accuracies that are comparable to the state of the art, with considerable running time improvements. When identifying 200 modules on a single sequence, BayesPairing 2 is over 100 times faster than its previous version, opening new doors for genome-wide applications.

Keywords: RNA structure prediction · RNA 3D modules · RNA modules identification in sequence

1 Introduction

RNAs use complex and well organized folding processes to support their many non-coding functions. The broad conservation of structures across species highlights the importance of this mechanism [14,35]. RNAs can operate using folding dynamics [25] or hybridization motifs [2]. Yet, many highly specific interactions need sophisticated three dimensional patterns to occur [11,13,15].

© Springer Nature Switzerland AG 2020
R. Schwartz (Ed.): RECOMB 2020, LNBI 12074, pp. 186–201, 2020.
https://doi.org/10.1007/978-3-030-45257-5_12

RNAs fold hierarchically [36]. First, Watson-Crick and Wobble base pairs are rapidly assembled into a secondary structure that determine the topology the RNA. Then, unpaired nucleotides form non-canonical base pairs interactions [16], stabilizing the loops while shaping the tertiary structure of the molecule. These non-canonical base pairing networks have thus been identified as critical components of the RNA architecture [4] and several catalogs of recurrent networks along with their characteristic 3D geometries are now available [7,10,12,27,28,30]. They act has structural organizers and ligand-binding centers [8] and we call them *RNA 3D modules*.

In contrast to well-established secondary structure prediction tools [20,22], we are still lacking efficient computational methods to leverage the information accumulated in the module databases. Software such as RMDetect [8], JAR3D [34] and our previous contribution BayesPairing 1 [32] have been released, but their precision and scalability remains a major bottleneck.

The significance of a module occurrence is typically assessed from recurrence: substructures that are found in distinct RNA structures are assumed to be functionally significant [30]. Based on this hypothesis, three approaches have been developed so far for the retrieval and scoring of 3D modules from sequence. The first one, RMDetect, takes advantage of Bayesian Networks to represent base pairing tendencies learned from sequence alignments. Candidate modules found in an input sequence are then scored with Bayesian probabilities. However, while showing excellent accuracy, RMDetect suffers from high computational costs, and minimal structure diversity among modules predicted [32]. Another option is JAR3D [34], which refined the graphical model-based scoring approach introduced by RMDetect and represents the state of the art for module scoring. However, it was not designed to maximize input sequence scanning efficiency and is limited in module diversity, only being applied to hairpin and internal loops. Finally, BayesPairing 1 [32], a recently introduced tool combining the Bayesian scoring of RMDetect to a regular expression based sequence parsing, is able to identify junction modules in input sequences and showed improved computational costs compared to RMDetect, which it was inspired from. Unfortunately, none of the aforementioned software can be used for the discovery of many RNA 3D modules in new sequences at the genome scale.

In this paper, we present BayesPairing 2, an efficient tool for high-throughput search of RNA modules in sequences. BayesPairing 2 analyzes the structural landscape of an input RNA sequence through secondary structure stochastic sampling and uses this information to identify candidate module insertion sites and select modules occurring in a favorable structural context. This pre-scoring stage enables us to dramatically reduce the number of putative matches and thus to (i) simultaneously search for multiple modules at once and (ii) eliminate false positives. BayesPairing 2 shows comparable performance to the state of the art while scaling gracefully with the number of modules searched. It also supports alignment search, a feature of RMDetect which could not be integrated in the BayesPairing 1 framework. All these improvements support potential applications at the genome scale (Fig. 1).

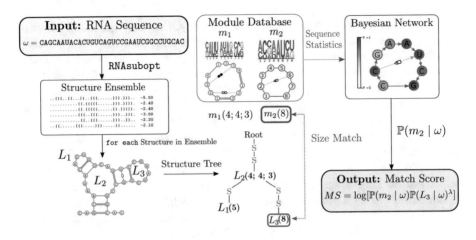

Fig. 1. The `BayesPairing 2` workflow addresses the identification of non-canonical 3D modules, *i.e.* arrangements of canonical and non canonical base pairs that are essential to the 3D architecture of RNAs. It takes as input either an RNA transcript or a multiple sequence alignment, possibly supplemented with a (shared) secondary structure, and returns an ordered list of occurrences for candidate modules. Its key idea is to match predicted secondary structure loops, highly likely to occur in thermodynamically-stable models, against a database of local modules learned from sequence data filtered for isostericity [19]. In this figure, we show the identification pipeline for one module on one structure of the ensemble. This is then repeated for all modules, for all structures.

2   Methods

Concepts and Model. A **non-canonical 3D module** consists in a set of non-canonical base pairs [17]. Modules occur within a **secondary structure loop**, consisting of one or several stretches of unpaired positions within an RNA transcript, also called **regions**, delimited by classic Watson-Crick/Wobble base pairs.

At the **thermodynamic equilibrium**, an RNA sequence w is expected to behave stochastically and adopt any of its secondary structure S, **compatible** with w with respect to canonical Watson-Crick/Wobble base pairing rules, with probability proportional to its **Boltzmann factor** [23]. The **Boltzmann probability** of a secondary structure S for an RNA sequence w is then

$$\mathbb{P}(S \mid w) = \frac{e^{-E_{S,w}/RT}}{\mathcal{Z}_w}$$

where $E_{S,w}$ represents the free-energy assigned to the (S, w) pair by the experimentally established Turner energy model [37], $\mathcal{Z}_w = \sum_{S'} e^{-E_{S',w}/RT}$ is the partition function [23], R is the Boltzmann constant and T the absolute temperature. By extension, the Boltzmann probability of a given loop to occur within

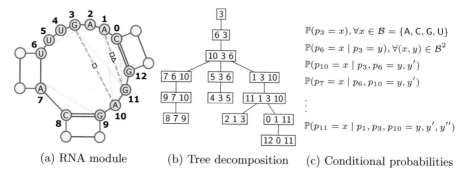

(a) RNA module   (b) Tree decomposition   (c) Conditional probabilities

Fig. 2. An RNA 3D module (a), here the three-way junction of the TPP riboswitch, represented in green, drawn in its structural context. Dashed and dotted lines respectively represent non-canonical base pairs and stacking interactions. A tree decomposition (b) of the module represents the dependencies between the module positions, leading to conditional probabilities (c), estimated from available sequence data

a sequence w is simply defined as

$$\mathbb{P}(\text{loop} \mid w) = \sum_{\substack{S \text{ compatible with } w \\ \text{loop} \in S}} \mathbb{P}(S \mid w).$$

In the current absence of thermodynamic data for non-canonical base-pairs and modules, we adopt a probabilistic approach, and model the sequence preferences associated with a module statistically as a **Bayesian network**, following Cruz *et al.* [8]. The structures of Bayesian networks are systematically derived from the base pairs occurring within **recurrent 3D motifs** [30]. Such motifs are typically mined within available 3D RNA structures in the PDB [5], and clustered geometrically.

Networks are then decomposed in a way that minimizes direct dependencies between individual positions of the module, while transitively preserving the emission probabilities. As illustrated in Fig. 2, we use a **tree decomposition** [6] of the network to minimize the maximum number of prior observations at each position, a strategy shared by instances of the junction tree methods [3]. Maximum likelihood conditional emission probabilities are then learned for each module using pseudo-counts.

The **emission probability** for the positions of a module m to be assigned to a nucleotide content A is then given by

$$\mathbb{P}(\text{assignment } A \mid \text{module } m) = \prod_{i \in m} \mathbb{P}(p_i = A_i \mid p_j = A_j \wedge p_{j'} = A_{j'} \wedge \ldots). \quad (1)$$

where $p_j, p_{j'}, \ldots$ represent the content of positions $j, j', \ldots$, the positions conditioning the content p_i of position i, as derived using the tree decomposition, and A_i represents the content of the i-th position in A. Using Bayes Theorem while assuming uniform priors for both assignments and modules (i.e.

$\mathbb{P}(m) = 1/|\mathcal{M}|, \mathbb{P}(A) = 1/4^{|m|}$), we obtain

$$\mathbb{P}(\text{module } m \mid \text{assignment } A) = \frac{\mathbb{P}(A \mid m) \times \mathbb{P}(m)}{\mathbb{P}(A)}$$

$$= \frac{4^{|m|} \prod_{i \in m} \mathbb{P}(p_i = A_i \mid p_j = A_j \wedge \ldots)}{|\mathcal{M}|}.$$

where $\mathcal{M}$ represents the set of admissible modules.

The final **match log-odds score** MS associated with a motif m being embedded within a given loop (*i.e.* at a given position) for an RNA sequence w is given by

$$\text{MS} = \lambda \log\left(\mathbb{P}(\text{loop} \mid w)\right) + \sum_{i \in m} \log\left(\mathbb{P}(p_i = A_i \mid p_j = A_j \wedge \ldots)\right)$$
$$+ |m| \log 4 - \log\left(|\mathcal{M}|\right) \tag{2}$$

where λ is a term that allows to control the weight of the structure and local sequence composition.

Algorithmic Considerations and Complexity. On an algorithmic level, for given sequence w and module m, we remark that it suffices to optimize for the first two terms of the above equations, the others being constant for a given module. A list of loops having highest Boltzmann probability $\mathbb{P}(\text{loop} \mid w)$ is first estimated from a statistical sample, generated using (non-redundant) stochastic backtrack [9,24,31]. The second term, *i.e.* the probability of the module content, is only evaluated for the loops that are compatible with the size constraints of the module, with tolerance for a size mismatch of up to one base per strand ($-\infty$ otherwise). Its evaluation uses conditional probabilities, learned from a tree-decomposition of the module, as described in Fig. 2. Matches featuring scores higher than a **cut-off** α are then reported as candidates.

The **overall complexity** of the method, when invoked with a module m and a transcript w of length n is in $\mathcal{O}(n^3 + kn \log n + \min(k, n^{2h(m)}) \times n \times |m|)$, where k denotes the number of sampled secondary structures and $h(m)$ is the total number of helices in m. It follows a sequence-agnostic precomputation in $\mathcal{O}(4^{w(m)} + |m| \times D)$, where $w(m)$ represents the tree-width of m, and D represents the overall size of the dataset used for training the model.

Remark that, while our reliance on sampling formally makes our method a heuristic in the context of optimizing the objective in Eq. (2), it must be noted that sampling provides a **statistically consistent** estimator for the probabilities of loops. Moreover, the probabilities associated with all possible loops could be computed exactly using constrained dynamic programming in time $\mathcal{O}(n^{3+2h(m)})$ [20].

Implementation. Secondary structures are non-redundantly sampled from the whole ensemble if the structure is not provided in the input, using RNAsubopt for a single sequence, or `RNAalifold` for a set of pre-aligned sequences [20, 24, 31]. Tree decompositions of modules are computed by the `htd` library [21] and conditional probabilities are learned using `pgmpy` [1]. `BayesPairing 2` is freely available as a downloadable software at (http://csb.cs.mcgill.ca/BP2).

Positioning Against Prior Work. Using stochastic sampling in `BayesPairing 2` allows to efficiently score all modules of a dataset in a single sequence search, unlike the previous version, which requires multiple regex searches on the sequence for each module. While searching structure-first improves the sensitivity, especially on modules without a strong sequence signal, it can add potential false positives, especially for small modules which appear a lot in secondary structures. This translates into more candidates scored, but scoring a candidate is much faster than scanning a sequence. Thus, `BayesPairing 2` is much more efficient when searching for many modules. In addition, the ability to sample with `RNAalifold` allows `BayesPairing 2` to take full advantage of aligned sequences.

3   Results

3.1   `Rna3Dmotif` Dataset

In order to assess the performance of `BayesPairing 2` on its own and in context with that of `BayesPairing 1`, we assembled a representative sequence-based dataset of local RNA 3D modules. We ran `Rna3Dmotif` on the non-redundant RNA PDB structure database [18]. Identified modules were then matched to `Rfam` family alignments via 3D structure positions. Sequences from these alignments were filtered to remove poorly aligned sequences, using isostericity substitution cutoffs ensuring that the extracted sequences could adopt their hypothesized structure. Modules matched to at least 35 sequences were added to the dataset. 75 modules, totaling 20 125 training sequences, were collected. To assess the presence and potential impact of **false positives (FP)** and **true negatives (TN)**, a negative dataset was assembled. To build this dataset, each sequence in the true positive dataset was shuffled while preserving its dinucleotide distribution. We assume motif occurrences to be homogeneous in length.

3.2   Validation on the `Rna3Dmotif` Dataset

Validating Searches on Sequences with Known Structure. A first aspect to validate is the ability of our method to retrieve the module when the native secondary structure is provided, ensuring the availability of a suitable loop for the module. For this test, the sequences were obtained from the positive dataset, and the structures accommodating their respective modules were generated with `RNAfold` hard constraint folding. As expected, structure-informed BP2 recovers every existing module.

Table 1. BayesPairing 2 module identification accuracy on Rna3Dmotif dataset

	F1 score	MCC	FDR	Sensitivity (top score)	Sensitivity (top 5 scores)
BayesPairing 2 performance	0.932	0.863	0.061	0.745	0.855

Joint Prediction of Secondary Structure Loops and Module Occurrences. To assess the performance of BayesPairing 2 on sequences of unknown structure, we performed two-fold cross-validation on 100 randomly sampled unique sequences (or on all sequences when fewer were available), for each module, amounting to a total of 6380 sequences. For each sequence-module pair, the candidate with highest score S through 20000 sampled structures was considered a **true positive (TP)** if its match score MS was above the score cutoff $T = -2.16$, and if its predicted position matched its real three-dimensional structure location. A sequence containing a module on which no accurate prediction was called above the cutoff was considered a **true negative (TN)**. We tested all λ values between 0 and 1 and cutoff values between -10 and 10, and found dataset-dependent optimal values of $\lambda = 0.35$ and a cutoff of -2.16 for this dataset. For the *top 5 scores* sensitivity, any correct prediction within the top 5 candidates could be considered a **TP**, whereas the *top score* test only accepted the highest score output. We also report the F1 score, the Matthews correlation coefficient (MCC), and the false discovery rate (FDR) associated with this cutoff. Formally the equations of those scores are:

$$F1 = \frac{(TP/(TP + FP))(TP/(TP + FN))}{(TP/(TP + FP)) + (TP/(TP + FN))} \qquad FDR = \frac{FP}{TP + FP}$$

$$MCC = \frac{TP \times TN - FP \times FN}{\sqrt{(TP + FP)(TP + FN)(TP + FP)(TN + FN)}}$$

Prediction Score Distribution and False Discovery Rate. We executed the same two-fold cross-validation experiment on the shuffled sequences described in Sect. 3.1. BayesPairing 2 found no hit on 92% of the 6380 sequences. It should be noted that it is not impossible for a shuffled sequence to contain a good hit for a module.

We obtained distributions of true and false hit scores from the cross-validation dataset. The score distributions, presented in Fig. 3a, are clearly distinct, and a score cutoff of -2.16 produced a false discovery rate of 0.061, as reported along with other common metrics in Table 1.

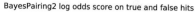

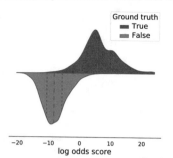

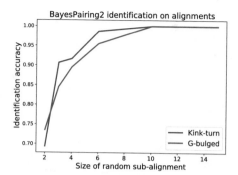

(a) Identification probabilistic scores output by BayesPairing 2 for 4500 true hits and 4500 false hits

(b) Prediction accuracy on sub-alignments for the kink-turn and g-bulged modules, by size of sub-alignment

Fig. 3. Evaluating BayesPairing 2 scores and accuracy.

Table 2. Rfam cross-family results for kink-turn (left) and G-bulged (right)

Trained	Identified on/with						Trained	Identified on/with			
Family	RF00162		RF02540		RF02541		*Family*	RF02540		RF02541	
Software	BP1	BP2	BP1	BP2	BP1	BP2	*Software*	BP1	BP2	BP1	BP2
RF00162	0.96	0.97	0.47	0.83	0.66	0.73	RF02540	0.98	1.0	0.91	0.98
RF02540	0.30	0.99	0.99	0.91	0.67	0.89	RF02541	0.82	0.99	0.93	0.99

(a) Kink-Turn	(b) G-bulged

3.3   Validation on Known Module Alignments from Rfam

Sequence Search. To complement our cross-validation experiments, we also tested BayesPairing 2 on Rfam alignments of the kink-turn and G-bulged internal loop modules. In these experiments, the modules were associated with their respective families through the Rfam motif database, then trained on one family and tested on the other. The results, for BayesPairing 1 and BayesPairing 2, are displayed in Tables 2a and b. We used standard parameters and selected the cutoffs associated to the same false discovery rate of 0.1 for both methods.

As observed in Sect. 3.2, BayesPairing 2 is slightly weaker at identifying modules with a strong sequence signal than BayesPairing 1, but considerably stronger when there is significant sequence variation as its signal appears to be more robust. This is particularly well illustrated by the capacity of BayesPairing 2 to identify the ribosomal kink-turn module on SAM riboswitch sequences. While the considerable sequence difference between the ribosome and riboswitch causes a sharp drop of 47% in BayesPairing 1 accuracy when predicting off-family, BayesPairing 2 only loses 25%.

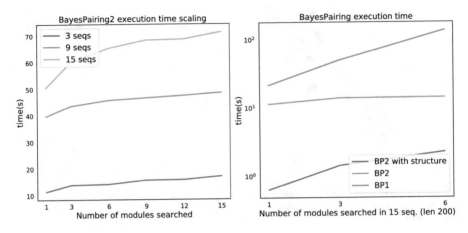

Fig. 4. Execution time of `BayesPairing 2`, as a function of numbers of modules and sequences (left), and compared to `BayesPairing 1` (right)

Alignment Search Improvement. Despite positive results in module identification on sequences taken from `Rfam`, sequence-based methods cannot fully take advantage of the common structure of an alignment. We show the relevance of including module identification on alignments in `BayesPairing 2` by improving the results presented in Sect. 3.3. If, instead of parsing individual sequences for modules, we parse randomly sampled sub-alignments, the predictions rise with the size of the sub-alignment until they reach 100%, up from 50 to 95% with sequence predictions by both software tools. Despite very low sample size (500 secondary structure sampled with `RNAalifold`), the alignment quickly outperforms the sequence predictions for all modules, on all tested families, as shown in Fig. 3b.

3.4   Time Benchmark

The execution time of `BayesPairing 2` was measured on 15 sequences (average size of ~200 nucleotides) containing a module each, with 5 hairpins, 5 internal loops and 5 multi-branched loops. We searched for 1, 3, 9 and 15 modules, and the execution time as a function of the sequence length and number of modules is displayed in Fig. 4. While the software typically requires 2–3 s to identify a module in a sequence of length 200, increasing the number of modules searched by a factor of fifteen only doubles its execution time.

Tests were executed on an Intel(R) Xeon(R) CPU E5-2667 @ 2.90 GHz, Ubuntu 16.0.4 with 23 cores, with a total physical memory of 792 gigabytes.

3.5   Comparison to the State of the Art

The first software to tackle the specific task of identifying 3D motifs in full RNA sequences was `RMDetect` (2011) [8], which showed good accuracy but was severely

Table 3. Performances of BayesPairing versions on Rna3Dmotif dataset

	F1 score	MCC	FDR	Sensitivity (1 candidate)	Sensitivity (5 candidates)
BP1	0.715	0.510	0.178	0.219	0.348
BP2	**0.932**	**0.863**	**0.061**	**0.745**	**0.855**

limited in the variety of motifs it could identify. BayesPairing 1 improved on this method by adding more flexibility and improving its search efficiency [32]. Another method, JAR3D, does not undertake full sequence searches but scores hairpin and internal loops against a database of models from the RNA 3D Motif Atlas. BayesPairing 2 can be adapted to fulfill the same task, and their purposes are close enough to be comparable. Because BayesPairing 1 has been shown to be a clear improvement on RMDetect, we focus our comparison on the former and JAR3D.

The good performances of BayesPairing 1 [32] relies on the assumption that the structural motif searched has a strong sequence signal. Indeed, the tool identifies motif location candidates through regular expressions. Thus, BayesPairing 1 struggles with motifs trained on a large number of distinct sequences with no dominant sequence pattern.

While it performed well on structure-based datasets with high sequence conservation, our Rfam-based dataset, with an average of 268 sequences from multiple Rfam families for each module, appears challenging for the method and is clearly outperformed by BayesPairing 2 on the dataset described in Sect. 3.1, as shown in Table 3. We also show in Fig. 4 that BayesPairing 2 scales much better in the number of modules searched.

JAR3D was also shown to outperform RMDetect in the identification of new variants of RNA 3D modules [40]. However, it does not perform a search on the input sequence, but only takes loops as input. As such, it executes a task that only accounts for a small proportion of BayesPairing 2's execution time. Indeed, scoring a loop against a model is very rapid, and both tools can score 10, 000

Table 4. BayesPairing 2 and JAR3D performances on hairpins (363 seq. in 33 loops), and internal loops (127 seq. in 28 loops) from the RNA 3D Motif Atlas.

Software	Average Identification TPR and FDR on RNA 3D Motif Atlas			
Loop type	Hairpin loops		Internal loops	
Software	TPR	FDR	TPR	FDR
BayesPairing 2	**0.9819**	**0.0020**	**1.00**	**0.0016**
JAR3D	0.9685	0.0509	0.957	0.0205

module candidates in less than 10 s, while the total runtime of BayesPairing 2 when searching for motifs in a single sequence of length 200 is greater than ~40 s. Therefore, we focus our comparison between BayesPairing 2 and JAR3D on true positive rate and false discovery rate, which contribute to the overall performance of both software.

In order to compare the software, we isolated the scoring component of BayesPairing 2, a function which takes as input a loop and a module and returns a match score between the two, the same input and output as JAR3D. We trained BayesPairing 2 on 51 motifs from the RNA 3D Motif Atlas, including 28 internal loops and 33 hairpin loops. Motifs which constituted full loops and only had occurrences of the same size, the two core assumptions of BayesPairing 2, were selected. Then, internal loops with fewer than three occurrences, and hairpin loops with fewer than 5 occurrences were removed from the dataset. True positive rates (TPR) were computed from predictions on RNA 3D Motif Atlas sequences. False discovery rates (FDR) were estimated from averaged predictions on 100 random sequences per true positive sequences (total 49000). Each random sequence was generated from the nucleotide distribution of the true positive sequences for that module. Default cutoffs were used. For BayesPairing 2, a cutoff of 3.5 was obtained by repeating the process presented in Sect. 3.2 after setting the weight of the secondary structure to 0, as the secondary structure is only considered in the context of the full sequence which is not part of the input for this specific task. The results are presented in Table 4. While the two software present comparable sensitivities, BayesPairing 2 achieves this high sensitivity with higher specificity.

4   Discussion

Applications. The most obvious application for an efficient and parallelizable motif identification framework is to parse sequences for local 3D structure signal. Modular approaches for RNA 3D structure construction like RNA-MoIP [29] have been shown to successfully take advantage of local tertiary structure information. In particular, RNA-MoIP leverages 3D module matches to select the most stable secondary structures to use as a scaffold for the full structure. Indeed, secondary structures that can accommodate known 3D modules are often more predictive of the real structure than those who cannot [8]. To this day, RMDetect, BayesPairing 1 and BayesPairing 2 are the only known full sequence probabilistic module identification tools to be able to identify hairpins, internal loops and junctions, which are key components of many well-known structures, namely several riboswitches. Of the three, BayesPairing 2 is the most scalable. This scalability is essential as many datasets include hundreds of modules [27,30], and this number will keep increasing as more structures are crystallized and mining methods improve.

While the tertiary structure signal encodes information that can be leveraged to build a full 3D structure, its implied functional significance can be taken advantage to refine tasks like sequence classification. Traditional methods for sequence classification include k-mer based techniques [38], as well as sequence and structure motifs [39], but those only use the sequence and secondary structure signals. 3D modules are highly complementary to those methods.

Identifying Multi-branched Loops in Sequences; Applications to Ribo-Switch Discovery. One of the distinctive characteristics of BayesPairing 2 is its ability to identify multi-branched loops. These motifs happen to be very common in riboswitches, in which they are often closely related to function, namely in the tyrosine pyrophosphate (TPP) riboswitch, the Cobalamin riboswitch, and the S-adenosyl methionine I (SAM-I) riboswitch [33]. We can use sequences from Rfam riboswitch families to train 3D module models, and then use those models to label new sequences as putative riboswitches.

The software also provides insight on the role of those of 3D modules in the folding dynamics of the riboswitch. Because BayesPairing 2 searches secondary structure ensembles for loops matching known structural modules, it can be used to observe, within the assumptions of the RNAfold library, how easily riboswitch sequences appear to fold into their junction. For instance, the TPP riboswitch's junction is very present in its Boltzmann ensemble, as its small (13 bases) three-way junction was correctly identified by our software on 81% of the sequences from the TPP Rfam family.

Because we could hypothesize the frequency of identification of a specific loop to be correlated with its size, it could be expected that the SAM-I riboswitch four-way junction, which counts 28 bases, would be identified less frequently. This is indeed the case as it was identified on 35% of the sequences of its family with a similar pipeline (Fig. 5).

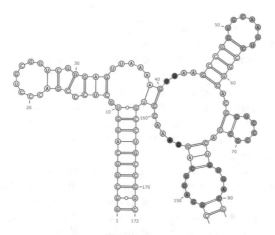

Fig. 5. The cobalamin riboswitch four-way junction (in yellow) in PDB structure 4GXY [26]. The adjacent structural motifs used to refine the structural search are highlighted. Bases within 3 Angstrom of the cobalamin molecule in the bound structure are indicated in bright red. Other colors highlight distinct modules.

The much smaller (17 bases) cobalamin riboswitch junction would then be expected to be found with a frequency somewhere in between 35% and 81%, based on this size assumption. Surprisingly, it was only successfully identified on 3.5% of the Rfam cobalamin family sequences.

However, interestingly, identifying small structural modules (two hairpins and one internal loop) around the junction with a first run of BayesPairing 2 and then using the position of those modules as constraints for a second run raises the frequency of identification of the multi-loop to 32%. The more adjacent motifs are found, the higher the identification confidence was observed to be. In contrast, applying the same method to the SAM riboswitch, or on shuffled cobalamin riboswitch sequences, does not leave to a significant improvement.

This difference in behavior between riboswitches could be rooted in different factors like co-transcriptional folding, RNA-RNA and RNA-protein interactions and/or the intrinsic difficulty of predicting riboswitch structural element with models learned from bound structures. However, the contrast between the constrained and unconstrained results in the cobalamin riboswitch tends to indicate that some, but not all multi-branched structure are strongly correlated with surrounding loops conformations.

Limitations and Future Work. Our approaches presents two main limitations. First, the assumption that motif occurrences have a consistent size is not a trivial one to make. For small modules, it is a reasonable assumption that the vast majority of occurrences will have the same size since adding or removing a base would have a large impact on the local 3D structure. However, for larger motifs, and especially junctions, the size constraint can prevent us from identifying some variants. This is something we alleviate in BayesPairing 2 by allowing imperfect matches, with a tolerated difference of up to one base per strand, but further work remains to be done to fully identify motifs bigger than 20 bases, for which this fuzzy matching might not be sufficient.

Second, a consequence of searching secondary structures before sequence is that in the rare cases when the sequence is better conserved than its secondary structure, the accuracy of the tool will suffer. It could however be argued that not overfitting to currently known sequences could be worth losing a bit of accuracy, although this can only be evaluated quantitatively as new structures and module occurrences become available, since the current structure datasets do not show sufficient sequence variability.

Interestingly, a large majority of the modules that cannot be predicted from sequence only by BayesPairing 2 occur in secondary structures that are never generated by RNAsubopt. In many of those cases, a base pair stacking was removed to allow the insertion of the module, at a considerable energy cost. We hypothesize that those small modifications, although not energetically favorable at the secondary structure level, are stabilized by 3D interactions which cannot be inferred from sequence. Going further with this hypothesis, differences in performances are then indicative of the stabilizing effect of non-canonical modules. This assumption could be tested in the future using coarse-grain molecular dynamics to correlate those two metrics.

The other notable limitation of the method is that the loop-based module definition used in our study does not allow the prediction of pseudoknots, nor canonical helices.

5   Conclusion

We presented `BayesPairing 2`, a software for efficient identification of RNA modules in sequences and alignments. `BayesPairing 2` strictly outperforms its previous version in execution time, search on provided secondary structures, and sequence search accuracy. It also appears to have complementary strengths to `JAR3D`, the state of the art for scoring. Finally, its structure-based approach brings a perspective on the place of the motif in the sequence's Boltzmann ensemble. This added context helps improve identification accuracy, but also the interpretation of the results, and can provide additional information about the role of a module in the folding process. Moreover, the time complexity improvement opens new doors for genome-wide sequence mining for local 3D structure patterns. As new RNA structures and sequences become available, more modules will be discovered, and `BayesPairing 2` is fast enough to take advantage of its customizability to contribute to filling the gap between secondary and tertiary structure prediction tool by associating a wide selection of RNA modules of interest to those new sequences.

Acknowledgements. The authors are greatly indebted to Anton Petrov for providing us with alignments between RNA PDB structures and `Rfam` families, which helped us match 3D modules to sequence alignments.

References

1. Ankan, A., Panda, A.: pgmpy: Probabilistic graphical models using python. In: Proceedings of the 14th Python in Science Conference (SCIPY 2015). Citeseer (2015)
2. Argaman, L., Altuvia, S.: fhlA repression by OxyS RNA: kissing complex formation at two sites results in a stable antisense-target RNA complex. J. Mol. Biol. **300**(5), 1101–1112 (2000)
3. Bach, F.R., Jordan, M.I.: Thin junction trees. In: Advances in Neural Information Processing Systems, pp. 569–576 (2002)
4. Beelen, R.H., Fluitsma, D.M., van der Meer, J.W., Hoefsmit, E.C.: Development of different peroxidatic activity patterns in pertoneal macrophages in vivo and in vitro. J. Reticuloendothel Soc. **25**(5), 513–523 (1979)
5. Berman, H.M., et al.: The protein data bank. Nucleic Acids Res. **28**, 235–242 (2000). https://doi.org/10.1093/nar/28.1.235
6. Bodlaender, H.L.: Dynamic programming on graphs with bounded treewidth. In: Lepistö, T., Salomaa, A. (eds.) ICALP 1988. LNCS, vol. 317, pp. 105–118. Springer, Heidelberg (1988). https://doi.org/10.1007/3-540-19488-6_110
7. Chojnowski, G., Walen, T., Bujnicki, J.M.: RNA bricks - a database of RNA 3D motifs and their interactions. Nucleic Acids Res. **42**, D123–D131 (2014). https://doi.org/10.1093/nar/gkt1084. Database issue

8. Cruz, J.A., Westhof, E.: Sequence-based identification of 3D structural modules in RNA with RMDetect. Nat. Methods **8**(6), 513–521 (2011). https://doi.org/10.1038/nmeth.1603

9. Ding, Y., Lawrence, C.E.: A statistical sampling algorithm for rna secondary structure prediction. Nucleic Acids Res. **31**, 7280–7301 (2003). https://doi.org/10.1093/nar/gkg938

10. Djelloul, M., Denise, A.: Automated motif extraction and classification in RNA tertiary structures. RNA **14**(12), 2489–2497 (2008). https://doi.org/10.1261/rna.1061108

11. Du, Z., Lind, K.E., James, T.L.: Structure of TAR RNA complexed with a Tat-TAR interaction nanomolar inhibitor that was identified by computational screening. Chem. Biol. **9**(6), 707–712 (2002)

12. Ge, P., Islam, S., Zhong, C., Zhang, S.: De novo discovery of structural motifs in RNA 3D structures through clustering. Nucleic Acids Res. **46**(9), 4783–4793 (2018). https://doi.org/10.1093/nar/gky139

13. Huck, L., et al.: Conserved tertiary base pairing ensures proper RNA folding and efficient assembly of the signal recognition particle Alu domain. Nucleic Acids Res. **32**(16), 4915–4924 (2004)

14. Kalvari, I., et al.: Rfam 13.0: shifting to a genome-centric resource for non-coding RNA families. Nucleic Acids Res. **46**(D1), D335–D342 (2017). https://doi.org/10.1093/nar/gkx1038

15. Lancaster, L., Lambert, N.J., Maklan, E.J., Horan, L.H., Noller, H.F.: The sarcin-ricin loop of 23S rRNA is essential for assembly of the functional core of the 50S ribosomal subunit. RNA **14**(10), 1999–2012 (2008)

16. Leontis, N.B., Westhof, E.: Geometric nomenclature and classification of RNA base pairs. RNA **7**(4), 499–512 (2001)

17. Leontis, N.B., Westhof, E.: Geometric nomenclature and classification of RNA base pairs. RNA (N.Y., NY) **7**, 499–512 (2001). https://doi.org/10.1017/s1355838201002515

18. Leontis, N.B., Zirbel, C.L.: Nonredundant 3D structure datasets for RNA knowledge extraction and benchmarking. In: Leontis, N., Westhof, E. (eds.) RNA 3D Structure Analysis and Prediction. Nucleic Acids and Molecular Biology, vol. 27, pp. 281–298. Springer, Heidelberg (2012). https://doi.org/10.1007/978-3-642-25740-7_13

19. Lescoute, A., Leontis, N.B., Massire, C., Westhof, E.: Recurrent structural RNA motifs, isostericity matrices and sequence alignments. Nucleic Acids Res. **33**, 2395–2409 (2005). https://doi.org/10.1093/nar/gki535

20. Lorenz, R., et al.: ViennaRNA package 2.0. Algorithms Mol. Biol. **6**, 26 (2011). https://doi.org/10.1186/1748-7188-6-26

21. mabseher: A small but efficient C++ library for computing (customized) tree and hypertree decompositions. https://github.com/mabseher/htd

22. Mathews, D.H.: RNA secondary structure analysis using RNAstructure. Curr. Protoc. Bioinform. **13**, 12.6.1–12.6.14 (2006). https://doi.org/10.1002/0471250953.bi1206s13

23. McCaskill, J.S.: The equilibrium partition function and base pair binding probabilities for RNA secondary structure. Biopolymers **29**, 1105–1119 (1990). https://doi.org/10.1002/bip.360290621

24. Michálik, J., Touzet, H., Ponty, Y.: Efficient approximations of RNA kinetics landscape using non-redundant sampling. Bioinform. (Oxford, Engl.) **33**, i283–i292 (2017). https://doi.org/10.1093/bioinformatics/btx269

25. Mustoe, A.M., Brooks, C.L., Al-Hashimi, H.M.: Hierarchy of RNA functional dynamics. Annu. Rev. Biochem. **83**, 441–466 (2014)
26. Peselis, A., Serganov, A.: Structural insights into ligand binding and gene expression control by an adenosylcobalamin riboswitch. Nat. Struct. Mol. Biol. **19**(11), 1182 (2012)
27. Petrov, A.I., Zirbel, C.L., Leontis, N.B.: Automated classification of RNA 3D motifs and the RNA 3D motif atlas. RNA **19**(10), 1327–1340 (2013). https://doi.org/10.1261/rna.039438.113
28. Popenda, M., et al.: RNA FRABASE 2.0: an advanced web-accessible database with the capacity to search the three-dimensional fragments within RNA structures. BMC Bioinform. **11**, 231 (2010). https://doi.org/10.1186/1471-2105-11-231
29. Reinharz, V., Major, F., Waldispühl, J.: Towards 3D structure prediction of large RNA molecules: an integer programming framework to insert local 3D motifs in RNA secondary structure. Bioinformatics **28**(12), i207–i214 (2012). https://doi.org/10.1093/bioinformatics/bts226
30. Reinharz, V., Soulé, A., Westhof, E., Waldispühl, J., Denise, A.: Mining for recurrent long-range interactions in RNA structures reveals embedded hierarchies in network families. Nucleic Acids Res. **46**(8), 3841–3851 (2018)
31. Rovetta, C., Michálik, J., Lorenz, R., Tanzer, A., Ponty, Y.: Non-redundant sampling and statistical estimators for RNA structural properties at the thermodynamic equilibrium (2019, under review). Preprint: https://hal.inria.fr/hal-02288811
32. Sarrazin-Gendron, R., Reinharz, V., Oliver, C.G., Moitessier, N., Waldispühl, J.: Automated, customizable and efficient identification of 3D base pair modules with BayesPairing. Nucleic Acids Res. **47**, 3321–3332 (2019)
33. Serganov, A., Nudler, E.: A decade of riboswitches. Cell **152**(1–2), 17–24 (2013)
34. Theis, C., Zirbel, C.L., Zu Siederdissen, C.H., Anthon, C., Hofacker, I.L., Nielsen, H., Gorodkin, J.: RNA 3D modules in genome-wide predictions of RNA 2D structure. PLoS ONE **10**(10), e0139900 (2015). https://doi.org/10.1371/journal.pone.0139900
35. Thiel, B.C., Ochsenreiter, R., Gadekar, V.P., Tanzer, A., Hofacker, I.L.: RNA structure elements conserved between mouse and 59 other vertebrates. Genes (Basel) **9**(8), 392 (2018)
36. Tinoco, I., Bustamante, C.: How RNA folds. J. Mol. Biol. **293**(2), 271–281 (1999). https://doi.org/10.1006/jmbi.1999.3001
37. Turner, D.H., Mathews, D.H.: NNDB: the nearest neighbor parameter database for predicting stability of nucleic acid secondary structure. Nucleic Acids Res. **38**, D280–D282 (2010). https://doi.org/10.1093/nar/gkp892
38. Wood, D.E., Salzberg, S.L.: Kraken: ultrafast metagenomic sequence classification using exact alignments. Genome Biol. **15**(3), R46 (2014)
39. Xue, C., Li, F., He, T., Liu, G.P., Li, Y., Zhang, X.: Classification of real and pseudo microrna precursors using local structure-sequence features and support vector machine. BMC Bioinform. **6**(1), 310 (2005)
40. Zirbel, C.L., Roll, J., Sweeney, B.A., Petrov, A.I., Pirrung, M., Leontis, N.B.: Identifying novel sequence variants of RNA 3D motifs. Nucleic Acids Res. **43**(15), 7504–7520 (2015). https://doi.org/10.1093/nar/gkv651

Lower Density Selection Schemes
via Small Universal Hitting Sets
with Short Remaining Path Length

Hongyu Zheng[ID], Carl Kingsford[ID], and Guillaume Marçais[(✉)][ID]

Computational Biology Department, Carnegie Mellon University,
Pittsburgh, PA 15213, USA
gmarcais@cs.cmu.edu
http://www.cs.cmu.edu/~gmarcais

Abstract. Universal hitting sets are sets of words that are unavoidable: every long enough sequence is hit by the set (i.e., it contains a word from the set). There is a tight relationship between universal hitting sets and minimizers schemes, where minimizers schemes with low density (i.e., efficient schemes) correspond to universal hitting sets of small size. Local schemes are a generalization of minimizers schemes which can be used as replacement for minimizers scheme with the possibility of being much more efficient. We establish the link between efficient local schemes and the minimum length of a string that must be hit by a universal hitting set. We give bounds for the remaining path length of the Mykkeltveit universal hitting set. Additionally, we create a local scheme with the lowest known density that is only a log factor away from the theoretical lower bound.

Keywords: de Bruijn graph · Minimizers · Universal hitting set · Depathing set

1 Introduction

We study the problem of finding *Universal Hitting Sets* [13] (UHS). A UHS is a set of words, each of length k, such that every long enough string (say of length L or longer) contains as a substring an element from the set. We call such a set a universal hitting set for parameters k and L. They are sets of unavoidable words, i.e., words that must be contained in any long strings, and we are interested in the relationship between the size of these sets and the length L.

More precisely, we say that a k-mer a (a string of length k) *hits* a string S if a appears as a substring of S. A set A of k-mers hits S if at least one k-mer of A hits S. A universal hitting set for length L is a set of k-mers that hits every string of length L. Equivalently, the *remaining path length* of a universal set is the length of the longest string that is not hit by the set ($L - 1$ here).

The study of universal hitting sets is motivated in part by the link between UHS and the common method of *minimizers* [14,15,17]. The minimizers method

© Springer Nature Switzerland AG 2020
R. Schwartz (Ed.): RECOMB 2020, LNBI 12074, pp. 202–217, 2020.
https://doi.org/10.1007/978-3-030-45257-5_13

is a way to sample a string for representative k-mers in a deterministic way by breaking a string into windows, each window containing w k-mers, and selecting in each window a particular k-mer (the "minimum k-mer", as defined by a preset order on the k-mers). This method is used in many bioinformatics software programs (e.g., [2,4–6,18]) to reduce the amount of computation and improve run time (see [11] for usage examples). The minimizers method is a family of methods parameterized by the order on the k-mers used to find the minimum. The *density* is defined as the expected number of sampled k-mers per unit length of sequence. Depending on the order used, the density varies.

In general, a lower density (i.e., fewer sampled k-mers) leads to greater computational improvements, and is therefore desirable. For example, a read aligner such a Minimap2 [7] stores all the locations of minimizers in the reference sequence in a database. It then finds all the minimizers in a read and searches in the database for these minimizers. The locations of these minimizers are used as seeds for the alignment. Using a minimizers scheme with a reduced density leads to a smaller database and fewer locations to consider, hence an increased efficiency, while preserving the accuracy.

There is a two-way correspondence between minimizers methods and universal hitting sets: each minimizers method has a corresponding UHS, and a UHS defines a family of *compatible* minimizers methods [9,10]. The remaining path length of the UHS is upper-bounded by the number of bases in each window in the minimizers scheme ($L \leq w + k - 1$). Moreover, the relative size of the UHS, defined as the size of UHS over the number of possible k-mers, provides an upper-bound on the density of the corresponding minimizers methods: the density is no more than the relative size of the universal hitting set. Precisely, $\frac{1}{w} \leq d \leq \frac{|U|}{\sigma^k}$, where d is the density, U is the universal hitting set, σ^k is the total number of k-mers on an alphabet of size σ, and w is the window length. In other words, the study of universal hitting sets with small size leads to the creation of minimizers methods with provably low density.

Local schemes [12] and forward schemes are generalizations of minimizers schemes. These extensions are of interest because they can be used in place of minimizers schemes while sampling k-mers with lower density. In particular, minimizers schemes cannot have density close to the theoretical lower bound of $1/w$ when w becomes large, while local and forward schemes do not suffer from this limitation [9]. Understanding how to design local and forward schemes with low density will allow us to further improve the computation efficiency of many bioinformatics algorithms.

The previously known link between minimizers schemes and UHS relied on the definition of an ordering between k-mers, and therefore is not valid for local and forward scheme that are not based on any ordering. Nevertheless, UHSs play a central role in understanding the density of local and forward schemes.

Our first contribution is to describe the connection between UHSs, local and forward schemes. More precisely, there are two connections: first between the density of the schemes and the relative size of the UHS, and second between the window size w of the scheme and the *remaining path length* of the UHS (i.e., the maximum length L of a string that does not contain a word from the UHS). This motivates our study of the relationship between the size of a universal hitting set U and the remaining path length of U.

There is a rich literature on unavoidable word sets (e.g., see [8]). The setting for UHS is slightly different for two reasons. First, we impose that all the words in the set U have the same length k, as a k-mer is a natural unit in bioinformatics applications. Second, the set U must hit any string of a given finite length L, rather than being unavoidable only by infinitely long strings.

Mykkeltveit [12] answered the question of what is the size of a minimum unavoidable set with k-mers by giving an explicit construction for such a set. The k-mers in the Mykkeltveit set are guaranteed to be present in any infinitely long sequence, and the size of the Mykkeltveit set is minimum in the sense that for any set S with fewer k-mers there is an infinitely long sequence that avoids S. On the other hand, the construction gives no indication on the remaining path length.

The DOCKS [13] and ReMuVal [3] algorithms are heuristics to generate unavoidable sets for parameters k and L. Both of these algorithms use the Mykkeltveit set as a starting point. In many practical cases, the longest sequence that does not contain any k-mer from the Mykkeltveit set is much larger than the parameter L of interest (which for a compatible minimizers scheme correspond to the window length). Therefore, the two heuristics extend the Mykkeltveit set in order to cover every L-long sequence. These greedy heuristics do not provide any guarantee on the size of the unavoidable set generated compared to the theoretical minimum size and are only computationally tractable for limited ranges of k and L.

Our second contribution is to give upper and lower bounds on the remaining path length of the Mykkeltveit sets. These are the first bounds on the remaining path length for minimum size sets of unavoidable k-mers.

Defining local or forward schemes with density of $O(1/w)$ (that is, within a constant factor of the theoretical lower bound) is not only of practical interest to improve the efficiency of existing algorithms, but it is also interesting for a historical reason. Both Roberts *et al.* [14] and Schleimer *et al.* [17] used a probabilistic model to suggest that minimizers schemes have an expected density of $2/(w + 1)$. Unfortunately, this simple probabilistic model does not correctly model the minimizers schemes outside of a small range of values for parameters k and w, and minimizers do not have an $O(1/w)$ density in general. Although the general question of whether a local scheme with $O(1/w)$ exists is still open, our third contribution is an almost optimal forward scheme with density of $O(\ln(w)/w)$ density. This is the lowest known density for a forward scheme, beating the previous best density of $O(\sqrt{w}/w)$ [9], and hinting that $O(1/w)$ might be achievable.

Understanding the properties of universal hitting sets and their many interactions with selection schemes (minimizers, forward and local schemes) is a crucial step toward designing schemes with lower density and improving the many algorithms using these schemes. In Sect. 2, we give an overview of the results, and in

Sect. 3 we give proofs sketches. Full proofs are available in the extended version of the paper on arXiv: https://arxiv.org/abs/2001.06550.

2   Results

2.1   Notation

Universal Hitting Sets. Consider a finite alphabet $\Sigma = \{0, \ldots, \sigma - 1\}$ with $\sigma \geq 2$ elements. If $a \in \Sigma$, a^k denotes the letter a repeated k times. We use Σ^k to denote the set of strings of length k on alphabet Σ, and call them k-mers. If S is a string, $S[n, l]$ denotes the substring starting at position n and of length l. For a k-mer $a \in \Sigma^k$ and an l-long string $S \in \Sigma^l$, we say "a hits S" if a appears as substring of S ($a = S[i, k]$ for some i). For a set of k-mers $A \subseteq \Sigma^k$ and $S \in \Sigma^l$, we say "A hits S" if there exists at least one k-mer in A that hits S. A set $A \subseteq \Sigma^k$ is a universal hitting set for length L if A hits every string of length L.

de Bruijn Graphs. Many questions regarding strings have an equivalent formulation with graph terminology using *de Bruijn graphs*. The de Bruijn graph $B_{\Sigma, k}$ on alphabet Σ and of order k has a node for every k-mer, and an edge (u, v) for every string of length $k + 1$ with prefix u and suffix is v. There are σ^k vertices and σ^{k+1} edges in the de Bruijn graph of order k.

There is a one-to-one correspondence between strings and paths in $B_{\Sigma, k}$: a path with w nodes corresponds to a string of $L = w + k - 1$ characters. A universal hitting set A corresponds to a *depathing set* of the de Bruijn graph: a universal hitting set for k and L intersects with every path in the de Bruijn graph with $w = L - k + 1$ vertices. We say "A is a (α, l)-UHS" if A is a set of k-mers that is a universal hitting set, with relative size $\alpha = |A|/\sigma^k$ and hits every walk of l vertices (and therefore every string of length $L = l + k - 1$).

A *de Bruijn sequence* is a particular sequence of length $\sigma^k + k - 1$ that contains every possible k-mer once and only once. Every de Bruijn graph is Hamiltonian and the sequence spelled out by a Hamiltonian tour is a de Bruijn sequence.

Selection Schemes. A *local scheme* [17] is a method to select positions in a string. A local scheme is parameterized by a *selection function* f. It works by looking at every w-mer of the input sequence S: $S[0, w], S[1, w], \ldots$, and selecting in each window a position according to the selection function f. The selection function selects a position in a window of length w, i.e., it is a function $f : \Sigma^w \to [0 : w - 1]$. The output of a forward scheme is a set of selected positions: $\{i + f(S[i, w]) \mid 0 \leq i < |S| - w\}$.

A *forward scheme* is a local scheme with a selection function such that the selected positions form a non-decreasing sequence. That is, if ω_1 and ω_2 are two consecutive windows in a sequence S, then $f(\omega_2) \geq f(\omega_1) - 1$.

A *minimizers scheme* is scheme where the selection function takes in the sequence of w consecutive k-mers and returns the "minimum" k-mer in the window (hence the name minimizers). The minimum is defined by a predefined

```
CACTGCTGTACCTCTTCT        CACTGCTGTACCTCTTCT        CACTGCTGTACCTCTTCT
```

```
      CACTGCT-----------         CACTGCT-----------         CACTGCT-----------
(a)   -ACTGCTG----------    (b)  -ACTGCTG----------    (c)  -ACTGCTG----------
      --CTGCTGT---------         --CTGCTGT---------         --CTGCTGT---------
      ---TGCTGTA--------         ---TGCTGTA--------         ---TGCTGTA--------
      ----GCTGTAC-------         ----GCTGTAC-------         ----GCTGTAC-------
      -----CTGTACC------         -----CTGTACC------         -----CTGTACC------
      ------TGTACCT-----         ------TGTACCT-----         ------TGTACCT-----
      -------GTACCTC----         -------GTACCTC----         -------GTACCTC----
      --------TACCTCT---         --------TACCTCT---         --------TACCTCT---
      ---------ACCTCTT--         ---------ACCTCTT--         ---------ACCTCTT--
      ----------CCTCTTC-         ----------CCTCTTC-         ----------CCTCTTC-
      -----------CTCTTCT        -----------CTCTTCT        -----------CTCTTCT
```

Fig. 1. (a) Example of selecting minimizers with $k = 3$, $w = 5$ and the lexicographic order (i.e., AAA < AAC < AAG < ... < TTT). The top line is the input sequence, each subsequent line is a 7-bases long window (the number of bases in a window is $w+k-1 = 7$) with the minimum 3-mer highlighted. The positions $\{1, 2, 5, 9, 10, 11\}$ are selected for a density $d = 6/(18-3+1) = 0.375$. (b) On the same sequence, an example of a selection scheme for $w = 7$ (and $k = 1$ because it is a selection scheme, hence the number of bases in a window is also w). The set of positions selected is $\{1, 6, 7, 8, 11, 13, 14\}$. This is not a forward scheme as the sequence of selected position is not non-decreasing. (c) A forward selection scheme for $w = 7$ with selected positions $\{1, 7, 8, 12, 13\}$. Like the minimizers scheme, the sequence of selected positions is non-decreasing.

order on the k-mers (e.g., lexicographic order) and the selection function is $f : \Sigma^{w+k-1} \to [0 : w - 1]$.

See Fig. 1 for examples of all 3 schemes. The local scheme concept is the most general as it imposes no constraint on the selection function, while a forward scheme must select positions in a non-decreasing way. A minimizers scheme is the least general and also selects positions in a non-decreasing way.

Local and forward schemes were originally defined with a function defined on a window of w k-mers, $f : \Sigma^{w+k-1} \to [0 : w - 1]$, similarly to minimizers. Selection schemes are schemes with $k = 1$, and have a single parameter w as the word length. While the notion of k-mer is central to the definition of the minimizers schemes, it has no particular meaning for a local or forward scheme: these schemes select positions within each window of a string S, and the sequence of the k-mers at these positions is no more relevant than sequence elsewhere in the window to the selection function.

There are multiple reasons to consider selection schemes. First, they are slightly simpler as they have only one parameter, namely the window length w. Second, in our analysis we consider the case where w is asymptotically large, therefore $w \gg k$ and the setting is similar to having $k = 1$. Finally, this simplified problem still provides information about the general problem of local schemes. Suppose that f is the selection function of a selection scheme, for any $k > 1$ we can define $g_k : \Sigma^{w+k-1} \to [0, w - 1]$ as $g_k(\omega) = f(\omega[0, w])$. That is, g_k is defined from the function f by ignoring the last $k - 1$ characters in a window. The functions g_k define proper selection functions for local schemes with parameter w and k, and because exactly the same positions are selected, the density of g_k

is equal to the density of f. In the following sections, unless noted otherwise, we use forward and local schemes to denote forward and local selection schemes.

Density. Because a local scheme on string S may pick the same location in two different windows, the number of selected positions is usually less than $|S|-w+1$. The *particular density* of a scheme is defined as the number of distinct selected positions divided by $|S| - w + 1$ (see Fig. 1). The *expected density*, or simply the *density*, of a scheme is the expected density on an infinitely long random sequence. Alternatively, the expected density is computed exactly by computing the particular density on any de Bruijn sequence of order $\geq 2w - 1$. In other words, a de Bruijn sequence of large enough order "looks like" a random infinite sequence with respect to a local scheme (see [10] and Sect. 3.1).

2.2 Main Results

The density of a local scheme is in the range $[1/w, 1]$, as $1/w$ corresponds to selecting exactly one position per window, and 1 corresponds to selecting every position. Therefore, the density goes from a low value with a constant number of positions per window (density is $O(1/w)$, which goes to 0 when w gets large), to a high with constant value (density is $\Omega(1)$) where the number of positions per window is proportional to w. When the minimizers and winnowing schemes were introduced, both papers used a simple probabilistic model to estimate the expected density to $2/(w + 1)$, or about 2 positions per window. Under this model, this estimate is within a constant factor of the optimal, it is $O(1/w)$.

Unfortunately, this simple model properly accounts for the minimizers behavior only when k and w are small. For large k—i.e., $k \gg w$—it is possible to create almost optimal minimizers scheme with density $\sim 1/w$. More problematically, for large w—i.e., $w \gg k$—and for all minimizer schemes the density becomes constant ($\Omega(1)$) [9]. In other words, minimizers schemes cannot be optimal or within a constant factor of optimal for large w, and the estimate of $2/(w + 1)$ is very inaccurate in this regime.

This motivates the study of forward schemes and local schemes. It is known that there exists forward schemes with density of $O(1/\sqrt{w})$ [9]. This density is not within a constant factor of the optimal density but at least shows that forward and local schemes do not have constant density like minimizers schemes for large w and that they can have much lower density.

Connection Between UHS and Selection Schemes. In the study of selection schemes, as for minimizers schemes, universal hitting sets play a central role. We describe the link between selection schemes and UHS, and show that the existence of a selection scheme with low density implies the existence of a UHS with small relative size.

Theorem 1. *Given a local scheme f on w-mers with density d_f, we can construct a $(d_f, w) - UHS$ on $(2w - 1)$-mers. If f is a forward scheme, we can construct a $(d_f, w) - UHS$ on $(w + 1)$-mers.*

Almost-Optimal Relative Size UHS for Linear Path Length. Conversely, because of their link to forward and local selection schemes, we are interested in universal hitting set with remaining path length $O(w)$. Necessarily a universal hitting hits any infinitely long sequences. On de Bruijn graphs, a set hitting every infinitely long sequences is a *decycling set*: a set that intersects with every cycle in the graph. In particular, a decycling set must contain an element in each of the cycles obtained by the rotation of the w-mers (e.g., cycle of the type $001 \rightarrow 010 \rightarrow 100 \rightarrow 001$). The number of these rotation cycles is known as the "necklace number" $N_{\sigma,w} = \frac{1}{n}\sum_{d|w} \varphi(d)\sigma^{w/d} = O(\sigma^w/w)$ [16], where $\varphi(d)$ is the Euler's totient function.

Consequently, the relative size of a UHS, which contains at least one element from each of these cycles, is lower-bounded by $O(1/w)$. The smallest previously known UHS with $O(w)$ remaining path length has a relative size of $O(\sqrt{w}/w)$ [9]. We construct a smaller universal hitting set with relative size $O(\ln(w)/w)$:

Theorem 2. *For every sufficiently large w, there is a forward scheme with density of $O(\ln(w)/w)$ and a corresponding $(O(\ln(w)/w), w)$-UHS.*

Remaining Path Length Bounds for the Mykkeltveit Sets. Mykkeltveit [12] gave an explicit construction for a decycling set with exactly one element from each of the rotation cycles, and thereby proved a long standing conjecture [16] that the minimal size of decycling sets is equal to the necklace number. Under the UHS framework, it is natural to ask what the remaining path length for Mykkeltveit sets is. Given that the de Bruijn graph is Hamiltonian, there exists paths of length exponential in w: the Hamiltonian tours have σ^w vertices. Nevertheless, we show that the remaining path length for Mykkeltveit sets is upper- and lower-bounded by polynomials of w:

Theorem 3. *For sufficiently large w, the Mykkeltveit set is a $(N_{\sigma,w}/\sigma^w, g(w))$-UHS, having the same size as minimal decycling sets, while $g(w) = O(w^3)$ and $g(w) > cw^2$ for some constant c.*

3   Methods and Proofs

Due to page limits, we provide proof sketches of the results. Full proofs are available in the extended paper on arXiv.

3.1   UHS from Selection Schemes

Contexts and Densities of Selection Schemes. We derive another way of calculating densities of selection schemes based on the idea of *contexts*.

Recall a local scheme is defined as a function $f : \Sigma^w \rightarrow [0, w-1]$. For any sequence S and scheme f, the set of selected locations are $\{f(S[i, w]) + i\}$ and the density of f on the sequence is the number of selected locations divided by $|S| - w + 1$. Counting the number of distinct selected locations is the same as counting the number of w-mers $S[i, w]$ such that f picks a new location from all previous w-mers. f can pick identical locations on two w-mers only if they overlap, so intuitively, we only need to look back $(w-1)$ windows to check if the position is already picked. Formally, f picks a new position in window $S[i, w]$ if and only if $f(S[i, w]) + i \neq f(S[i - d, w]) + (i - d)$ for all $1 \leq d \leq w - 1$.

For a location i in sequence S, the context at this location is defined as $c_i = S[i - w + 1, 2w - 1]$, a $(2w - 1)$-mer whose last w-mer starts at i. Whether f picks a new position in $S[i, w]$ is entirely determined by its context, as the conditions only involve w-mers as far back as $S[i - w + 1, w]$, which are included in the context. This means that instead of counting selected positions in S, we can count the contexts c satisfying $f(c[w - 1, w]) + w - 1 \neq f(c[j, w]) + j$ for all $0 \leq j \leq w - 2$, which are the contexts such that f on the last w-mer of c picks a new location. We denote by $\mathcal{C}_f \subset \Sigma^{2w-1}$ the set of contexts that satisfy this condition.

The expected density of f is computed as the number of selected positions over the length of the sequence for a random sequence, as the sequence becomes infinitely long. For a sufficiently long random sequence ($|S| \gg w$), the distribution of its contexts converges to a uniform random distribution over $(2w - 1)$-mers. Because the distribution of these contexts is exactly equal to the uniform distribution on a circular de Bruijn S sequence of order at least $2w - 1$, we can calculate the expected density of f as the density of f on S, or as $|\mathcal{C}_f|/\sigma^{2w-1}$.

UHS from Selection Schemes. The set $\mathcal{C}_f$ over $(2w - 1)$-mers is the UHS needed for Theorem 1. $\mathcal{C}_f$ is a UHS with remaining path length of at most $w - 1$ as in a walk of length w of $(2w - 1)$-mers, the first and last w-long window do not share a k-mer. Therefore, one of the contexts among these w must pick a new location and by definition belongs to $\mathcal{C}_f$.

When f is a forward scheme, to determine if a new location is picked in a window, looking back one window is sufficient. This is because if we do not pick a new location, we have to pick the same location as in last window. This means context with two w-mers, or as a $(w + 1)$-mer, is sufficient, and our other arguments involving contexts still hold. Combining the pieces, we prove the following theorem:

Theorem 1. *Given a local scheme f on w-mers with density d_f, we can construct a $(d_f, w) - UHS$ on $(2w - 1)$-mers. If f is a forward scheme, we can construct a $(d_f, w) - UHS$ on $(w + 1)$-mers.*

3.2 Forbidden Word Depathing Set

Construction and Path Length. In this section, we construct a set that is a $(O(\ln(w)/w), w) - UHS$.

Definition 1 (Forbidden Word UHS). *Let $d = \lfloor \log_\sigma(w/\ln(w)) \rfloor - 1$. Define $\mathcal{F}_{\sigma,w}$ as the set of w-mers that satisfies either of the following clauses: (1) 0^d is the prefix of x (2) 0^d is not a substring of x.*

We assume that w is sufficiently large such that $d \geq 1$. The longest remaining path after removing $\mathcal{F}_{\sigma,w}$ is $w - d$ as a w-mer m either does not have 0^d as a substring (then by clause 2, m is in $\mathcal{F}_{\sigma,w}$), or 0^d is a substring of m and in at most $(w-d)$ steps 0^d would become the prefix of current w-mer (and in $\mathcal{F}_{\sigma,w}$ by clause 1). The number of w-mer satisfying clause 1 is $\sigma^{w-d} = O(\ln(w)\sigma^w/w)$.

For the rest of this section, we focus on counting w-mers satisfying clause 2 in Definition 1, that is, the number of w-mers not containing 0^d.

Number of w-mers Not Containing 0^d. We construct a finite state machine (FSM) that recognizes 0^d as follows. The FSM consists of $d + 1$ states labeled "0" to "d", where "0" is the initial state and "d" is the terminal state. The state "i" with $0 \leq i \leq d - 1$ means that the last i characters were 0 and $d - i$ more zeroes are expected to match 0^d. The terminal state "d" means that we have seen a substring of d consecutive zeroes. If the machine is at non-terminal state "i" and receives the character 0, it moves to state "$i + 1$", otherwise it moves to state "0"; once the machine reaches state "d", it remains in that state forever.

Now, assume we feed a random w-mer to the finite state machine. The probability that the machine does not reach state "d" for the input w-mer is the relative size of the set of w-mer satisfying clause 2. Denote $p_k \in \mathbb{R}^d$ such that $p_k(j)$ is the probability of feeding a random k-mer to the machine and ending up in state "j", for $0 \leq j < d$ (note that the vector does not contain the probability for the terminal state "d"). The answer to our problem is then $\|p_w\|_1 = \sum_{i=0}^{d-1} p_w(i)$, that is, the sum of the probabilities of ending at a non-terminal state.

Define $\mu = 1/\sigma$. Given that a randomly chosen w-mer is fed into the FSM, i.e., each base is chosen independently and uniformly from Σ, the probabilities of transition in the FSM are: "i" $\rightarrow$ "$i + 1$" with probability μ, "i" $\rightarrow$ "0" with probability $1 - \mu$. The probability matrix to not recognize 0^d is a $d \times d$ matrix, as we discard the row and column associated with terminal state "d":

$$
A_d = \begin{bmatrix} 1-\mu & 1-\mu & \dots & 1-\mu & 1-\mu \\ \mu & 0 & \dots & 0 & 0 \\ 0 & \mu & \dots & 0 & 0 \\ \vdots & \vdots & \ddots & \vdots & \vdots \\ 0 & 0 & \dots & \mu & 0 \end{bmatrix}_{d \times d} = \begin{bmatrix} (1-\mu)\mathbf{1}_{d-1}^T & 1-\mu \\ \mu I_{d-1} & \mathbf{0}_{d-1} \end{bmatrix}
$$

Starting with $p_0 = (1, 0, \dots, 0) \in \mathbb{R}^d$ as initially no sequence has been parsed and the machine is at state "0" with probability 1, we can compute the probability vector p_w as $p_w = A_d p_{w-1} = A_d^w p_0$.

Bounding $\|p_w\|_1$. To bound $\|p_w\|_1$, we find bounds on the largest eigenvalue λ_0 of A_d and of the norm of its eigenvector ν_0. For $\lambda \neq \mu$, the characteristic polynomial is $p_{A_d}(\lambda) = (-1)^d(\lambda^{d+1} - \lambda^d - \mu^{d+1} + \mu^d)/(\lambda - \mu)$. The largest eigenvalue by norm is real and satisfies $\lambda_0 \leq 1 - \mu^{d+1}$, and its eigenvector satisfies $\|\nu_0\|_1 \leq 3$. Because of the special relation between λ_0 and p_0, and the choice of d, we have $\|p_w\|_1 = \|A_d^w p_0\|_1 < \lambda_0^w \|\nu_0\|_1 < 3(1 - \mu^{d+1})^w = O(1/w)$.

These lemma implies that the relative size for the set $\mathcal{F}_{\sigma,w}$ is dominated by the w-mers satisfying clause 1 of Definition 1 and $\mathcal{F}_{\sigma,w}$ is of relative size $O(\ln(w)/w)$. This completes the proof that $\mathcal{F}_{\sigma,w}$ is a $(O(\ln(w)/w), w) - UHS$.

3.3 Construction of the Mykkeltveit Sets

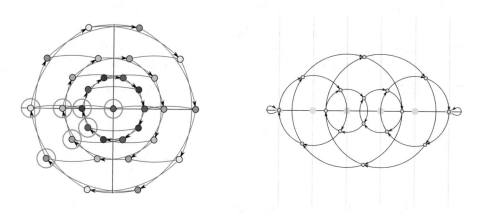

Fig. 2. (a) Mykkeltveit embedding of the de Bruijn graph of order 5 on the binary alphabet. The nodes of a conjugacy class have the same color and form a circle (there is more than one class per circle). The pure rotations are represented by the red edges. A non-pure rotation $S_a(x)$ is a red edge followed by a horizontal shift (blue edge). The set of nodes circled in gray is the Mykkeltveit set. (b) Weight-in embedding of the same graph. Multiple w-mers map to the same position in this embedding and each circle represent a conjugacy class. The gray dots on the horizontal axis are the w centers of rotations and the vertical gray lines going through the centers separate the space in sub-regions of interest.

In this section, we construct the Mykkeltveit set $\mathcal{M}_{\sigma,w}$ and prove some important properties of the set. We start with the definition of the Mykkeltveit embedding of the de Bruijn graph.

Definition 2 (Modified Mykkeltveit Embedding). *For a w-mer x, its embedding in the complex plane is defined as $P(x) = \sum_{i=0}^{w-1} x_i r_w^{i+1}$, where r_w is a wth root of unity, $r_w = e^{2\pi i/w}$.*

Intuitively, the position of a w-mer x is defined as the following center of mass. The w roots of unity form a circle in the complex plane, and a weight equal to the value of the base x_i is set at the root r_w^{i+1}. The position of x is the center of mass of these w points and associated weights. Originally, Mykkeltveit defined the embedding with weight r_w^i [12]. This extra factor of r_w in our modified embedding rotates the coordinate and is instrumental in the proof.

Define the *successor function* $S_a(x) = x_1 x_2 \cdots x_{w-1} a$, where $a \in \Sigma$. The successor function gives all the neighbors of x in the de Bruijn graph. A *pure rotation* of x is the particular neighbor $R(x) = S_{x_0}(x)$, i.e., the sequence of $R(x)$ is a left rotation of x.

We focus on a particular kind of cycle in the de Bruijn graph. A *pure cycle* in the de Bruijn graph, also known as *conjugacy class* is the sequence of w-mers obtained by repeated rotation: $(x, R(x), R^2(x), \ldots)$. Each pure cycle consists of w distinct w-mers, unless $x_0 x_1 \cdots x_{w-1}$ is periodic, and in this case the size of the cycle is equal to its shortest period.

The embeddings from pure rotations satisfy a curious property:

Lemma 1 (Rotations and Embeddings). *$P(R(x))$ on the complex plane is $P(x)$ rotated clockwise around origin by $2\pi/w$. $P(S_a(x))$ is $P(R(x))$ shifted by $\delta = a - x_0$ on the real part, with the imaginary part unchanged.*

Proof. By Definition 2 and the definition of successor function $S_a(x)$:

$$P(S_a(x)) = \sum_{i=0}^{w-1} (S_a(x))_i r_w^{i+1}$$
$$= \sum_{i=0}^{w-2} x_{i+1} r_w^{i+1} + a r_w^{w-1+1}$$
$$= r_w^{-1} \sum_{i=0}^{w-1} x_i r_w^{i+1} + (a - x_0)$$
$$= r_w^{-1} P(x) + \delta$$

Note that for pure rotations $\delta = 0$, and $r_w^{-1} P(x)$ is exactly $P(x)$ rotated clockwise by $2\pi/w$. $\square$

The range for δ is $[-\sigma + 1, \sigma - 1]$. In particular, δ can be negative. In a pure cycle, either all w-mer satisfy $P(x) = 0$, or they lie equidistant on a circle centered at origin. Figure 2(a) shows the embeddings and pure cycles of 5-mers. It is known that we can partition the set of all w-mers into $N_{\sigma,k}$ disjoint pure cycles. This means any decycling set that breaks every cycle of the de Bruijn graph will be at least this large. We now construct our proposed depathing set with this idea in mind.

Definition 3 (Mykkeltveit Set). *We construct the Mykkeltveit set $\mathcal{M}_{\sigma,w}$ as follows. Consider each conjugacy class, we will pick one w-mer from each of them by the following rule:*

1. *If every w-mer in the class embeds to the origin, pick an arbitrary one.*
2. *If there is one w-mer x in the class such that $\operatorname{Re}(P(x)) < 0$ and $\operatorname{Im}(P(x)) = 0$ (on the negative real axis), pick that one.*
3. *Otherwise, pick the unique w-mer x such that $\operatorname{Im}(P(x)) < 0$ and $\operatorname{Im}(P(R(x))) > 0$. Intuitively, this is the w-mer in the cycle right below the negative real axis.*

3.4 Upper Bounding the Remaining Path Length in Mykkeltveit Sets

In this section, we show the remaining path after removing $\mathcal{M}_{\sigma,w}$ is at most $O(w^3)$ long. This polynomial bound is a stark contrast to the number of remaining vertices after removing the Mykkeltveit set—i.e., $\sigma^w - N_{\sigma,w} \sim (1 - \frac{1}{w})\sigma^w$, which is exponential in w. Our main argument involves embedding a w-mer to point in the complex plane, similar to Mykkeltveit's construction.

Weight-in Embedding. Let $W(x) = \sum_{i=0}^{w-1} x_i$ be the *weight* of the w-mer x. The weight satisfies $0 \le W(x) \le W_{\max} = w(\sigma - 1)$. The weight of a successor is $W(S_a(x)) = W(x) + \delta$ where $\delta = a - x_0$ as in Lemma 1.

We defined the *weight-in* embedding of w-mer x by $Q(x) = P(x) - W(x)$. In this embedding, the position of the successor $S_a(x)$ is obtained by a rotation of angle $-2\pi/w$ around the point $(-W(x), 0)$: $Q(S_a(x)) = P(S_a(x)) - W(S_a(x)) = r_w^{-1}(Q(x) + W(x)) - W(x)$.

Figure 2(b) shows the weight-in embedding of a de Bruijn graph. The set $\mathcal{C}_{\sigma,w} = \{(-j, 0) \mid 0 \le j \le W_{\max}\}$ is the set of all the possible center of rotations, and is shown by large gray dots on Fig. 2(b). Because all the w-mer in a given conjugacy class have the same weight, say j, the conjugacy classes form a circle around a particular center $(-j, 0)$.

Embedding Mykkeltveit Set. Consider a w-mer x and a successor $S_a(x)$ such that their embeddings $Q(x)$ and $Q(S_a(x))$ are both in the upper half-plane (positive imaginary part). Because $Q(S_a(x))$ is obtained by a rotation around a point $(W(x), 0)$ on the x-axis from $Q(x)$, it has a larger real part (moving to the right). Similarly, if both $Q(x)$ and $Q(S_a(x))$ are both in the lower half-plane, the real part will decrease. Hence, any long enough walk in the de Bruijn graph travels to the right on the upper half-plane, then cross to the lower half-plane and travels to the left, to eventually return to the upper half-plane.

Because the imaginary part of $Q(x)$ and $P(x)$ are identical, by definition of the Mykkeltveit set, every edge going from the lower half-plane to the upper half-plane has a node in the set. Therefore, $\mathcal{M}_{\sigma,w}$ hits every long enough path, and, by symmetry of the de Bruijn graph, to upper-bound the longest remaining path it suffices to upper-bound the longest path in the upper half-plane.

Relaxations. We consider the following problem: what is the longest sequence of complex numbers $(z_0, z_1, \dots)$ that stays in the upper half-plane, where z_i is obtained from z_{i-1} by a rotation around one of the centers from $\mathcal{C}_{\sigma,w}$? This is a relaxation of the original problem in the sense that any valid walk in the de Bruijn graph in the upper half-plane is a valid sequence of complex numbers in the relaxed formulation, but the converse is not true.

This problem is broken down in different regions of the complex plane (see Fig. 2):

1. If z_i is to the right of all rotation centers (i.e., $\mathrm{Re}(z_i) > 0$), then any rotation will decrease the polar angle by at least $2\pi/w$, and after at most $w/4$ steps the sequence leaves the upper-half plane.
2. If z_i is to the left of all rotation centers (i.e., $\mathrm{Re}(z_i) < -W_{\max}$), then the situation is similar as in the previous case and after at most $w/4$ steps the sequence must satisfy $\mathrm{Re}(z) \ge -W_{\max}$.

3. If z_i is in between two rotation centers (i.e., $-j \leq \mathrm{Re}(z_i) < -j + 1$ for some $1 \leq j \leq W_{\max}$), let's define L as an an upper-bound on the number of steps before the sequence reach the next region (i.e., $\mathrm{Re}(z_{i+L}) \geq -j + 1$).

Because there are $W_{\max} = O(w)$ regions in case 3, the total number of steps is upper bounded by $O(w + Lw)$. The extended paper gives a simple argument to bound $L = O(w^3)$ for a total bound of $O(w^4)$, then a refined and longer potential-based argument that yields $L = O(w^2)$, for a final bound of $O(w^3)$.

3.5   Lower Bounding the Remaining Path Length in Mykkeltveit Sets

We provide here a constructive proof of the existence of a $\Omega(w^2)$ long path in the de Bruijn graph after removing $\mathcal{M}_{\sigma,w}$. Since all w-mers in $\mathcal{M}_{\sigma,w}$ satisfy $\mathrm{Im}(P(x)) \leq 0$, a path satisfying $\mathrm{Im}(P(x)) > 0$ at every step is guaranteed to avoid $\mathcal{M}_{\sigma,w}$ and our construction will satisfy this criteria. It suffices to prove the theorem for binary alphabet as the path constructed will also be a valid path in a graph with larger alphabet. We present the constructions for even w here.

We need an alternative view of w-mers in this section, close to a shift register. Imagine a paper ring with w slots, labelled tag 0 to tag $w - 1$ with content $y = y_0 y_1 \cdots y_{w-1}$, and a pointer initially at 0. The w-mer from the ring is $y_j y_{j+1} \cdots y_{w-1} y_0 \cdots y_{j-1} = y[j, w - j] \cdot y[0, j]$, assuming pointer is at tag j. A pure rotation $R(x)$ on the ring is simply moving the pointer one base forward, and an impure one $S_a(x)$ is to write a to y_j before moving the pointer forward.

Let $w = 2m$. We create $\lceil w/8 \rceil$ ordered quadruples of tags taken modulo w: $Q_j = \{a - j, a + j, b - j, b + j\}$ where $j \in [1, \lceil w/8 \rceil]$, $a = m - 1$, and $b = w - 1$. In each quadruple Q_j, the set of associated root of unity r_w^{i+1} for the 4 tags are of form $\{-e^{-i\theta}, -e^{i\theta}, e^{-i\theta}, e^{i\theta}\}$, adding up to 0. Consequently, changing y_k for each k in Q_j from 1 to 0 does not change the resulting embedding. The strategy consists of creating "pseudo-loops": start from a w-mer, rotate it a certain number of times and switch the bit of the w-mer corresponding to the index in a quadruple to 0 to return to almost the starting position (the same position in the plane but a different w-mer with lower weight).

More precisely, the initial w-mer x is all ones but x_{w-1} set to zero, with paper ring content $y = x$ and pointer at tag 0. The resulting w-mer satisfies $P(x) = -1$. The sequence of operations is as follows. First, do a pure rotation on x. Then, for each quadruple Q_j from $j = 1$ to $j = \lceil w/8 \rceil$, we perform the following actions on x: pure rotations until the pointer is at tag $a - j$, impure rotation S_0, pure rotations until the pointer is at tag $a + j$, impure rotation S_0, pure rotations until pointer is at tag $b - j$, impure S_0, pure rotations until pointer is at tag $b + j$, impure S_0.

Each round involves exactly $w + 1$ rotations since the last step is to an impure rotation S_0 at tag $b + j$ which increases by one between quadruple Q_j and Q_{j+1}. The total length of the path over all Q_i is at least cw^2 for some constant c. Figure 3 shows an example of quadruples and a generated long path that fits in the upper halfplane.

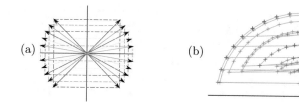

Fig. 3. (a) For $w = 40$, each set of 4 arrows of the same color represent a quadruple set of root of unity. There are a total of 5 sets. They were crafted so that the 4 vector in each set cancel out. (b) The path generated by these quadruple sets. The top circle of radius 1 is traveled many times (between tags r_1 and r_2 in each quadruple), as after setting the 4 bits to 0, the w-mer has the same norm as the starting point.

The correctness proof, alongside with the construction for odd w (by approximating movements from $w' = 2w$) can be found in the full paper.

4   Discussion

Relationship of UHS and Selection Schemes. Our construction of a UHS of relative size $O(\ln(w)/w)$ and remaining path length w also implies the existence of a forward selection scheme with density $O(\ln(w)/w)$, only a $\ln(w)$ factor away from the lower bound on density achievable by forward scheme and local schemes.

Unfortunately this construction does not apply for arbitrary UHS. In general, given a UHS with relative size d and remaining path length w, it is still unknown how to construct a forward or local scheme with density $O(d)$. As described in Sect. 3.1, we can construct a UHS from a scheme by taking the set $\mathcal{C}_f$ of contexts that yields new selections. But it is not always possible to go the other way: there are universal hitting sets that are not equal to a set of contexts $\mathcal{C}_f$ for any function f.

We are thus interested in the following questions. Given a UHS U with relative size d, is it possible to create another UHS U' from U that has the same relative size d and correspond to a local scheme (i.e., there exists f such that $U' = \mathcal{C}_f$)? If not, what is the smallest price to pay (extra density compared to relative size of the UHS) to derive a local scheme from UHS U?

Existence of "Perfect" Selection Schemes. One of the goal in this research is to confirm or deny the existence of asymptotically "perfect" selection schemes with density of $1/w$, or at least $O(1/w)$. Study of UHS might shed light on this problem. If such perfect selection scheme exists, asymptotic perfect UHS defined as $(O(1/w), w)$-UHS would exist. On the other hand, if we denied existence of an asymptotic perfect UHS, this would imply nonexistence of "perfect" forward selection scheme with density $O(1/w)$.

Remaining Path Length of Minimum Decycling Sets. There is more than one decycling set of minimum size (MDS) for given w. The Mykkeltveit [12] set is one possible construction, and a construction based on very different ideas is given in Champarnaud *et al.* [1]. The number of MDS is much larger than the two sets obtained by these two methods. Empirically, for small values of w, we can exhaustively search all the MDS on the binary alphabet: for $2 \leq w \leq 7$ the number of MDS is respectively 2, 4, 30, 28, 68 288 and 18 432.

While experiments suggest the longest remaining path in a Mykkeltveit depathing set defined in the original paper is around $\Theta(w^3)$, matching our upper bound, we do not know if such bound is tight across all possible minimal decycling sets. The Champarnaud set seems to have a longer remaining path than the Mykkeltveit set, although it is unknown if it is within a constant factor, bounded by a polynomial of w of different degree, or is exponential. More generally, we would like to know what is the range of possible remaining path lengths as a function of w over the set of all MDSs.

Funding Information. This work was partially supported in part by the Gordon and Betty Moore Foundation's Data-Driven Discovery Initiative through Grant GBMF4554 to C.K., by the US National Science Foundation (CCF-1256087, CCF-1319998) and by the US National Institutes of Health (R01GM122935).

Conflict of Interests. C.K. is a co-founder of Ocean Genomics, Inc. G.M. is V.P. of software development at Ocean Genomics, Inc.

References

1. Champarnaud, J.M., Hansel, G., Perrin, D.: Unavoidable sets of constant length. Int. J. Algebra Comput. **14**(2), 241–251 (2004). https://doi.org/10.1142/S0218196704001700

2. Chikhi, R., Limasset, A., Medvedev, P.: Compacting de Bruijn graphs from sequencing data quickly and in low memory. Bioinformatics **32**(12), i201–i208 (2015). https://doi.org/10.1093/bioinformatics/btw279. https://academic.oup.com/bioinformatics/article/32/12/i201/2289008/Compacting-de-Bruijn-graphs-from-sequencing-data

3. DeBlasio, D., Gbosibo, F., Kingsford, C., Marçais, G.: Practical universal k-mer sets for minimizer schemes. In: Proceedings of the 10th ACM International Conference on Bioinformatics, Computational Biology and Health Informatics, Niagara Falls, NY, USA, BCB 2019, pp. 167–176. ACM, New York (2019). https://doi.org/10.1145/3307339.3342144. http://doi.acm.org/10.1145/3307339.3342144

4. Deorowicz, S., Kokot, M., Grabowski, S., Debudaj-Grabysz, A.: KMC 2: fast and resource-frugal k-mer counting. Bioinformatics **31**(10), 1569–1576 (2015). https://doi.org/10.1093/bioinformatics/btv022. http://bioinformatics.oxfordjournals.org/content/31/10/1569

5. Grabowski, S., Raniszewski, M.: Sampling the suffix array with minimizers. In: Iliopoulos, C., Puglisi, S., Yilmaz, E. (eds.) SPIRE 2015. LNCS, vol. 9309, pp. 287–298. Springer, Cham (2015). https://doi.org/10.1007/978-3-319-23826-5_28

6. Jain, C., Dilthey, A., Koren, S., Aluru, S., Phillippy, A.M.: A fast approximate algorithm for mapping long reads to large reference databases. In: Sahinalp, S.C. (ed.) RECOMB 2017. LNCS, vol. 10229, pp. 66–81. Springer, Cham (2017). https://doi.org/10.1007/978-3-319-56970-3_5

7. Li, H., Birol, I.: Minimap2: pairwise alignment for nucleotide sequences. Bioinformatics **34**(18), 3094–3100 (2018). https://doi.org/10.1093/bioinformatics/bty191. https://academic.oup.com/bioinformatics/article/34/18/3094/4994778

8. Lothaire, M., Lothaire, M.: Algebraic Combinatorics on Words, vol. 90. Cambridge University Press, Cambridge (2002)

9. Marçais, G., DeBlasio, D., Kingsford, C.: Asymptotically optimal minimizers schemes. Bioinformatics **34**(13), i13–i22 (2018). https://doi.org/10.1093/bioinformatics/bty258. https://academic.oup.com/bioinformatics/article/34/13/i13/5045769

10. Marçais, G., Pellow, D., Bork, D., Orenstein, Y., Shamir, R., Kingsford, C.: Improving the performance of minimizers and winnowing schemes. Bioinformatics **33**(14), i110–i117 (2017). https://doi.org/10.1093/bioinformatics/btx235. https://academic.oup.com/bioinformatics/article/33/14/i110/3953951

11. Marçais, G., Solomon, B., Patro, R., Kingsford, C.: Sketching and sublinear data structures in genomics. Ann. Rev. Biomed. Data Sci. **2**(1), 93–118 (2019). https://doi.org/10.1146/annurev-biodatasci-072018-021156

12. Mykkeltveit, J.: A proof of Golomb's conjecture for the de Bruijn graph. J. Comb. Theory Ser. B **13**(1), 40–45 (1972). https://doi.org/10.1016/0095-8956(72)90006-8. http://www.sciencedirect.com/science/article/pii/0095895672900068

13. Orenstein, Y., Pellow, D., Marçais, G., Shamir, R., Kingsford, C.: Compact universal k-mer hitting sets. In: Frith, M., Storm Pedersen, C.N. (eds.) WABI 2016. LNCS, vol. 9838, pp. 257–268. Springer, Cham (2016). https://doi.org/10.1007/978-3-319-43681-4_21

14. Roberts, M., Hayes, W., Hunt, B.R., Mount, S.M., Yorke, J.A.: Reducing storage requirements for biological sequence comparison. Bioinformatics **20**(18), 3363–3369 (2004). https://doi.org/10.1093/bioinformatics/bth408

15. Roberts, M., Hunt, B.R., Yorke, J.A., Bolanos, R.A., Delcher, A.L.: A preprocessor for shotgun assembly of large genomes. J. Comput. Biol. **11**(4), 734–752 (2004). https://doi.org/10.1089/cmb.2004.11.734

16. Golomb, S.W.: Nonlinear shift register sequences. In: Shift Register Sequences, pp. 110–168. World Scientific, September 2014. https://doi.org/10.1142/9789814632010_0006. http://www.worldscientific.com/doi/abs/10.1142/9789814632010_0006

17. Schleimer, S., Wilkerson, D.S., Aiken, A.: Winnowing: local algorithms for document fingerprinting. In: Proceedings of the 2003 ACM SIGMOD International Conference on Management of Data, SIGMOD 2003, pp. 76–85. ACM (2003). https://doi.org/10.1145/872757.872770

18. Ye, C., Ma, Z.S., Cannon, C.H., Pop, M., Yu, D.W.: Exploiting sparseness in de novo genome assembly. BMC Bioinform. **13**, S1 (2012). https://doi.org/10.1186/1471-2105-13-S6-S1. http://www.biomedcentral.com/1471-2105/13/S6/S1/abstract

Short Papers

Strain-Aware Assembly of Genomes from Mixed Samples Using Flow Variation Graphs

Jasmijn A. Baaijens[1,2], Leen Stougie[1,3,4], and Alexander Schönhuth[1,4,5(✉)]

[1] Centrum Wiskunde & Informatica, Amsterdam, The Netherlands
alexander.schoenhuth@cwi.nl
[2] Harvard Medical School, Boston, MA, USA
[3] Vrije Universiteit, Amsterdam, The Netherlands
[4] INRIA-Erable, Lyon, France
[5] Utrecht University, Utrecht, The Netherlands

Extended Abstract

Introduction. The goal of strain-aware genome assembly is to reconstruct all individual haplotypes from a mixed sample at the strain level and to provide abundance estimates for the strains. Given that the use of a reference genome can introduce significant biases, de novo approaches are most suitable for this task. So far, reference-genome-independent assemblers have been shown to reconstruct haplotypes for mixed samples of limited complexity and genomes not exceeding 10000 bp in length. This renders such approaches applicable to viral quasispecies, but one cannot use them for bacterial sized genomes. In experiments presented here, we notice that even reference-dependent approaches tend to struggle with bacterial sized genomes.

We present VG-Flow, a de novo approach that enables full-length haplotype reconstruction from pre-assembled contigs of complex mixed samples. Our method increases contiguity of the input assembly and, at the same time, it performs haplotype abundance estimation. VG-Flow is the first approach to require polynomial, and not exponential runtime in terms of the underlying graphs. Since runtime increases only linearly in the length of the genomes in practice, it enables the reconstruction also of genomes that are longer by orders of magnitude, thereby establishing the first de novo solution to strain-aware full-length genome assembly applicable to bacterial sized genomes.

Methods. The methodical novelty that underlies VG-Flow's advances is to derive *flow variation graphs* from the (common) variation graphs that one constructs from the input contigs. General variation graphs [2,3] derived from input contigs had been presented in earlier work as a means for overcoming linear reference induced biases and aiming at the reconstruction of full-length strain-level haplotypes [1]. We introduce the concept of a flow variation graph and cast the relevant computational problem in terms of this graph, which renders the problem polynomial-time solvable for the first time.

© Springer Nature Switzerland AG 2020
R. Schwartz (Ed.): RECOMB 2020, LNBI 12074, pp. 221–222, 2020.
https://doi.org/10.1007/978-3-030-45257-5_14

Our approach consists of five steps, depicted in Fig. 1. As input it takes a data set of next-generation sequencing reads and a collection of strain-specific contigs assembled from the data. The final output is presented as a *genome variation graph* capturing all haplotypes present, along with the estimated relative abundances. While the already efficient or practially feasible steps (1) and (5) correspond to prior work [1], steps (2), (3) and (4) are novel, and replace the exponential-runtime procedure presented earlier.

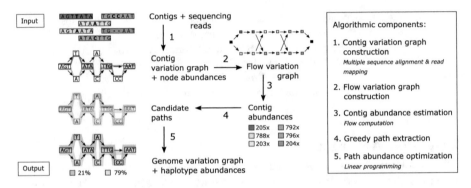

Fig. 1. Algorithm overview

Results. We demonstrate that VG-flow scales approximately linearly in genome size in practice, which allows to process mixtures of genomes that are longer on orders of magnitude. In this, VG-Flow presents the first comprehensive solution to assembling haplotypes from mixed samples at the strain level, also for small bacterial genomes and samples of considerably increased complexity. Benchmarking experiments show that our method outperforms state-of-the-art approaches on mixed samples from viral genomes in terms of assembly accuracy as well as abundance estimation. Experiments on longer, bacterial sized genomes demonstrate that VG-Flow is the only current approach that can reconstruct full-length haplotypes from mixed samples at the strain level in human-affordable runtime.

A full version of this paper is available at https://doi.org/10.1101/645721 and the software can be downloaded from https://bitbucket.org/jbaaijens/vg-flow.

References

1. Baaijens, J., Van der Roest, B., Köster, J., Stougie, L., Schönhuth, A.: Full-length de novo viral quasispecies assembly through variation graph construction. Bioinformatics **35**(24), 5086–5094 (2019). https://doi.org/10.1093/bioinformatics/btz443
2. Garrison, E., et al.: Variation graph toolkit improves read mapping by representing genetic variation in the reference. Nat. Biotechnol. **36**, 875–879 (2018). https://doi.org/10.1038/nbt.4227
3. Paten, B., Novak, A., Eizenga, J., Garrison, E.: Genome graphs and the evolution of genome inference. Genome Res. **27**(5), 665–676 (2017). https://doi.org/10.1101/gr.214155.116

Spectral Jaccard Similarity: A New Approach to Estimating Pairwise Sequence Alignments

Tavor Z. Baharav[1], Govinda M. Kamath[2], David N. Tse[1],
and Ilan Shomorony[3(⊠)]

[1] Department of Electrical Engineering, Stanford University, Stanford, CA, USA
{tavorb,dntse}@stanford.edu
[2] Microsoft Research New England, Cambridge, MA, USA
gokamath@microsoft.com
[3] Department of Electrical and Computer Engineering, University of Illinois,
Urbana-Champaign, Urbana, IL, USA
ilans@illinois.edu

A key step in many genomic analysis pipelines is the identification of regions of similarity between pairs of DNA sequencing reads. This task, known as pairwise sequence alignment, is a heavy computational burden, particularly in the context of third-generation long-read sequencing technologies, which produce noisy reads. This issue is commonly addressed via a two-step approach: first, we filter pairs of reads which are likely to have a large alignment, and then we perform computationally intensive alignment algorithms only on the selected pairs. The *Jaccard similarity* between the set of k-mers of each read can be shown to be a proxy for the alignment size, and is usually used as the filter. This strategy has the added benefit that the Jaccard similarities don't need to be computed exactly, and can instead be efficiently estimated through the use of *min-hashes*. This is done by hashing all k-mers of a read and computing the minimum hash value (the min-hash) for each read. For a randomly chosen hash function, the probability that the min-hashes are the same for two distinct reads is precisely their k-mer Jaccard similarity (JS). Hence, one can estimate the Jaccard similarity by computing the fraction of min-hash collisions out of the set of hash functions considered.

However, when the k-mer distribution of the reads being considered is significantly non-uniform, Jaccard similarity is no longer a good proxy for the alignment size. In particular, genome-wide GC biases and the presence of common k-mers increase the probability of a min-hash collision, thus biasing the estimate of alignment size provided by the Jaccard similarity. In this work, we introduce a min-hash-based approach for estimating alignment sizes called *Spectral Jaccard Similarity* which naturally accounts for an uneven k-mer distribution in the reads being compared. The Spectral Jaccard Similarity is computed by considering a min-hash collision matrix (where rows correspond to pairs of reads and columns correspond to different hash functions), removing an offset, and performing a *singular value decomposition*. The leading left singular vector provides the Spectral

---

T. Z. Baharav and G. M. Kamath—Contributed equally and listed alphabetically.

© Springer Nature Switzerland AG 2020
R. Schwartz (Ed.): RECOMB 2020, LNBI 12074, pp. 223–225, 2020.
https://doi.org/10.1007/978-3-030-45257-5_15

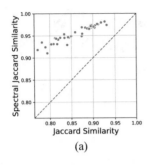
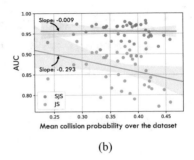

(a)   (b)

Fig. 1. SJS has uniformly higher area under the ROC curve for experiments on 40 PacBio bacterial datasets from the NCTC library [1]. In these experiments, Spectral Jaccard Similarity and Jaccard Similarity were used to filter pairs of reads with an overlap of at least 30%. SJS was computed based on 1000 hash functions, while the Jaccard similarity was computed exactly. (a) We show the AUC values using Daligner alignments as ground truth. (b) The higher the min-hash collision probability is, the worse both methods perform, indicating a "harder" dataset. However, the performance of the SJS filter degrades less than that of the JS filter.

Jaccard Similarity (SJS) for each pair of reads, while the corresponding right singular vector can be understood as a measure of the "unreliability" of each hash function. Intuitively, a hash function that assigns low values to common k-mers is more unreliable for estimating alignment size, since it is more likely to create spurious min-hash collisions. Implicitly, this approach leads to a kind of weighted Jaccard similarity, where the weight of different hash functions is learned from the dataset.

In Fig. 1(a), we demonstrate improvements in performance of the Spectral Jaccard Similarity based filters over Jaccard Similarity based filters on 40 datasets of PacBio reads from the NCTC collection.

Given a dataset with average read length L and a value of k, we define the *mean collision probability over the dataset* as the probability that a randomly selected read from the dataset has a min-hash collision with a bag of L randomly sampled k-mers from the k-mer distribution of the data set. This gives us a measure of the hardness of the dataset to compute pairwise alignments. In Fig. 1(b) we show that the performance of both Spectral Jaccard Similarity and Jaccard Similarity degrades as the collision probability of the dataset increases, but the Spectral Jaccard Similarity degrades much more gracefully.

In order to address the increased computational complexity associated with computing the Spectral Jaccard Similarities, we develop an approximation to SJS that can be computed with a single matrix-vector product, instead of a full singular value decomposition. Empirically, we show that the loss in performance due to this approximation is negligible on most real datasets.

The full version of the manuscript is available at [2]. The code is available at https://github.com/TavorB/spectral_jaccard_similarity.

References

1. Parkhill, J., et al.: National collection of type cultures (NCTC)-3000. https://www.sanger.ac.uk/resources/downloads/bacteria/nctc/
2. Baharav, T., Kamath, G.M., Tse, D.N., Shomorony, I.: Spectral Jaccard Similarity: a new approach to estimating pairwise sequence alignments. BioRxiv (2019). https://doi.org/10.1101/800581. https://www.biorxiv.org/content/early/2019/10/10/800581

MosaicFlye: Resolving Long Mosaic Repeats Using Long Reads

Anton Bankevich[✉] and Pavel Pevzner

Department of Computer Science and Engineering,
University of California at San Diego, La Jolla, San Diego, USA
abankevich@ucsd.edu

Although long-read sequencing technologies opened a new era in genome assembly, the problem of resolving *unbridged* repeats (i.e., repeats that are not spanned by any reads) such as long segmental duplications in the human genome remains largely unsolved, making it a major obstacle towards achieving the goal of complete genome assemblies. Thus, long-read assemblers currently face the same repeat-resolution challenge that short-read assemblers faced a decade ago (albeit at a different scale of repeat lengths).

Long error-prone reads and short accurate reads have their strengths and weaknesses, e.g., short reads may resolve some long repeats that are difficult to resolve with long reads. For example, diverged copies of a long repeat (e.g., copies differing by 3%) often don't share k-mers (for a typical k-mer size used in short-read assemblers) and thus are automatically resolved by the de Bruijn graph-based assemblers. In contrast, long-read assemblers face difficulties resolving such repeats since repeat copies with a 3% divergence are difficult to distinguish using the error-prone reads with error rates exceeding 10%. Thus, long-read assemblers trade the ability to resolve the unbridged but divergent repeat copies for the ability to resolve bridged repeats.

Vollger et al. 2019 used variations between repeat copies for reconstructing all copies of a divergent repeat based on its consensus. However, segmental duplications (SDs) have a *mosaic structure* (Pu et al. 2018) that prevents the utilization of a single consensus sequence as a template for aligning all repeat copies. Such difficult cases were not considered in Vollger et al. 2019. Our mosaicFlye algorithm reconstructs each copy of a mosaic repeat based on differences between various copies.

Given a parameter k, we define the *genome graph* by representing each chromosome of length n as a path on $n − k + 1$ vertices. The *de Bruijn graph DB(Genome, k)* is constructed by "gluing" identical k-mers in the genome graph. Given a read-set *Reads* sampled from *Genome*, one can view each read as a "mini-chromosome" and construct the de Bruijn graph of the resulting genome (Pevzner et al. 2004). Since the resulting graph *DB(Reads, k)* encodes all errors in reads, it is much more complex than the graph *DB(Genome, k)*. Various short-read assembly tools transform the graph *DB (Reads, k)* into the assembly graph that approximates *DB(Genome, k)* (Pevzner et al. 2004). However, constructing an accurate assembly graph from long error-prone reads is a challenging problem.

© Springer Nature Switzerland AG 2020
R. Schwartz (Ed.): RECOMB 2020, LNBI 12074, pp. 226–228, 2020.
https://doi.org/10.1007/978-3-030-45257-5_16

Kolmogorov et al. 2019 developed a Flye assembler that solves this problem by making some concessions. Flye constructs the repeat graph of long reads with the goal to approximate the de Bruijn graph *DB(Genome, k)* in the case of a large *k*, e.g., *k* = 1500. Since this task proved to be difficult in the case of error-prone reads, the Flye assembler collapses similar (rather than only identical as in the de Bruijn graph) *k*-mers in the genome graph into a single vertex in the repeat graph. To construct the repeat graph of a genome, Flye generates all local self-alignments of the genome against itself that have divergence below a threshold *d%*. Pevzner et al. 2004 defined the repeat graph *RG(Genome, k, d)* as the graph obtained from the genome graph by collapsing all aligned positions (vertices) into a single vertex.

Kolmogorov et al. 2019 defined the repeat graph *RG(Reads, k, d)* similarly to *RG (Genome, k, d)* by applying the same approach to a "genome" formed by all reads (each read is viewed as a "mini-chromosome"). They further described how to construct *RG (Reads, k, d)* in the case when *d* is not too small (e.g., exceeds 5%) and demonstrated that *RG(Reads, k, d)* approximates *RG(Genome, k, d)*. However, although the problem of constructing the repeat graph from long error-prone reads has been solved, it remains unclear how to construct the de Bruijn graph from such reads. Solving this problem is arguably one of the most pressing needs in the long-read assembly since it would result in assemblies of the same quality as assemblies of long error-free reads. mosaicFlye uses variations between various copies of a mosaic repeat for resolving these copies and thus untangling the repeat graph of reads *RG(Reads,k,d)*. We show that mosaicFlye improves assemblies of the human genome as well as bacterial genomes and meta-genomes. Below we describe only one application of mosaiFlye to a long segmental duplication (SD) on human chromosome 6 that has only two unassembled regions: the centromere and a 300 kb long SD. The mosaicFlye code is available at https://antonbankevich.github.io/mosaic/. The full paper is available at https://biorxiv.org/cgi/content/short/2020.01.15.908285v1.

Recently, the Telomere-To-Telomere consortium (Miga et al. 2019) initiated an effort to generate a complete assembly of the human genome from ultralong ONT reads by assembling them using Canu (Koren et al. 2017) and Flye assemblers. We applied mosaicFlye to the only unresolved SD on chromosome 6. Since two copies of this SD flank the centromere, the chromosome 6 can be represented as *ARCRB*, where *A* and *B* refer to the arms of the chromosome, *R* refers to the unresolved SD, and *C* refers to the centromere. Canu incorrectly resolved the repeat *R* (resulting in a misassembly *ARB*, while Flye represented the repeat *R* as a single edge in the assembly graph (no misassembly). mosaicFlye recruited 3500 reads from the repeat *R* and resolved this repeat by leveraging the small divergence in the repeat copies and several insertions of size 1–3 kb that distinguish two copies of this repeat. Thus, combining the Flye assembly with the mosaicFlye repeat resolution resulted in a complete assembly of both arms of chromosome 6.

References

Kolmogorov, M., Yuan, J., Lin, Y., Pevzner, P.A.: Assembly of long, error-prone reads using repeat graphs. Nat. Biotechnol. **37**, 540 (2019)

Koren, S., et al.: Canu: scalable and accurate long-read assembly via adaptive *k*-mer weighting and repeat separation. Genome Res. **27**, 722–736 (2017)

Miga, K.H., et al.: Telomere-to-telomere assembly of a complete human X chromosome, bioRxiv (2019). https://doi.org/10.1101/735928

Pevzner, P.A., Tang, H., Tesler, G.: De novo repeat classification and fragment assembly. Genome Res. **14**, 1786–1796 (2004)

Pu, L., Lin, Y., Pevzner, P.A.: Detection and analysis of ancient segmental duplications in mammalian genomes. Genome Res. **28**, 901–909 (2018)

Vollger, M.R., et al.: Long-read sequence and assembly of segmental duplications. Nat. Methods **16**, 88–94 (2019)

Bayesian Non-parametric Clustering of Single-Cell Mutation Profiles

Nico Borgsmüller[1,2], Jose Bonet[3,4], Francesco Marass[1,2],
Abel Gonzalez-Perez[3,4], Nuria Lopez-Bigas[3,5], and Niko Beerenwinkel[1,2(✉)]

[1] Department of Biosystems Science and Engineering, ETH Zürich,
Basel, Switzerland
niko.beerenwinkel@bsse.ethz.ch
[2] SIB, Swiss Institute of Bioinformatics, Basel, Switzerland
[3] Institute for Research in Biomedicine (IRB Barcelona), The Barcelona Institute
of Science and Technology, Barcelona, Spain
[4] Research Program on Biomedical Informatics, Universitat Pompeu Fabra,
Barcelona, Spain
[5] Institució Catalana de Recerca i Estudis Avançats (ICREA), Barcelona, Spain

Cancer is an evolutionary process characterized by the accumulation of mutations that drive tumor initiation, progression, and treatment resistance. The underlying evolutionary process can follow different modes but ultimately leads to multiple coexisting cell populations differing in their genotype. This genomic heterogeneity, known as intra-tumor heterogeneity (ITH), poses major challenges for cancer treatment as treatment-surviving populations are likely to cause cancer recurrence. The high resolution of single-cell DNA sequencing (scDNA-seq) offers great potential to resolve ITH by distinguishing clonal populations based on their accumulated mutations. However, technical limitations of scDNA-seq, such as high error rates and a large proportion of missing values introduced through necessary DNA amplification processes, complicate this task. Generic clustering algorithms, such as centroid- or density-based methods, do not account for these characteristics and are therefore unsuitable for scDNA-seq data. Hence, a variety of tailored methods was introduced recently, varying in the main objective, model choice, and inference. For example, tools like SCITE [1] and SPhyR [2] focus on resolving the phylogenetic relationship among cells and in doing so can also provide clusters and genotypes. SCG [3] employs a parametric model and variational inference to predict clonal composition and genotypes. SiCloneFit [4] infers the phylogenetic relation and clonal composition jointly. However, approaches that use a tree structure are computationally expensive due to the difficulties of tree search.

Here we introduce BnpC, a novel non-parametric method to determine the clonal composition and genotypes of scDNA-seq data, which is applicable to data sets with thousands of cells. As a prior for the unknown number of clusters, BnpC employs a representation of the Dirichlet Process known as a Chinese Restaurant Process (CRP). The joint posterior space of all parameters is explored by a Markov chain Monte Carlo sampling scheme consisting of a mixture of Gibbs sampling, a modified non-conjugate split-merge move, and Metropolis-Hastings

N. Borgsmüller, J. Bonet—These authors contributed equally to this work.

R. Schwartz (Ed.): RECOMB 2020, LNBI 12074, pp. 229–230, 2020.
https://doi.org/10.1007/978-3-030-45257-5_17

updates. BnpC can either sample clusters and genotypes from the posterior distribution, or provide a point estimate that can be used in downstream analysis.

We compared our method with the state-of-the-art methods SCG and SiCloneFit on synthetic and biological data. The simulated data sets contained 1250, 2500, 5000, and 10000 cells with 200, 350, and 500 mutations each. On large data sets ($\geq$5000 cells), we were not able to run SCG due to memory requirements larger than 128 GB. The runtime of SiCloneFit varied widely and we were not able to run it for more than 10 steps on most of the data sets within a 24 h time limit. On the smaller data sets, SCG and BnpC achieved similar clustering accuracy, whereas SiCloneFit performed less accurately in a comparable runtime. For genotype prediction, BnpC outperformed the other two methods in all cases. BnpC's runtime scaled nearly linearly with data size and optimal results on the largest data sets could be obtained after only two hours. On biological data, our method did not only reproduce previous findings for three different data sets [5–7] but also identified additional clones missed in the original analysis but confirmed by additional experiments in Patient 4 [5] and Patient CRC0907 [6]. For patient 9 [7], we were able to recapitulate the previous results without requiring the manual pre-processing step conducted in the original analysis.

In summary, our model and inference scheme result in robust and scalable predictions of clonal composition and genotype. Thus far, BnpC is the only method to provide accurate clonal genotypes on large data sets within a reasonable time. Besides the biological insights and the potential clinical benefit, the inferred clusters and genotypes can be used to reduce data size to facilitate downstream analyses. As scDNA-seq data size continues to grow due to biotechnological progress, scalable and accurate inference, as provided by BnpC, will be increasingly relevant. A preprint of the full paper is available at https://doi.org/10.1101/2020.01.15.907345. BnpC is freely available under MIT license at https://github.com/cbg-ethz/BnpC.

References

1. Jahn, K., Kuipers, J., Beerenwinkel, N.: Tree inference for single-cell data. Genome Biol. **17**(1), 86 (2016)
2. El-Kebir, M.: SPhyR: tumor phylogeny estimation from single-cell sequencing data under loss and error. Bioinformatics **34**(17), i671–i679 (2018)
3. Roth, A., et al.: Clonal genotype and population structure inference from single-cell tumor sequencing. Nat. Methods **13**(573), 573–576 (2016)
4. Zafar, H., et al.: Siclonefit: Bayesian inference of population structure, genotype, and phylogeny of tumor clones from single-cell genome sequencing data. Genome Res. **29**(11), 1847–1859 (2019)
5. Gawad, C., Koh, W., Quake, S.R.: Dissecting the clonal origins of childhood acute-lymphoblastic leukemia by single-cell genomics. PNAS **111**(50), 17947–17952 (2014)
6. Wu, H., et al.: Evolution and heterogeneity of non-hereditary colorectal cancer revealed by single-cell exome sequencing. Oncogene **36**(20), 2857–2867 (2017)
7. McPherson, A., et al.: Divergent modes of clonal spread and intraperitoneal mixing in high-grade serous ovarian cancer. Nat. Genet. **48**, 758–767 (2016)

PaccMann^{RL}: Designing Anticancer Drugs From Transcriptomic Data via Reinforcement Learning

Jannis Born[1,2](✉) ⓘ, Matteo Manica[1] ⓘ, Ali Oskooei[1] ⓘ, Joris Cadow[1] ⓘ, and María Rodríguez Martínez[1] ⓘ

[1] IBM Research Zurich, Zurich, Switzerland
{jab,tte,osk,dow,mrm}@zurich.ibm.com
[2] Machine Learning for Computational Biology Lab, ETH Zurich, Zurich, Switzerland

Motivation. The pharmaceutical industry has experienced a significant productivity decline: Less than 0.01% of drug candidates obtain market approval, with an estimated 10–15 years until market release and costs that range between one [2] to three billion dollars per drug [3]. With the advent of deep generative models in computational chemistry, de-novo drug design is undergoing an unprecedented transformation. While state-of-the-art deep learning approaches have shown potential in generating compounds with desired chemical properties, they entirely disregard the biomolecular characteristics of the target disease.

Approach. Here, we introduce the first molecule generator that can be driven through a disease context deemed to represent the target environment in which the drug has to act. Showcased at the daunting task of de-novo anticancer drug discovery, our generative model is demonstrated being capable of tailoring anticancer candidate compounds for a specific biomolecular profile. Using a RL framework, the transcriptomic profiles (bulk RNA-seq) of cancer cells are used as a context for the generation of candidate molecules. Our molecule generator combines two separately pretrained variational autoencoders (VAEs). The first VAE encodes transcriptomic profiles into a smooth, latent space which in turn is used to condition a second VAE to generate novel molecular structures (represented as SMILES sequences) on the given transcriptomic profile. The generative process is optimized through PaccMann [1], a previously developed drug sensitivity prediction model, to obtain effective anticancer compounds for the given context (see Fig. 1).

Results. Starting from a molecule generator that has been pretrained on bioactive compounds from ChEMBL but never exposed to anticancer drugs, we demonstrate how the de-novo generation can be biased towards compounds with high predicted inhibitory effect against individual cell lines or specific cancer

---

J. Born, M. Manica, A. Oskooei—Equal Contributions.

R. Schwartz (Ed.): RECOMB 2020, LNBI 12074, pp. 231–233, 2020.
https://doi.org/10.1007/978-3-030-45257-5_18

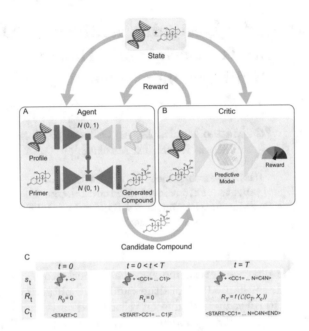

Fig. 1. **Graphical abstract of the proposed framework for designing anti-cancer compounds against specific cancer profiles.** Treating conditional molecular generation in an actor-critic-setting. (A) The actor is composed of a biomolecular profile VAE (top row) and a molecule VAE (bottom row) that are combined to obtain a conditional molecule generator. Both VAEs as well as the predictive critic (B) are pretrained independently. To sample compounds against a cancer site, a biomolecular profile is encoded into the latent space of the first VAE. The latent embedding is decoded (see (C)) through the molecular decoder to yield a candidate compound. The proposed compound is then evaluated through the critic, a multimodal drug sensitivity prediction model that ingests the compound and the target profile of interest. (A) aims for maximizing the reward given by (B), so that over time, (A) will learn to produce candidate compounds with higher predicted efficacy against the given cancer profile.

sites. We provide initial verification of our approach by investigating candidate drugs generated against specific cancer types and find the highest structural similarity to existing compounds with known efficacy against these cancer types. We envision our approach to transform in silico anticancer drug design by leveraging the biomolecular characteristics of the disease in order to increase success rates in lead compound discovery.

Availability. The full preprint is available at https://arxiv.org/abs/1909.05114. The omics data (TCGA) and the molecular (ChEMBL) used to train the two VAEs and can be found on https://ibm.box.com/v/paccmann-pytoda-data. All code to reproduce the experiments is available on https://github.com/PaccMann/ with a detailed example at https://github.com/PaccMann/paccmann_rl. To assess the critic, see [1].

References

1. Manica, M., et al.: Toward explainable anticancer compound sensitivity prediction via multimodal attention-based convolutional encoders. Molecular Pharmaceutics (2019). ACS Publications
2. Scannell, J.W., Blanckley, A., Boldon, H., Warrington, B.: Diagnosing the decline in pharmaceutical R&D efficiency. Nat. Rev. Drug Discovery **11**(3), 191 (2012)
3. Schneider, G.: Mind and machine in drug design. Nat. Mach. Intell. **1**, 128–130 (2019)

CluStrat: A Structure Informed Clustering Strategy for Population Stratification

Aritra Bose[1,2], Myson C. Burch[2], Agniva Chowdhury[3], Peristera Paschou[4(✉)], and Petros Drineas[2]

[1] Computational Genomics, IBM T.J. Watson Research Center, Yorktown Heights, NY, USA
a.bose@ibm.com

[2] Computer Science Department, Purdue University, West Lafayette, IN, USA
{bose6,mcburch,pdrineas}@purdue.edu

[3] Department of Statistics, Purdue University, West Lafayette, IN, USA
agniva@purdue.edu

[4] Department of Biological Sciences, Purdue University, West Lafayette, IN, USA
ppaschou@purdue.edu

Genome-wide association studies (GWAS) have been extensively used to estimate the signed effects of trait-associated alleles. One of the key challenges in GWAS are confounding factors, such as population stratification, which can lead to spurious genotype-trait associations. Recent independent studies [1,8,10] failed to replicate the strong evidence of previously reported signals of directional selection on height in Europeans in the UK Biobank cohort, and attributed the loss of signal to cryptic relatedness in populations. Population structure causes genuine genetic signals in causal variants to be mirrored in numerous non-causal loci due to *linkage disequilibrium (LD)* [3], resulting in spurious associations. Thus, it is important to account for LD in the computation of the distance matrix [6]. One way to account for the LD structure is to use the squared Mahalanobis distance [5]. Here, we present CluStrat, a stratification correction algorithm for complex population structure that leverages the LD-induced distances between individuals. It performs agglomerative hierarchical clustering using the Mahalanobis distance based *Genetic Relationship Matrix (GRM)* which captures the population-level covariance of the genotypes. Thereafter, we apply sketching-based randomized ridge regression on the clusters and perform a meta-analysis to obtain the association statistics.

With the growing size of data, computing and storing the genome wide GRM is a non-trivial task. We get around this overhead by computing the Mahalanobis distance between two vectors efficiently without storing or inverting the covariance matrix, but instead computing the corresponding rank-k leverage and cross-leverage scores. We compute the rank-k Mahalanobis distance with respect to the top k-left singular vectors of the genotype matrix, thus making the computation feasible for UK Biobank-scale datasets using methods such as TeraPCA [2] to approximate the left singular vectors accurately and efficiently.

Supported by NSF IIS 1715202 and NFS DMS 1760353 awarded to PD and PP.

© Springer Nature Switzerland AG 2020
R. Schwartz (Ed.): RECOMB 2020, LNBI 12074, pp. 234–236, 2020.
https://doi.org/10.1007/978-3-030-45257-5_19

We test CluStrat on a large simulation study of arbitrarily-structured, admixed sub-populations by generating 100 GWAS datasets (with 1,000 individuals genotyped on one million genetic markers) from a quantitative trait model (and it's equivalent binary trait) based on previous work [9]. We simulated 30 different scenarios, varying proportions of true genetic effect and admixture and compared it's performance to standard population structure correction approaches such as EIGENSTRAT [7], GEMMA [11], and EMMAX [4]. We identified two to three-fold more true causal variants when compared to the above methods for almost all scenarios, while trading off for a slightly higher spurious associations, but, far less than the uncorrected Armitage trend χ^2 test. Applying CluStrat on WTCCC2 Parkinson's disease (PD) data with a p-value threshold set to 10^{-7}, we identified loci mapped to a host of genes known to be associated with PD such as BACH2, MAP2, NR4A2, SLC11A1, UNC5C to name a few. In summary, CluStrat highlights the advantages of biologically relevant distance metrics, such as the Mahalanobis distance, which seems to capture the cryptic interactions within populations in the presence of LD better than the Euclidean distance. Of independent interest is a simple, but not necessarily well-known, connection between the regularized Mahalanobis distance-based GRM and the leverage and cross-leverage scores of the genotype matrix. CluStrat source code and user manual is available at: https://github.com/aritra90/CluStrat and the full version is available at https://doi.org/10.1101/2020.01.15.908228.

References

1. Berg, J.J., Harpak, A., Sinnott-Armstrong, N., et al.: Reduced signal for polygenic adaptation of height in UK Biobank. eLife **8**, e39725 (2019)
2. Bose, A., Kalantzis, V., Kontopoulou, E.M., et al.: TeraPCA: a fast and scalable software package to study genetic variation in tera-scale genotypes. Bioinformatics **35**, 3679–3683 (2019)
3. Ewens, W.J., Spielman, R.S.: The transmission/disequilibrium test: history, subdivision, and admixture. Am. J. Hum. Genet. **57**(2), 455 (1995)
4. Kang, H.M., Sul, J.H., Service, S.K., et al.: Variance component model to account for sample structure in genome-wide association studies. Nat. Genet. **42**(4), 348 (2010)
5. Mahalanobis, P.C.: On the generalized distance in statistics. National Institute of Science of India (1936)
6. Mathew, B., Léon, J., Sillanpää, M.J.: A novel linkage-disequilibrium corrected genomic relationship matrix for SNP-heritability estimation and genomic prediction. Heredity **120**(4), 356 (2018)
7. Price, A.L., Patterson, N.J., Plenge, R.M., et al.: Principal components analysis corrects for stratification in genome-wide association studies. Nat. Genet. **38**(8), 904 (2006)
8. Sohail, M., Maier, R.M., Ganna, A., et al.: Polygenic adaptation on height is overestimated due to uncorrected stratification in genome-wide association studies. eLife **8**, e39702 (2019)
9. Song, M., Hao, W., Storey, J.D.: Testing for genetic associations in arbitrarily structured populations. Nat. Genet. **47**(5), 550 (2015)

10. Uricchio, L.H., Kitano, H.C., Gusev, A., et al.: An evolutionary compass for detecting signals of polygenic selection and mutational bias. Evol. Lett. **3**(1), 69–79 (2019)
11. Zhou, X., Stephens, M.: Genome-wide efficient mixed-model analysis for association studies. Nat. Genet. **44**(7), 821 (2012)

PWAS: Proteome-Wide Association Study

Nadav Brandes[1]([✉]) [iD], Nathan Linial[1] [iD], and Michal Linial[2] [iD]

[1] School of Computer Science and Engineering,
The Hebrew University of Jerusalem, Jerusalem, Israel
nadav.brandes@mail.huji.ac.il, nati@cs.huji.ac.il
[2] Department of Biological Chemistry, The Hebrew University of Jerusalem,
Jerusalem, Israel
michall@cc.huji.ac.il

Abstract. Over the last two decades, Genome-Wide Association Study (GWAS) has become a canonical tool for exploratory genetic research, generating countless gene-phenotype associations. Despite its accomplishments, several limitations and drawbacks still hinder its success, including low statistical power and obscurity about the causality of implicated variants. We introduce PWAS (Proteome-Wide Association Study), a new method for detecting protein-coding genes associated with phenotypes through protein function alterations. PWAS aggregates the signal of all variants jointly affecting a protein-coding gene and assesses their overall impact on the protein's function using machine-learning and probabilistic models. Subsequently, it tests whether the gene exhibits functional variability between individuals that correlates with the phenotype of interest. By collecting the genetic signal across many variants in light of their rich proteomic context, PWAS can detect subtle patterns that standard GWAS and other methods overlook. It can also capture more complex modes of heritability, including recessive inheritance. Furthermore, the discovered associations are supported by a concrete molecular model, thus reducing the gap to inferring causality. To demonstrate its applicability for a wide range of human traits, we applied PWAS on a cohort derived from the UK Biobank ($\sim$330K individuals) and evaluated it on 49 prominent phenotypes. 23% of the significant PWAS associations on that cohort (2,998 of 12,896) were missed by standard GWAS. A comparison between PWAS to existing methods proves its capacity to recover causal protein-coding genes and highlighting new associations with plausible biological mechanism.

Keywords: GWAS · Machine learning · Protein function · UK Biobank · Recessive heritability

1 Background

GWAS is limited by a number of crucial factors. Insufficient statistical power is often caused, partly, by the large number of tested variants across the genome, especially when rare variants of small effect sizes are involved. Due to Linkage Disequilibrium (LD) and population stratification, even when a genomic locus is robustly implicated with a phenotype, pinning the exact causal variant(s) is a convoluted task. To arrive at more interpretable, actionable discoveries, some methods seeks to implicate genes

© Springer Nature Switzerland AG 2020
R. Schwartz (Ed.): RECOMB 2020, LNBI 12074, pp. 237–239, 2020.
https://doi.org/10.1007/978-3-030-45257-5_20

directly, by carrying the association tests at the level of annotated functional elements in the first place. The most commonly used gene-level method is SKAT [1], which aggregates the signal across an entire genomic region, be it a gene or any other functional entity. Another approach, recently explored by methods such as PrediXcan [2] and TWAS [3], tests whether the studied phenotypes correlate with gene expression levels predicted from genetic variants.

2 Methods

A natural enhancement to existing gene-based association studies would be a protein-centric method that considers the effects of genetic variants on the functionality of genes, rather than their abundance (be it at the transcript or protein level). We present PWAS: Proteome Wide Association Study (Fig. 1). PWAS is based on the premise that causal variants in coding regions affect phenotypes by altering the biochemical functions of the genes' protein products (Fig. 1a). PWAS takes the same inputs as GWAS (Fig. 1b): (i) called genotypes of m variants across n individuals, (ii) a vector of n phenotype values (binary or continuous), and (iii) a covariate matrix for the n individuals with possible confounders (e.g. sex, age, principal components, batch). By exploiting a rich proteomic knowledgebase, a pre-trained machine-learning model [4] estimates the extent of damage caused to each of the k proteins in the human proteome, as a result of the m observed variants, for each of the n individuals (typically $k \ll m$). To calculate these effect scores, PWAS considers any variant that affects the coding-regions of genes (e.g. missense, nonsense, frameshift). The per-variant damage predictions are aggregated at the gene level, in view of each sample's genotyping, generating protein function *effect score* matrices. PWAS generates two such matrices, reflecting either a *dominant* or a *recessive* effect on phenotypes. Intuitively, the dominant effect score is intended to express the probability of at least one hit damaging the protein function, while the recessive score attempts to express the probability of at least two damaging hits. PWAS identifies significant associations between the phenotype values to the effect score values (comprising columns of the matrices, each representing a distinct protein-coding gene), while accounting for the provided

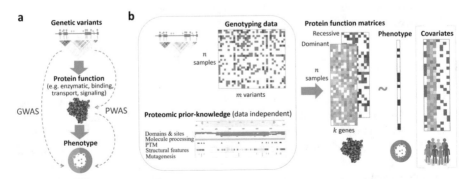

Fig. 1. The PWAS framework: (a) Assumed causal model. (b) The method's workflow.

covariates. Each gene can be tested by the dominant model, the recessive model, or a generalized model that uses both the dominant and recessive values.

3 Results

Using a dataset derived from the UK Biobank [5] ($\sim$330K samples), we tested PWAS on 49 prominent phenotypes (including major metabolic, psychiatric and autoimmune diseases, various cancer types, blood tests, and a variety of physical measurements such as height and BMI). Altogether, PWAS discovered 12,896 gene-phenotype associations, only 5,338 of which (41%) contain a GWAS-significant non-synonymous variant within the gene's coding region. In other words, although PWAS considers the same set of variants, in 59% of the associations it is able to recover an aggregated signal that is overlooked by GWAS when considering each of the variants individually. Even when considering all the variants in proximity of the gene to account for LD (up to 500,000 bp to each side of the coding region), 2,998 of the 12,896 PWAS associations (23%) are still missed by GWAS. We further compared PWAS to SKAT, the most commonly used method for detecting genetic associations at the gene level. Importantly, whereas SKAT attempts to recover all existing genetic associations, PWAS focuses specifically on protein-coding genes that are associated with a phenotype through protein function. We found PWAS to be superior to SKAT in the number of discovered associations for most phenotypes.

4 Conclusion

Like other gene-based approaches, PWAS benefits from a reduced burden of multiple-testing correction and is able to uncover associations spread across multiple variants. In addition, it provides concrete functional interpretations for the protein-coding genes it discovers, and can determine whether the proposed causal effect appears to be dominant, recessive or some mixture of the two (e.g. additive). PWAS is an open-source Python project, available as a command-line tool at https://github.com/nadavbra/pwas. A preliminary of the manuscript is available on bioRxiv at https://doi.org/10.1101/812289.

References

1. Wu, M.C., Lee, S., Cai, T., et al.: Rare-variant association testing for sequencing data with the sequence kernel association test. Am. J. Hum. Genet. **89**, 82–93 (2011)
2. Gamazon, E.R., Wheeler, H.E., Shah, K.P., et al.: A gene-based association method for mapping traits using reference transcriptome data. Nat. Genet. **47**, 1091 (2015)
3. Gusev, A., Ko, A., Shi, H., et al.: Integrative approaches for large-scale transcriptome-wide association studies. Nat. Genet. **48**, 245 (2016)
4. Brandes, N., Linial, N., Linial, M.: Quantifying gene selection in cancer through protein functional alteration bias. Nucleic Acids Res. **47**, 6642–6655 (2019)
5. Sudlow, C., Gallacher, J., Allen, N., et al.: UK Biobank: an open access resource for identifying the causes of a wide range of complex diseases of middle and old age. PLoS Med. **12**, e1001779 (2015)

Estimating the Rate of Cell Type Degeneration from Epigenetic Sequencing of Cell-Free DNA

Christa Caggiano[1], Barbara Celona[2], Fleur Garton[3], Joel Mefford[1],
Brian Black[2,4], Catherine Lomen-Hoerth[5], Andrew Dahl[6],
and Noah Zaitlen[1(✉)]

[1] Department of Neurology, University California Los Angeles,
Los Angeles, CA, USA
nzaitlen@ucla.edu
[2] Cardiovascular Research Institute, University of California San Francisco,
San Francisco, CA, USA
[3] Institute for Molecular Biosciences, University of Queensland, Brisbane, Australia
[4] Department of Biochemistry and Biophysics, University of California,
San Francisco, USA
[5] Department of Neurology, University of California San Francisco,
San Francisco, CA, USA
[6] Section of Genetic Medicine, Department of Medicine, University of Chicago,
Chicago, IL, USA

Cells die at different rates as a function of disease state, age, environmental exposure, and behavior [8,10]. Knowing the rate at which cells die is a fundamental scientific question, with direct translational applicability. A quantifiable indication of cell death could facilitate disease diagnosis and prognosis, prioritize patients for admission into clinical trials, and improve evaluations of treatment efficacy and disease progression [1,4,14,16]. Circulating cell-free DNA (cfDNA) in the bloodstream originates from dying cells and is a promising non-invasive biomarker for cell death.

To understand what drives changes in the biology of people with disease, we can decompose the cfDNA mixture into the cell types from which the cfDNA originates. This can give a non-invasive picture of cell death, which can be used to characterize an individual's disease, or health, at a particular moment. While each cell type has the same DNA sequence, which does not give us information on where a cfDNA fragment arises from, DNA methylation is cell type informative [7]. Subsequently, there is a rich literature of cell type decomposition approaches using DNA methylation, often focusing on estimating the contribution of immunological cell types to whole blood [2,3,11,12].

Recent work has attempted to use cfDNA methylation patterns to decompose tissue of origin for cfDNA [5,6,9,13]. These approaches, however, do not address some of the unique challenges of cfDNA, or were designed for different purposes. Several were designed for reference and input data from a methylation chip, which are high coverage and have relatively low noise. Since cfDNA is only present in the blood in small amounts, as an onerous amount of blood must be extracted from a patient to get the required amount of input DNA for

© Springer Nature Switzerland AG 2020
R. Schwartz (Ed.): RECOMB 2020, LNBI 12074, pp. 240–242, 2020.
https://doi.org/10.1007/978-3-030-45257-5_21

methylation chips, which may not be practical for clinical use. While increasingly powerful, these approaches can not provide biomarker discovery, especially for unknown cell types, or comprehensive deconvolution. In this work, we turn to using whole genome bisulfite sequencing (WGBS) to assess the methylation of cfDNA. Unlike methylation chips that target specific genomic locations, WGBS covers the entire genome, typically resulting in lower coverage per-site, and increased noise relative to chip data. Current methods are ill-equipped to handle such noise in either the reference or input.

Here, we develop a method to accurately estimate the relative abundances of cell types contributing to cfDNA. We leverage the distinct DNA methylation profile of each cell type throughout the body. Decomposing the cfDNA mixture is difficult, as fragments from relevant cell types may only be present in a small amount. We propose an algorithm, CelFiE, that estimates cell type proportion from both whole genome cfDNA input and reference data. CelFiE accommodates low coverage data, does not rely on CpG site curation, and estimates contributions from multiple unknown cell types that are not available in reference data.

We show in realistic simulations that we can accurately estimate known and unknown cell types even at low coverage and relatively few sites. We also can estimate rare cell types that contribute to only a small fraction of the total cfDNA. Decomposition of real WGBS complex mixtures demonstrates that CelFiE is robust to several violations of our model assumptions. Specifically, the real data allow: correlation across regions and between cell types, read counts that are drawn from heavy-tailed distributions, and reference samples that are actually heterogeneous mixtures of many cell types. Additionally, we develop an approach for unbiased site selection.

Finally, we apply CelFiE to two real cfDNA data sets. First, we apply CelFiE to cfDNA extracted from pregnant women and non-pregnant women. Since placenta is not expected in non-pregnant women, this data enables validation for our method. CelFiE estimates a large placenta component specifically in pregnant women ($p = 9.1 \times 10^{-5}$). We also apply CelFiE to cfDNA from amyotrophic lateral sclerosis (ALS) patients and age matched controls. Specifically, CelFiE estimates increased skeletal muscle component in the cfDNA of ALS patients ($p = 2.6 \times 10^{-3}$), which is consistent with muscle impairment characterizing ALS. Currently, there are no established biomarkers for ALS. Subsequently, it is difficult to monitor disease progression and efficiently evaluate treatment response [15]. CfDNA provides an opportunity to measure cell death in ALS that could fill these gaps. Differences in the cfDNA of ALS patients is a novel observation, and along with successful decomposition of cfDNA from pregnant women, demonstrates that CelFiE has the potential to meaningfully decompose cfDNA in realistic conditions. CfDNA decomposition has broad translational potential for understanding the biology of cell death, and in applications such as quantitative biomarker discovery or in the non-invasive monitoring and diagnosis of disease.

The full version of the paper is available at https://www.biorxiv.org/content/10.1101/2020.01.15.907022v1.

242   C. Caggiano et al.

References

1. Bowser, R., Turner, M.R., Shefner, J.: Biomarkers in amyotrophic lateral sclerosis: opportunities and limitations. Nat. Rev. Neurol. **7**, 631–8 (2011)
2. Houseman, E.A., Molitor, J., Marsit, C.J.: Reference-free cell mixture adjustments in analysis of DNA methylation data. Bioinformatics **30**, 1431–1439 (2014)
3. Houseman, E.A., et al.: DNA methylation arrays as surrogate measures of cell mixture distribution. BMC Bioinformatics **13**, 86 (2012). https://doi.org/10.1186/1471-2105-13-86
4. Joka, D., et al.: Prospective biopsy-controlled evaluation of cell death biomarkers for prediction of liver fibrosis and nonalcoholic steatohepatitis. Hepatology **55**, 455–64 (2012)
5. Lehmann-Werman, R., et al.: Identification of tissue-specific cell death using methylation patterns of circulating DNA. Proc. Natl. Acad. Sci. **113**, E1826–E1834 (2016)
6. Liu, X., et al.: Comprehensive DNA methylation analysis of tissue of origin of plasma cell-free DNA by methylated CpG tandem amplification and sequencing (MCTA-Seq). Clin. Epigenetics **11**, 93 (2019)
7. Lokk, K., et al.: DNA methylome profiling of human tissues identifies global and tissue-specific methylation patterns. Genome Biol. **15**, r54 (2014). https://doi.org/10.1186/gb-2014-15-4-r54
8. Meier, P., Finch, A., Evan, G.: Apoptosis in development. Nature **407**, 796–801 (2000)
9. Moss, J., et al.: Comprehensive human cell-type methylation atlas reveals origins of circulating cell-free DNA in health and disease. Nat. Commun. **9**, 1–12 (2018)
10. Nagata, S.: Apoptosis by death factor. Cell **88**, 355–65 (1997)
11. Rahmani, E., Schweiger, R., Shenhav, L., Eskin, E., Halperin, E.: A Bayesian framework for estimating cell type composition from DNA methylation without the need for methylation reference. In: Sahinalp, S.C. (ed.) RECOMB 2017. LNCS, vol. 10229, pp. 207–223. Springer, Cham (2017). https://doi.org/10.1007/978-3-319-56970-3_13
12. Rahmani, E., et al.: Sparse PCA corrects for cell-type heterogeneity in epigenome-wide association studies. Nat. Methods **13**, 443–445 (2016)
13. Snyder, M.W., Kircher, M., Hill, A.J., Daza, R.M., Shendure, J.: Cell-free DNA comprises an in vivo nucleosome footprint that informs its tissues-of-origin. Cell **164**, 57–68 (2016)
14. Turner, M.R., et al.: Mechanisms, models and biomarkers in amyotrophic lateral sclerosis. Amyotroph. Lateral Scler. Frontotemporal Degener. **14**, 19–32 (2013)
15. Verber, N.S., et al.: Biomarkers in motor neuron disease: a state of the art review. Front. Neurol. **10**, 291 (2019)
16. Vila, M., Przedborski, S.: Targeting programmed cell death in neurodegenerative diseases. Nat. Rev. Neurosci. **4**, 365–375 (2003)

Potpourri: An Epistasis Test Prioritization Algorithm via Diverse SNP Selection

Gizem Caylak[1] and A. Ercument Cicek[1,2(✉)]

[1] Department of Computer Engineering, Bilkent University, 06800 Ankara, Turkey
cicek@cs.bilkent.edu.tr
[2] Computational Biology Department, Carnegie Mellon University, Pittsburgh,
PA 15213, USA

Extended Abstract

Genome-wide association studies (GWAS) have been an important tool for susceptibility gene discovery in genetic disorders and investigating the interplay among multiple loci has played an imporatant role. Such interactions between two or more loci is called epistasis and it has a major role in complex genetic traits.

Exhaustive identification of interacting loci, even just pairs, is potentially intractable for large GWAS [1]. Moreover, such an approach lacks statistical power due to multiple hypothesis testing. Another approach is to reduce the search space by filtering pairs based on a type of statistical threshold. However this apporach does not follow a biological reasoning and tends to detect interactions that are in linkage disequilibrium (LD). On another track, incorporating biological background and testing the SNP pairs that are annotated has also proven useful. Yet, this approach requires most SNPs to be discarded as many are quite far away from any gene to be associated. Moreover, this introduces a literature bias in the selections of the algorithms. A rather more popular approach is to prioritize the tests to be performed rather than discarding pairs from the search space and controlling for Type-I error. In this approach, the user can keep performing tests in the order specified by the algorithm until a desired number of significant pairs are found. The idea is to provide the user with a manageable number of true positives (statistically significant epistatic pairs) while minimizing the number of tests to ensure statistical power. Despite using various heuristics, all methods still have high false discovery rates.

Linkage disequilibrium is an important source of information for epistasis prioritization algorithms. Two SNPs that appear to be interacting statistically, might not be biologically meaningful if they are on the same haplotype block. In an orthogonal study, Yilmaz et al. propose a feature (SNP) selection algorithm which avoids LD for better phenotype prediction [2]. Authors show that while looking for a small set of loci (i.e., 100) that is the most predictive of a continuous phenotype, selecting SNPs that are far away from each other, results in better predictive power. This method, SPADIS, is designed for feature selection for multiple regression. As the SNP set it generates contains diverse and

R. Schwartz (Ed.): RECOMB 2020, LNBI 12074, pp. 243–244, 2020.
https://doi.org/10.1007/978-3-030-45257-5_22

complementary SNPs, it results in better R^2 values by covering more biological functions.

Inspired by this idea, we conjecture that selecting pairs of SNPs from genomic regions that (i) harbor individually informative SNPs, and (ii) are diverse in terms of genomic location would avoid LD better and yield more functionally complementing and more epistatic SNP pairs compared to the current state of the art since no other algorithm exploits this information. We propose a new method that for the first time incorporates the genome location diversity with the population coverage density. Specifically, our proposed method, *Potpourri*, maximizes a submodular set function to select a set of genomic regions (i) that include SNPs which are individually predictive of the cases, and (ii) that are topologically distant from each other on the underlying genome. Epistasis tests are performed for pairs across these regions, such that pairs that densely co-cover the case cohort are given priority.

We validate our hypothesis and show that Potpourri is able to detect statistically significant and biologically meaningful epistatic SNP pairs. We perform extensive tests on three Wellcome Trust Case Control Consortium (WTCCC) GWA studies and compare our method with the state-of-the-art LINDEN algorithm [1]. First, we *guide* LINDEN by pruning its search space using Potpourri-selected-SNPs to show that (i) it is possible to significantly improve the precision (from 0.003 up to 0.302) and (ii) that our diversification approach is sound. Then, we show that the ranking of the diverse SNPs by the co-coverage of the case cohort further improves the prediction power and the precision (up to 0.652 in the selected setting). Potpourri drastically reduces the number of hypothesis tests to perform (from ∼380k to ∼15k), and yet is still able to detect more epistatic pairs with similar significance levels in all there GWA studies considered. The running time is also cut by 4 folds in the selected settings. Another problem with the current techniques is the biological interpretation of the obtained epistatic pairs. Once the most significant SNP pairs returned are in the non-coding regions and are too isolated to be associated with any gene, the user can hardly make sense of such a result despite statistical significance. We investigate the advantage of promotion of SNPs falling into regulatory and non-coding regions for testing and propose three techniques. We show that these techniques further improve the precision (up to 0.8) with similar number of epistatic pairs detected. Potpourri is available at http://ciceklab.cs.bilkent.edu.tr/potpourri and the preprint is available at https://www.biorxiv.org/content/10.1101/830216v3. This research has been supported by TUBITAK grant #116E148 to AEC.

References

1. Cowman, T., Koyutürk, M.: Prioritizing tests of epistasis through hierarchical representation of genomic redundancies. Nucleic Acids Res. **45**(14), e131–e131 (2017)
2. Yilmaz, S., Tastan, O., Cicek, E.: SPADIS: an algorithm for selecting predictive and diverse SNPs in GWAS. IEEE/ACM Trans. Comput. Biol. Bioinf. (2019). https://doi.org/10.1109/TCBB.2019.2935437

Privacy-Preserving Biomedical Database Queries with Optimal Privacy-Utility Trade-Offs

Hyunghoon Cho[1]([⊠]), Sean Simmons[1,4], Ryan Kim[2], and Bonnie Berger[1,3,4]([⊠])

[1] Broad Institute of MIT and Harvard, Cambridge, MA 02142, USA
hhcho@broadinstitute.org
[2] Harvard University, Cambridge, MA 02138, USA
[3] CSAIL, MIT, Cambridge, MA 02139, USA
[4] Department of Mathematics, MIT, Cambridge, MA 02139, USA
bab@mit.edu

1 Introduction

Sharing data across research groups is an essential driver of biomedical research. In particular, biomedical databases with interactive query-answering systems allow users to retrieve information from the database using restricted types of queries. For example, medical data repositories allow researchers designing clinical studies to query how many patients in the database satisfy certain criteria, a workflow known as *cohort discovery*. In addition, genomic "beacon" services allow users to query whether or not a given genetic variant is observed in the database, a workflow we refer to as *variant lookup*. While these systems aim to facilitate the sharing of aggregate biomedical insights without divulging sensitive individual-level data, they can still leak private information about the individuals through the query answers. To address these privacy concerns, existing studies have proposed to perturb query results with a small amount of noise in order to reduce sensitivity to underlying individuals [1,2]. However, these existing efforts either lack rigorous guarantees of privacy or introduce an excessive amount of noise into the system, limiting their effectiveness in practice.

2 Methods

Here, we build upon recent advances in differential privacy (DP) to introduce query-answering systems with formal privacy guarantees while ensuring that the query results are as accurate as theoretically possible. We newly propose to leverage the truncated α-geometric mechanism (α-TGM), previously developed for a limited class of count queries in a theoretical context [3], to limit disclosure risk in both cohort discovery and variant lookup workflows. We show that α-TGM, combined with a post-processing step performed by the user, provably maximizes the expected utility (encompassing accuracy) of the system for a

---

H. Cho and S. Simmons—Equal-contributions.

R. Schwartz (Ed.): RECOMB 2020, LNBI 12074, pp. 245–247, 2020.
https://doi.org/10.1007/978-3-030-45257-5_23

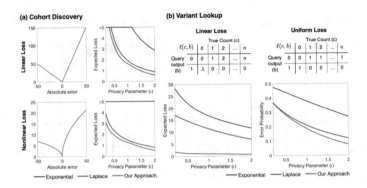

Fig. 1. Our approach improves the utility of medical cohort discovery and genomic variant lookup with differential privacy. We compared the performance of our optimal differential privacy mechanisms to the exponential and Laplace mechanisms for (a) cohort discovery and (b) variant lookup queries with different choices of loss functions. We used uniform prior for cohort discovery and a real-world query distribution from the ExAC database as prior for variant lookup. Overall, our mechanisms achieve consistently higher accuracy of query results compared to existing approaches.

broad range of user-defined notions of utility and for any desired level of privacy. Notably, the optimality of α-TGM was previously known for only symmetric utility functions, which are insufficient for workflows we typically encounter in biomedical databases [1]. We extend this result to a more general class of utility functions including *asymmetric* functions, thereby answering an open question posed in the original publication of α-TGM [3]. Moreover, we demonstrate that α-TGM can be transformed to obtain an optimal DP mechanism for the variant lookup problem, a novel result enabled by our generalized notion of utility.

3   Results

We compared the effectiveness of our proposed approach for cohort discovery and variant lookup queries to the exponential and Laplace mechanisms proposed by earlier studies [1,2]. Our mechanisms consistently improved the accuracy of query results across all values of privacy parameter ϵ and different types of loss functions (Fig. 1). Furthermore, we observed that using an appropriate prior can further increase query accuracy and that our improved queries can be used to enhance downstream analyses, such as association tests. Given the optimality of our schemes, our results illustrate the theoretical boundaries of leveraging DP for mitigating privacy risks in biomedical query-answering systems. A full version of this work is available on bioRxiv.[1]

---

[1] https://doi.org/10.1101/2020.01.16.909010.

References

1. Vinterbo, S.A., et al.: Protecting count queries in study design. J. Am. Med. Inform. Assoc. **19**, 750–757 (2012)
2. Raisaro, J.L., et al.: MedCo: enabling secure and privacy-preserving exploration of distributed clinical and genomic data. IEEE/ACM Trans. Comput. Biol. Bioinf. **16**(4), 1328–1341 (2018)
3. Ghosh, A., et al.: Universally utility-maximizing privacy mechanisms. SIAM J. Comput. **41**(6), 1673–1693 (2012)

Iterative Refinement of Cellular Identity from Single-Cell Data Using Online Learning

Chao Gao[1] and Joshua D. Welch[1,2(✉)]

[1] Department of Computational Medicine and Bioinformatics,
University of Michigan, Ann Arbor, MI, USA
gchao@umich.edu, welchjd@umich.edu
[2] Department of Computer Science and Engineering, University of Michigan,
Ann Arbor, MI, USA

Abstract. Recent experimental advances have enabled high-throughput single-cell measurement of gene expression, chromatin accessibility and DNA methylation. We previously employed integrative non-negative matrix factorization (iNMF) to jointly align multiple single-cell datasets (X_i) and learn interpretable low-dimensional representations using dataset-specific (V_i) and shared metagene factors (W) and cell factor loadings (H_i). We developed an alternating nonnegative least squares (ANLS) algorithm to solve the iNMF optimization problem [2]:

$$\min_{\substack{W,V_i,H_i \geq 0 \\ i \in 1,..,N}} \sum_{i=1}^{N} \|X_i - (W + V_i)H_i\|_F^2 + \lambda \sum_{i=1}^{N} \|V_iH_i\|_F^2 \qquad (1)$$

The resulting metagenes and cell loadings provide a principled, quantitative definition of cellular identity and how it varies across biological contexts. However, datasets exceeding 1 million cells are now widely available, creating computational barriers to scientific discovery. For instance, it is no longer feasible to use the entire available datasets as inputs to implement standard pipelines on a personal computer with limited memory capacity. Moreover, there is a need for an algorithm capable of iteratively refining the definition of cellular identity as efforts to create a comprehensive human cell atlas continually sequence new cells.

To address these challenges, we developed an online learning algorithm for integrating massive and continually arriving single-cell datasets. The key innovation that makes it possible to perform online learning [1] is to optimize a "surrogate function", given by (2), that asymptotically converges to the same solution as the original iNMF objective (1). We can then perform matrix factorization in an online fashion by iteratively minimizing the expected cost $\hat{f}_t(H, W, V)$ as new data points x_t (or points randomly sampled from a large fixed training set) arrive.

$$\hat{f}_t(W, V_1, ..., V_N, H_1, ..., H_N) = \frac{1}{t}\sum_{i=1}^{t} \|x_i - (W + V_{d_i})h_{d_i}\|_F^2 + \lambda \|V_{d_i}h_{d_i}\|_F^2 \qquad (2)$$

where d_i indicates which dataset the ith data point belongs to.

© Springer Nature Switzerland AG 2020
R. Schwartz (Ed.): RECOMB 2020, LNBI 12074, pp. 248–250, 2020.
https://doi.org/10.1007/978-3-030-45257-5_24

Intuitively, this strategy allows online learning because it expresses a formula for incorporating a new observation x_t given the factorization result $(H^{(t-1)}, W^{(t-1)}, V^{(t-1)})$ for previously seen data points. Thus, we can iterate over the data points one-by-one or in "mini-batches"—and also rapidly update the factorization when new data points arrive. We also derived a novel hierarchical alternating least squares (HALS) algorithm for iNMF and incorporated it into an efficient online algorithm.

Our online approach accesses the training data as mini-batches, decoupling memory usage from dataset size and allowing on-the-fly incorporation of new data as it is generated. Because the online algorithm does not require all of the data on each iteration (only a single data point or fixed-size mini-batch), we used the rhdf5 package to load each mini-batch from disk on the fly. For a mini-batch size of 5,000 cells, we found that reading each mini-batch from disk added minimal overhead (less than 0.35 s per iteration).

We first investigated the performance of the online learning algorithm on aligning the single-cell RNA-seq datasets. For this purpose, we benchmarked the online learning algorithm with three other batch algorithms, ANLS, HALS, and the multiplicative update method, using several datasets from different tissues, including human PBMCs, human pancreas, and the adult mouse brain. By evaluating each method on the training and testing sets, we confirmed that the online implementation of iNMF converges much more quickly using a fraction of the memory required for the batch implementation, without sacrificing solution quality. We also found that the time required for the online iNMF algorithm to converge does not grow steadily with increasing dataset size, but stabilizes once the number of cells crosses a minimum size threshold. Thus, the advantage of the online algorithm compared to the batch algorithm increases with dataset size. Our new approach enables factorization of 691962 single cells from 9 regions of the mouse brain on a standard laptop in ~20 min using about 500 MB of RAM. In comparison, we estimate that carrying out the same analysis using our previous batch iNMF approach would have taken 4–5 h and required over 40 GB of RAM.

Secondly, we demonstrated that our online algorithm allows for iterative refinement of cell identity from continually arriving datasets. We constructed a cell atlas of the mouse motor cortex by iteratively incorporating 6 single-cell RNA-seq datasets generated by the BRAIN Initiative Cell Census Network over a period of two years. We found that by accessing the data in a true online fashion—seeing each cell only once during training—we could accurately integrate the datasets as they arrived. In order to assess the robustness of this approach, we investigated how both random data ordering and random initialization affect the results. We found that different orders of dataset arrival have only minimal effect on the final objective function value: different orders resulted in a standard deviation of 1.2% in the final iNMF objective, compared to 0.2% from random initializations. Furthermore, the clustering results derived from online iNMF are stable even under different orders of dataset arrival, with an adjusted Rand index of (ARI) 0.7—only slightly larger than the variation from different random initializations (ARI of 0.74). We also attempted to incorporate single-nucleus ATAC-seq data in true online fashion after processing the 6 scRNA-seq datasets, but found that the alignment quality is lower than when we process all of the modalities at once.

In summary, our new online learning algorithm shows several promising characteristics in different contexts. Given multiple finite single-cell datasets, our approach offers rapid convergence speed without sacrificing solution quality. It also obviates the need to recompute results each time additional cells are sequenced and allows for processing of datasets too large to fit in memory. Most importantly, it facilitates continual refinement of cell identity as new single-cell datasets from different biological contexts and data modalities are generated chronologically. We also found that our newly developed batch HALS algorithm can converge nearly as rapidly as online iNMF on smaller datasets, though it does not offer the benefits of memory usage independent of dataset size or iterative incorporation of new data. We anticipate that online iNMF will prove increasingly useful for assembling a comprehensive atlas of cell types and states as single-cell dataset sizes increase.

Full text: Preprint version of the full manuscript is available at https://www. biorxiv.org/content/10.1101/2020.01.16.909861v2.

References

1. Mairal, J., Bach, F., Ponce, J., Sapiro, G.: Online learning for matrix factorization and sparse coding. J. Mach. Learn. Res. **11**(Jan), 19–60 (2010)
2. Welch, J.D., Kozareva, V., Ferreira, A., Vanderburg, C., Martin, C., Macosko, E.Z.: Single-cell multi-omic integration compares and contrasts features of brain cell identity. Cell **177**(7), 1873–1887 (2019)

A Guided Network Propagation Approach to Identify Disease Genes that Combines Prior and New Information

Borislav H. Hristov, Bernard Chazelle, and Mona Singh[(✉)]

Department of Computer Science and Lewis-Sigler Institute for Integrative Genomics,
Princeton University, Princeton, USA
mona@cs.princeton.edu

Summary. A major challenge in biomedical data science is to identify the causal genes underlying complex genetic diseases. Despite the massive influx of genome sequencing data, identifying disease-relevant genes remains difficult as individuals with the same disease may share very few, if any, genetic variants. Protein-protein interaction networks provide a means to tackle this heterogeneity, as genes causing the same disease tend to be proximal within networks. Previously, network propagation approaches have spread "signal" across the network from either known disease genes *or* genes that are newly putatively implicated in the disease. Here we introduce a general framework that considers both sources of data within a network context. Specifically, we use prior knowledge of disease-associated genes to guide random walks initiated from genes that are newly identified as perhaps disease-relevant. In rigorous, large-scale testing across 24 cancer types, we demonstrate that our approach for integrating both prior and new information not only better identifies cancer driver genes than using either source of information alone but also readily outperforms other state-of-the-art network-based approaches. To demonstrate the versatility of our approach, we also apply it to genome-wide association data to identify genes functionally relevant for several complex diseases. Overall, our work suggests that guided network propagation approaches that utilize both prior and new data are a powerful means to identify disease genes.

Methods. At a high level, our approach uKIN (**u**sing **K**nowledge **I**n **N**etworks) propagates new information across a network, while using prior information to guide this propagation. While our approach is generally applicable, we focus on the case of propagating information across biological networks in order to find disease genes. In the scenario of uncovering cancer genes, prior information comes from the set of known cancer genes, and new information corresponds to those genes that are found to be somatically mutated across patient tumors. For other complex diseases, new information may arise from (say) genes weakly associated with a disease via GWAS studies or found to have *de novo* mutations in a patient population of interest. As we expect the genes that are actually disease-relevant to be proximal in the network to each other and to the previously known set of disease genes, we spread the signal from the newly implicated genes biasing it to

© Springer Nature Switzerland AG 2020
R. Schwartz (Ed.): RECOMB 2020, LNBI 12074, pp. 251–252, 2020.
https://doi.org/10.1007/978-3-030-45257-5_25

move towards genes that are closer to the known disease genes. We accomplish this by performing random walks with restarts (RWRs), where with probability α, the walk jumps back to one of the starting genes. That is, α controls the extent to which we use new versus prior information, where higher values of α weigh the new information more heavily. With probability $1 - \alpha$, the walk moves to a neighboring node, but instead of moving from one gene to one of its neighbors uniformly at random as is typically done, the probability instead is higher for neighbors that are closer to the prior knowledge set of genes. We show that the walk has a stationary distribution that can be computed numerically. The genes whose nodes have high scores are most frequently visited and, therefore, are more likely disease-relevant as they are close to both the mutated starting nodes as well as to already known disease genes.

Results. We apply our method to 24 different TCGA cancer types. First, we show that uKIN successfully integrates prior knowledge and new information across all cancer types by comparing uKIN's performance when using both prior and new knowledge (RWRs with $\alpha = 0.5$), to versions of uKIN using either only new information ($\alpha = 1$) or only prior information ($\alpha = 0$). Second, we demonstrate that uKIN readily outperforms frequency-based methods such as MutSigCV 2.0 as well as other state-of-the-art network-based methods. Our results are consistent across both networks that we tested (HPRD and Biogrid), showing the robustness of our method with respect to the underlying network. Further, we demonstrate that using more accurate prior knowledge such as cancer-type specific driver genes yields better performance. We examine the genes that are highly ranked by uKIN and observe that they display diverse mutational rates. We uncover potential oncogenic roles of several infrequently mutated novel genes. Finally, we showcase uKIN's versatility by applying it to GWAS data for three complex diseases: age-related macular degeneration, amyotrophic lateral sclerosis and epilepsy.

Conclusion. We present uKIN, a network propagation method that incorporates both existing knowledge as well as diverse types of new information, and demonstrate that it is highly effective in uncovering disease genes. As our knowledge of disease-associated genes continues to grow and be refined, and as new experimental data becomes more abundant, we expect that the power of uKIN for accurately prioritizing disease genes will continue to increase. uKIN can be freely downloaded at: http://compbio.cs.princeton.edu/ukin/ and the full paper can be found at https://arxiv.org/abs/2001.06135.

A Scalable Method for Estimating the Regional Polygenicity of Complex Traits

Ruth Johnson[1], Kathryn S. Burch[2], Kangcheng Hou[2], Mario Paciuc[3], Bogdan Pasaniuc[2,4,5,6(✉)], and Sriram Sankararaman[1,2,5,6(✉)]

[1] Department of Computer Science, University of California Los Angeles, Los Angeles, CA 90095, USA
sriram@cs.ucla.edu
[2] Bioinformatics Interdepartmental Program, University of California Los Angeles, Los Angeles, CA 90095, USA
[3] Department of Statistics, Rice University, Houston, TX 77005, USA
[4] Department of Pathology and Laboratory Medicine, David Geffen School of Medicine, University of California Los Angeles, Los Angeles, CA 90095, USA
[5] Department of Human Genetics, David Geffen School of Medicine, University of California Los Angeles, Los Angeles, CA 90095, USA
[6] Department of Computational Medicine, David Geffen School of Medicine, University of California Los Angeles, Los Angeles, CA 90095, USA

A key question in human genetics is understanding the proportion of SNPs modulating a particular phenotype or the proportion of susceptibility SNPs for a disease, termed *polygenicity*. Previous studies have observed that complex traits tend to be highly polygenic, opposing the previous belief that only a handful of SNPs contribute to a trait [1–4]. Beyond these genome-wide estimates, the distribution of polygenicity across genomic regions as well as the genomic factors that affect regional polygenicity remain poorly understood.

A reason for this gap is that methods for estimating polygenicity utilize SNP effect sizes from GWAS. However, due to LD and noise from the regression performed in GWAS, all effect sizes estimated from GWAS are non-zero, but not every SNP is truly a susceptibility SNP. Estimating polygenicity from GWAS while accounting for LD requires fully conditioning on the "susceptibility status" of every SNP and explicitly enumerating all possible configurations of susceptibility SNPs. This creates an exponential search space of 2^M, where M is the number of SNPs, which is intractable even when analyses are within small regions in the genome.

To circumvent the large computational bottleneck, existing methods that estimate polygenicity from GWAS do not explicitly condition on the susceptibility status of every SNP [5]. We expect these methods to lead to a downward bias when estimating polygenicity since only partially modeling the LD structure prevents these methods from fully re-capturing SNPs' effects that have been spread throughout the region due to LD. At a regional level, we expect the impact of a bias to be more pronounced since underestimating the polygenicity in a

B. Pasaniuc and S. Sankararaman—Equally contributed to this work.

R. Schwartz (Ed.): RECOMB 2020, LNBI 12074, pp. 253–254, 2020.
https://doi.org/10.1007/978-3-030-45257-5_26

region with a small fraction of susceptibility SNPs can be the difference between estimating the absence of susceptibility SNPs or the presence of only a few.

In this work, we propose a statistical framework, Bayesian EstimAtion of Variants in Regions (BEAVR), which relies on (MCMC) to estimate the regional polygenicities of a complex trait from GWAS while fully modeling the correlation structure due to LD. We present a fully generative Bayesian statistical model that estimates, for a given region, the regional polygenicity. Our model inherently allows for a variety of genetic architectures as it does not make prior assumptions on the number of susceptibility SNPs for a trait. A straightforward implementation of the MCMC sampler still presents a computational bottleneck since each iteration of the sampler is $\mathcal{O}(M^2)$ due to the full conditioning on each SNP, where M is the number of SNPs. To address this, we introduce a new inference algorithm that leverages the intuition that the majority of SNPs are not susceptibility SNPs. The runtime of our algorithm is $\mathcal{O}(MK)$ for M SNPs and K susceptibility SNPs, where the number of susceptibility SNPs is typically $K << M$, allowing us to perform efficient inference that scales approximately linearly with the number of SNPs analyzed in a region. Through comprehensive simulations and an analysis of BMI, eczema, and high cholesterol from the UK Biobank [6], we show that our method can accurately estimate regional polygenicity across many settings and provides insight into the genetic architecture for a variety of traits that is consistent with previous knowledge as well as provides novel information about the physical distribution of susceptibility SNPs across the genome.

The full version of the paper can be accessed at:
http://biorxiv.org/content/10.1101/2020.01.15.908095v1.

References

1. Yang, J., et al.: Common SNPs explain a large proportion of the heritability for human height. Nat. Genet. **42**(7), 565 (2010)
2. Stahl, E.A., et al.: Bayesian inference analyses of the polygenic architecture of rheumatoid arthritis. Nat. Genet. **44**(5), 483 (2012)
3. Loh, P.R., et al.: Contrasting genetic architectures of schizophrenia and other complex diseases using fast variance-components analysis. Nat. Genet. **47**(12), 1385 (2015)
4. Moser, G., Lee, S.H., Hayes, B.J., Goddard, M.E., Wray, N.R., Visscher, P.M.: Simultaneous discovery, estimation and prediction analysis of complex traits using a bayesian mixture model. PLoS Genet. **11**(4), e1004969 (2015)
5. Zhang, Y., Qi, G., Park, J.-H., Chatterjee, N.: Estimation of complex effect-size distributions using summary-level statistics from genome-wide association studies across 32 complex traits. Nat. Genet. **50**(9), 1318 (2018)
6. Bycroft, C., et al.: The UK biobank resource with deep phenotyping and genomic data. Nature **562**(7726), 203 (2018)

Efficient and Accurate Inference of Microbial Trajectories from Longitudinal Count Data

Tyler A. Joseph[1], Amey P. Pasarkar[1], and Itsik Pe'er[1,2,3($\boxtimes$)]

[1] Department of Computer Science, Columbia University, New York, NY 10027, USA
{tjoseph,itsik}@cs.columbia.edu
[2] Department of Systems Biology, Columbia University, New York, NY 10032, USA
[3] Data Science Institute, Columbia University, New York, NY 10027, USA

Introduction. The human body is home to trillions of microbial cells that play an essential role in health and disease [2]. The gut microbiome, for instance, is responsible for a variety of normal physiological processes such as the regulation of immune response and breakdown of xenobiotics [3]. Disturbances in gut communities have been associated with several diseases, notably obesity [7] and colitis [8]. Moreover, changes to the vaginal microbiome during pregnancy are associated with risk of preterm birth [4]. Consequently, investigating the human microbiome can provide insight into biological processes and the etiology of disease.

The recently completed second phase of the Human Microbiome Project [9] has highlighted the relationship between dynamic changes in the microbiome and disease, motivating new microbiome study designs based on longitudinal sampling and 16S rRNA gene sequencing. Nonetheless, analysis of 16S datasets faces multiple domain-specific challenges. First, 16S datasets are inherently compositional [5]: they only contain information about the relative proportions of taxa in a sample. In addition, technical noise, such as uneven amplification during PCR, can produce read counts that differ substantially from the underlying community structure [6]. In particular, species near the detection threshold may fail to appear in a sample, necessitating a distinction between a biological zero—where a species is absent in the community—from a technical zero where it drops below the detection threshold [1]. Finally, the number of taxa and time points in a sample may be large, requiring methods that scale to high dimensional data.

Methods. To address these challenges, we propose LUMINATE (LongitUdinal Microbiome INference And zero deTEction), a fast and accurate method for inferring relative abundances from noisy read count data. LUMINATE takes as input time-stamped read counts from multiple samples, and an optional a list of external perturbations. It outputs denoised relative abundances and posterior probabilities of biological zeros. Our contribution is two-fold. First, we reformulate the problem of posterior inference in a state-space model as an optimization problem with special structure using variational inference, allowing for efficient inference. Second, we propose a novel approach to differentiate between biological zeros and technical zeros.

© Springer Nature Switzerland AG 2020
R. Schwartz (Ed.): RECOMB 2020, LNBI 12074, pp. 255–256, 2020.
https://doi.org/10.1007/978-3-030-45257-5_27

We benchmarked LUMINATE in comparison to three other models representing the current state of the art. We evaluated how well each model reconstructed community trajectories by simulating ground truth trajectories with varying amounts of sequencing noise. We then performed simulations to evaluate how each model scales with increasing number of observed time points. We further used simulations to assess LUMINATE's ability to distinguish biological from technical zeros. Finally, we demonstrated the utility of LUMINATE on real data by using estimated relative abundances to infer the parameters of a dynamical system, leading to more accurate predictions of community trajectories.

Results. Our results demonstrate that LUMINATE is as accurate or better than the current state of the art. LUMINATE further constitutes a significant advancement as it runs orders of magnitude faster, facilitating previously infeasible scale-up of analysis to datasets with multiple longitudinal samples and many observed taxa. Furthermore, LUMINATE accurately detects biological zeros, providing insightful interpretation, rather than just inference, while also simplifying downstream parameter fitting.

Code availability: https://github.com/tyjo/luminate.

bioRχiv: https://doi.org/10.1101/2020.01.10.902163.

References

1. Äijö, T., Müller, C.L., Bonneau, R.: Temporal probabilistic modeling of bacterial compositions derived from 16s rRNA sequencing. Bioinformatics **34**(3), 372–380 (2017)
2. Cho, I., Blaser, M.J.: The human microbiome: at the interface of health and disease. Nat. Rev. Genet. **13**(4), 260 (2012)
3. Clemente, J.C., Ursell, L.K., Parfrey, L.W., Knight, R.: The impact of the gut microbiota on human health: an integrative view. Cell **148**(6), 1258–1270 (2012)
4. DiGiulio, D.B., et al.: Temporal and spatial variation of the human microbiota during pregnancy. Proc. Natl. Acad. Sci. **112**(35), 11060–11065 (2015)
5. Gloor, G.B., Macklaim, J.M., Pawlowsky-Glahn, V., Egozcue, J.J.: Microbiome datasets are compositional: and this is not optional. Front. Microbiol. **8**, 2224 (2017)
6. Kuczynski, J., et al.: Experimental and analytical tools for studying the human microbiome. Nat. Rev. Genet. **13**(1), 47 (2012)
7. Ley, R.E., Turnbaugh, P.J., Klein, S., Gordon, J.I.: Microbial ecology: human gut microbes associated with obesity. Nature **444**(7122), 1022 (2006)
8. Manichanh, C., et al.: Reduced diversity of faecal microbiota in crohn's disease revealed by a metagenomic approach. Gut **55**(2), 205–211 (2006)
9. Proctor, L.M., et al.: The integrative human microbiome project. Nature **569**(7758), 641–648 (2019)

Identifying Causal Variants by Fine Mapping Across Multiple Studies

Nathan LaPierre[1], Kodi Taraszka[1], Helen Huang[2], Rosemary He[3],
Farhad Hormozdiari[6], and Eleazar Eskin[1,4,5(✉)]

[1] Department of Computer Science, UCLA, Los Angeles, CA, USA
eeskin@cs.ucla.edu
[2] Department of Ecology and Evolutionary Biology, UCLA, Los Angeles, CA, USA
[3] Department of Mathematics, UCLA, Los Angeles, CA, USA
[4] Department of Human Genetics, UCLA, Los Angeles, CA, USA
[5] Department of Computational Medicine, UCLA, Los Angeles, CA, USA
[6] Harvard T.H. Chan School of Public Health, Boston, MA, USA

Genome-Wide Association Studies (GWAS) have successfully identified numerous genetic variants associated with a variety of complex traits in humans. However, most of these associated variants are not causal, and are simply in Linkage Disequilibrium (LD) with the true causal variants. This problem is addressed by statistical "fine mapping" methods, which attempt to prioritize a small subset of variants for further testing while accounting for LD structure [1]. CAVIAR [2] introduced a widely-adopted Bayesian approach that accounted for uncertainty in association statistics using a multivariate normal (MVN) model and allowed for potentially multiple causal SNPs at a locus. There is growing interest in improving fine-mapping by leveraging information from multiple studies. One example of this is trans-ethnic fine mapping, which can significantly improve fine mapping power and resolution by leveraging the distinct LD structures in each population. However, existing methods either assume a single causal SNP at each locus or do not explicitly model heterogeneity, limiting their power.

In this abstract, we present MsCAVIAR, a novel method that addresses these challenges. We retain the Bayesian MVN framework of CAVIAR while introducing a novel approach to explicitly account for the heterogeneity of effect sizes between studies using a Random-Effects (RE) model. MsCAVIAR takes as input the association statistics for SNPs at the same locus in multiple studies and the linkage disequilibrium (LD) structure between variants obtained from in-sample genotyped data. MsCAVIAR computes and outputs a minimal-sized "causal set" of SNPs that, with probability at least ρ, contains *all* causal SNPs.

By our definition of a causal set, every causal SNP must be contained in the set with high probability, but not every SNP in the set needs to be causal. Concretely, each SNP can be assigned a binary causal status: 1 for causal or 0 for non-causal. So long as none of the SNPs outside of the causal set are set to 1, the assignments are compatible with our definition of a causal set. We can represent these causal status assignments in a binary vector with one entry for each SNP denoting its causal status; we call such a vector a "configuration"

N. LaPierre and K. Taraszka—These authors contributed equally.

R. Schwartz (Ed.): RECOMB 2020, LNBI 12074, pp. 257–258, 2020.
https://doi.org/10.1007/978-3-030-45257-5_28

and denote it as C. For each configuration C compatible with the causal set, we compute its (posterior) probability in a Bayesian manner: the probability of a configuration of SNPs being causal given the association statistics can be computed by modeling a prior probability for that configuration and a likelihood function for the association statistics given the assumed causal SNPs from C.

The overall likelihood function can be decomposed into a product over the likelihood function for each study, since we assume that the studies are independent. More specifically, we assume that there is a true global effect size for a SNP over all possible populations, around which the effect sizes for that SNP in different studies are independently drawn according to a heterogeneity variance parameter. This allows MsCAVIAR to model the fact that effect sizes of a SNP across different studies are related, but not equal. Because we expect the summary statistics to be a function of their LD with the causal SNPs, the parameters of the likelihood function for each study are different, assuming the studies have different LD patterns. By computing the product over the likelihood of each study, we account for all of their LD patterns in determining the likelihood over all the studies. The posterior probability for a causal set is then computed by summing the posterior probabilities of all compatible configurations.

In order to evaluate MsCAVIAR and compare its performance with other methods, we performed a simulation study. We chose two chromosome regions from the 1000 Genomes project [3], one with low LD (20% of the SNPs have LD equal or higher than 0.5), and one with higher LD (80% of the SNPs are in association equal or higher than 0.5). We combined the two populations for a total of 1000 simulations and set the posterior probability threshold to 0.95. We then implanted 3 causal SNPs in each locus and simulated genomewide-significant association statistics for each of those SNPs, calculating the association statistics of the non-causal SNPs by their LD with causal SNPs. In addition to MsCAVIAR, we tested another trans-ethnic fine mapping method, PAINTOR [4], as well as running CAVIAR [2] on each population individually. We found that, while all methods were well-calibrated, in that their accuracy was equal to or greater than 0.95, MsCAVIAR achieved better fine-mapping resolution, e.g. it returned smaller causal sets on average.

MsCAVIAR is freely available at: https://github.com/nlapier2/MsCAVIAR.

The full paper is available at: https://www.biorxiv.org/content/10.1101/2020.01.15.908517v1.

References

1. Schaid, D.J., Chen, W., Larson, N.B.: From genome-wide associations to candidate causal variants by statistical fine-mapping. Nat. Rev. Genet. **19**, 491–504 (2018)
2. Hormozdiari, F., Kostem, E., Kang, E.Y., Pasaniuc, B., Eskin, E.: Identifying causal variants at loci with multiple signals of association. Genetics **198**, 497–508 (2014)
3. Genomes Project Consortium, et al.: A global reference for human genetic variation. Nature, **526**(7571), 68 (2015)
4. Kichaev, G., Pasaniuc, B.: Leveraging functional-annotation data in trans-ethnic fine-mapping studies. Am. J. Hum. Genet. **97**(2), 260–271 (2015)

MONN: A Multi-objective Neural Network for Predicting Pairwise Non-covalent Interactions and Binding Affinities Between Compounds and Proteins

Shuya Li[1], Fangping Wan[1], Hantao Shu[1], Tao Jiang[2,3], Dan Zhao[1(✉)], and Jianyang Zeng[1,4(✉)]

[1] Institute for Interdisciplinary Information Sciences, Tsinghua University, Beijing, China
{zhaodan2018,zengjy321}@tsinghua.edu.cn
[2] Department of Computer Science and Engineering, University of California, Riverside, CA, USA
[3] Bioinformatics Division, BNRIST/Department of Computer Science and Technology, Tsinghua University, Beijing, China
[4] MOE Key Laboratory of Bioinformatics, Tsinghua University, Beijing, China

Extended Abstract

Background. Computational approaches for inferring the mechanisms of compound-protein interactions (CPIs) can greatly facilitate drug development. Recently, although a number of deep learning based methods have been proposed to predict binding affinities of CPIs and attempt to capture local interaction sites in compounds and proteins through neural attentions, they still lack a systematic evaluation on the interpretability of the identified local features [1–3]. In this work, we constructed the *first* benchmark dataset containing the pairwise inter-molecular non-covalent interactions for more than 10,000 compound-protein pairs. Our comprehensive evaluation suggested that current neural attention based approaches have difficulty in automatically capturing the accurate local non-covalent interactions between compounds and proteins.

Method. Motivated by the above observation, we developed a multi-objective neural network, called MONN, to effectively learn both local atom-level interactions (*i.e.*, pairwise non-covalent interactions) and global binding strengths (*i.e.*, affinities) between compounds and proteins.

MONN is a structure-free model that takes only graph representations of compounds and primary sequences of proteins as input. A graph convolution

This work was supported in part by the National Natural Science Foundation of China [61872216, 81630103, 31900862], the Turing AI Institute of Nanjing and the Zhongguancun Haihua Institute for Frontier Information Technology.
S. Li and F. Wan—These authors contributed equally to this work.

R. Schwartz (Ed.): RECOMB 2020, LNBI 12074, pp. 259–260, 2020.
https://doi.org/10.1007/978-3-030-45257-5_29

network with a warp unit [4] is employed to capture both local features for atoms of a compound and a global representation for the whole compound. In addition, a convolution neural network (CNN) is used to effectively extract features from the local contexts of residues along a protein sequence. Then, MONN learns the pairwise non-covalent interactions between atoms of the compound and residues of the protein from labels derived from the high-quality 3D structure of the compound-protein complex. Finally, MONN integrates information from both the compound and the protein to predict their binding affinity. During this process, the predicted pairwise non-covalent interactions are also incorporated to enable information sharing between the components of both molecules to benefit the prediction of binding affinity.

Results. Comprehensive evaluation demonstrated that while the previous neural attention based approaches trained using only binding affinity labels fail to exhibit satisfactory interpretability results, MONN can successfully predict non-covalent interactions between compounds and proteins from our benchmark dataset. The generalization ability of MONN was further validated by an additional test dataset constructed from the Protein Data Bank (PDB) [5].

Moreover, MONN can outperform the state-of-the-art baseline methods in predicting compound-protein binding affinities on our constructed benchmark dataset. Additional tests on a much larger dataset suggested that MONN can achieve better performance than the baseline methods even without extra supervision from structural data.

These results suggested that MONN can offer a powerful tool in predicting binding affinities of compound-protein pairs and also provide useful insights into understanding the molecular mechanisms of compound-protein interactions, which thus can greatly advance the drug discovery process.

Availability

Source code and data: https://github.com/lishuya17/MONN.

Full-text preprint: https://www.biorxiv.org/content/10.1101/2019.12.30.89151 5v1.

References

1. Tsubaki, M., Tomii, K., Sese, J.: Compound-protein interaction prediction with end-to-end learning of neural networks for graphs and sequences. Bioinformatics **35**(2), 309–318 (2018)
2. Gao, K.Y., Fokoue, A., Luo, H., Iyengar, A., Dey, S., Zhang, P.: Interpretable drug target prediction using deep neural representation. In: IJCAI, pp. 3371–3377 (2018)
3. Karimi, M., Wu, D., Wang, Z., Shen, Y.: DeepAffinity: interpretable deep learning of compound-protein affinity through unified recurrent and convolutional neural networks. Bioinformatics **35**, 3329–3338 (2019)
4. Ishiguro, K., Maeda, S.I., Koyama, M.: Graph warp module: an auxiliary module for boosting the power of graph neural networks. arXiv preprint arXiv:1902.01020, (2019)
5. Berman, H.M., et al.: The protein data bank. Nucleic Acids Res. **28**(1), 235–242 (2000)

Evolutionary Context-Integrated Deep Sequence Modeling for Protein Engineering

Yunan Luo[1], Lam Vo[2], Hantian Ding[1], Yufeng Su[1], Yang Liu[1],
Wesley Wei Qian[1], Huimin Zhao[2], and Jian Peng[1($\boxtimes$)]

[1] Department of Computer Science, University of Illinois at Urbana-Champaign,
Urbana, USA
jianpeng@illinois.edu

[2] Department of Chemical and Biomolecular Engineering,
University of Illinois at Urbana-Champaign, Urbana, USA

1 Introduction

Protein engineering seeks to design proteins with improved or novel functions. Compared to rational design and directed evolution approaches, machine learning-guided approaches traverse the fitness landscape more effectively and hold the promise for accelerating engineering and reducing the experimental cost and effort. A critical challenge here is whether we are capable of predicting the function or fitness of unseen protein variants. By learning from the sequence and large-scale screening data of characterized variants, machine learning models predict functional fitness of sequences and prioritize new variants that are very likely to demonstrate enhanced functional properties, thereby guiding and accelerating rational design and directed evolution. While existing generative models and language models have been developed to predict the effects of mutation and assist protein engineering, the accuracy of these models is limited due to their unsupervised nature of the general sequence contexts they captured that is not specific to the protein being engineered. The full paper describing ECNet is available on bioRxiv at https://doi.org/10.1101/2020.01.16.908509.

2 Methods

We developed ECNet, a supervised deep learning model that guides protein engineering by predicting protein fitness from the sequence. We constructed a sequence representation that incorporated the local evolutionary context specific to the protein to be engineered. This representation explicitly encodes the residue interdependencies of all residue pairs in the sequence, which informs our prediction model to quantify the effects of mutations – especially higher-order mutations – in the sequence. We further incorporated with global evolutionary context from an LM model trained on large sequence databases to model the semantic grammar within protein sequences as well as other structure and stability relevant contexts. Finally, a recurrent neural network model, trained on the

© Springer Nature Switzerland AG 2020
R. Schwartz (Ed.): RECOMB 2020, LNBI 12074, pp. 261–263, 2020.
https://doi.org/10.1007/978-3-030-45257-5_30

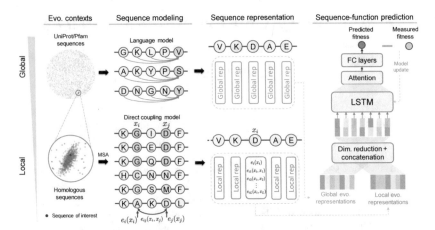

Fig. 1. Overview of ECNet. Our method, ECNet, integrates global and local evolutionary contexts to represent the protein sequence of interest. First, a language model is used to learn global semantic-rich global sequence representations from the protein sequence databases such as UniProt or Pfam. Next, a direct coupling analysis model is used to capture the dependencies between residues in protein sequences, which encodes the local evolutionary context. The global and local evolutionary representations are then combined as sequence representations and used as the input of a deep learning model that predicts the fitness of proteins. Quantitative fitness data measured by deep mutagenesis scans (DMS) are used to supervise the training of the deep learning model. (MSA: multiple sequence alignment; Dim. reduction: dimensionality reduction; LSTM: long short-term memory network; FC layers: fully-connected layers; Evo. contexts: evolutionary contexts; Evo. representations: evolutionary representations.)

fitness data of screened variants, is used for the sequence-to-function modeling with both representations.

3   Results

We performed multiple benchmarking experiments to assess the ability of ECNet in predicting the functional fitness from protein sequences. Through extensive benchmark experiments, we showed that our method outperforms existing methods, including unsupervised methods and supervised methods using handcrafted features, on ~50 deep mutagenesis scanning and random mutagenesis datasets, demonstrating its potential of guiding and expediting protein engineering. We also assessed ECNet's performance on predicting the fitness of higher-order variants when lower-order data were used for model training. We collected the fitness measurements of both single and double mutants of six proteins from previous DMS studies. We then trained our prediction model using single mutants data only and tested its performance on double mutants. The model achieved Spearman correlation ranging from 0.73 to 0.94 for the six proteins and outperformed the Supervised LM and the EVmutation baselines, suggesting its generalizability to the prediction of higher-order variants from low-order data.

Acknowledgments. J.P. acknowledges the support from the Sloan Research Fellowship and the NSF CAREER Award. Y. Luo acknowledges the support from the CompGen Fellowship.

Log Transformation Improves Dating
of Phylogenies

Uyen Mai[1(✉)] and Siavash Mirarab[2]

[1] Computer Science and Engineering, UC San Diego, San Diego, CA 92093, USA
umai@eng.ucsd.edu
[2] Electrical and Computer Engineering, UC San Diego, San Diego, CA 92093, USA
smirarab@ucsd.edu

Keywords: Time tree · Divergence time estimation · Phylogenetic
dating · Molecular dating · Non-convex optimization

Extended Abstract

The level of divergence between species represented by sequence data is a function of unknown time and mutation rates. Therefore, sequence data do not reveal exact timing of evolutionary events, and inferred phylogenies often have branch lengths estimated in the unit of the expected number of substitutions. Dating a phylogeny is the process of translating branch lengths from this unit to time unit. Such a process requires soft or hard constraints for the timing of *some* nodes and infers the divergence times of the remaining nodes. Dating is crucial for understanding evolutionary processes [2] and is necessary in many downstream applications of phylogenetics and phylodynamics. Many methods have been developed for phylogenetic dating, but none is universally accepted [3,5]. While some of these methods assume a parametric model for mutation rates and use maximum likelihood or Bayesian inference (e.g., [1,4,8,9]), other non-parametric methods rely on optimization problems based on assumed properties of the distribution of the rates (e.g., [4,6–8]). Parametric methods tend to work well for correct models but can be sensitive to model misspecification.

We introduce a non-parametric dating method called LogDate. We define mutation rates necessary to compute time unit branch lengths as the product of a single global rate and a set of rate multipliers, one per branche. Let μ_i be the rate specific to a branch i, then $\mu_i = \mu \nu_i$ where μ is the global rate and ν_i is the rate multiplier for branch i. We find the global rate μ and all rate multipliers such that the log-transformed rate multipliers have the minimum variance, subject to the constraints defined by the calibration points or sampling times. We further weight the terms of the objective function to define weighted LogDate (wLogDate) as the solution to:

$$\mathrm{argmin}_\nu \sum_{i=1}^{2n-1} \sqrt{\hat{b}_i} \log^2(\nu_i) \quad \text{subject to sampling point constraints} \quad (1)$$

© Springer Nature Switzerland AG 2020
R. Schwartz (Ed.): RECOMB 2020, LNBI 12074, pp. 264–265, 2020.
https://doi.org/10.1007/978-3-030-45257-5_31

where $\hat{b}_i$ is the inferred mutation unit branch length (plus a fixed pseudo-count). This formulation gives us a constrained optimization problem that can be solved efficiently using standard numerical techniques. Our main insight is the realization that log transformation of the rate multipliers results in more accurate dates. Our objective function, unlike related methods Least-Squares Dating [8] and Langley-Fitch [4], assigns symmetrical penalties for increased or decreased rates, a property which we refer to as "symmetry of ratios". Moreover, the Log-Date method corresponds to the ML estimate under a specific statistical model.

We compare wLogDate to several alternatives, including computationally expensive Bayesian methods. Through simulation of varied clock and tree models, wLogDate often infers more accurate node ages than other methods. Improvements are most visible under the hardest conditions with high variance clock models and inter-host tree models. We test wLogDate on three biological datasets of 892 sequences from H1N1 2009 pandemic, 904 sequences from San Diego HIV, and 1610 sequences from West African Ebola epidemic. For each dataset, LogDate computes the time tree in minutes and estimates tMRCA close to the reported literature.

LogDate is open-source and available at https://github.com/uym2/LogDate. The preprint version of the paper can be found at https://www.biorxiv.org/content/10.1101/2019.12.20.885582v1.

References

1. Drummond, A.J., Ho, S.Y.W., Phillips, M.J., Rambaut, A.: Relaxed phylogenetics and dating with confidence. PLoS Biol. **4**(5), e88 (2006). https://doi.org/10.1371/journal.pbio.0040088
2. Forest, F.: Calibrating the tree of life: fossils, molecules and evolutionary timescales. Ann. Botany **104**(5), 789–794 (2009). https://doi.org/10.1093/aob/mcp192
3. Kumar, S., Hedges, S.B.: Advances in time estimation methods for molecular data. Mol. Biol. Evol. **33**(4), 863–869 (2016). https://doi.org/10.1093/molbev/msw026
4. Langley, C.H., Fitch, W.M.: An examination of the constancy of the rate of molecular evolution. J. Mol. Evol. **3**(3), 161–177 (1974). https://doi.org/10.1007/BF01797451
5. Rutschmann, F.: Molecular dating of phylogenetic trees: a brief review of current methods that estimate divergence times. Divers. Distrib. **12**(1), 35–48 (2006). https://doi.org/10.1111/j.1366-9516.2006.00210.x
6. Sanderson, M.J., Shaffer, H.B.: Troubleshooting molecular phylogenetic analyses. Annu. Rev. Ecol. Syst. **33**(1), 49–72 (2002)
7. Tamura, K., Battistuzzi, F.U., Billing-Ross, P., Murillo, O., Filipski, A., Kumar, S.: Estimating divergence times in large molecular phylogenies. Proc. Natl. Acad. Sci. **109**(47), 19333–19338 (2012). https://doi.org/10.1073/pnas.1213199109
8. To, T.H., Jung, M., Lycett, S., Gascuel, O.: Fast dating using least-squares criteria and algorithms. Syst. Biol. **65**(1), 82–97 (2015). https://doi.org/10.1093/sysbio/syv068
9. Volz, E.M., Frost, S.D.W.: Scalable relaxed clock phylogenetic dating. Virus Evol. **3**(2) (2017). https://doi.org/10.1093/ve/vex025

Reconstructing Genotypes in Private Genomic Databases from Genetic Risk Scores

Brooks Paige[1,2] , James Bell[1] , Aurélien Bellet[3] , Adrià Gascón[1,4],
and Daphne Ezer[1,4,5(✉)]

[1] The Alan Turing Institute, London, UK
[2] Department of Computer Science, UCL, London, UK
[3] Inria, Rocquencourt, France
[4] University of Warwick, Coventry, UK
[5] Department of Biology, University of York, York, UK
daphne.ezer@york.ac.uk

Genomic researchers are already aware that some forms of aggregate data from their databases should not be released publicly, because there is a risk that an attacker may be able to determine whether a particular individual is a member of the database (a membership inference attack). These kinds of aggregate statistics about the frequency or presence/absence of a particular SNP might be useful to release to the broader research community, but it is not an essential output of the research process. However, the main research findings—i.e. the SNPs associated with the trait of interest and their strength of association—are essential to publish since the entire purpose of these genomic research projects is to uncover the relationship between genetic variants and phenotypic traits. Here, we demonstrate that it is possible to recover the SNPs of individuals in the database (a reconstruction attack), using Genetic Risk Scores (GRS), a common research output. GRS models describe the relationship between a particular phenotype of interest and particular SNPs. First, a reduced set of SNPs is selected; then, this reduced set is used as the independent variables in a linear regression analysis.

We begin by investigating a simple scenario: two GWAS studies for the same trait are performed on the same set of N SNPs and M participants, except that the second study includes one extra individual. We assume that both studies publish the coefficients associated with the GRS models that they infer as part of the analysis ($\hat{\beta}_M$ and $\hat{\beta}_{M+1}$). Our approach centres on the use of the vector we define as d_1,

$$d_1 \triangleq K(\hat{\beta}_{M+1} - \hat{\beta}_M) = C\phi_0, \tag{1}$$

where K is a matrix in which K_{ii} corresponds to the frequency of SNP i in the private database and $K_{i,j}$ corresponds to the frequency of SNP i and j occurring simultaneously in the same individual. C is a scalar value and ϕ_0 is the vector of genotypes of the individual who was found in the first study, but not the

---

Full article available here: https://doi.org/10.1101/2020.01.15.907808.

R. Schwartz (Ed.): RECOMB 2020, LNBI 12074, pp. 266–268, 2020.
https://doi.org/10.1007/978-3-030-45257-5_32

second. Here we assume that ϕ_0 is binary, having a value of 1 if the individual has the SNP and 0 otherwise. If K is known precisely, then ϕ_0 is known with perfect accuracy, because the vector d_1 will contain 0 at positions in which ϕ_0 is 0 (and $1/C$ otherwise). Often, K will not be known, but can be estimated based on public databases. We have developed a custom expectation-maximisation algorithm that predicts ϕ_0 based on an estimated $\hat{K}$. We have experimentally demonstrated that this performs better than a baseline, using the Cornell Dog Database [1].

We additionally consider the case where m additional individuals have been included in the second study, yielding a new GRS model $\hat{\beta}_{M+m}$ including these $M + m$ participants. The analog to Eq. (1) for multiple individuals is:

$$d_m \triangleq K(\hat{\beta}_{M+m} - \hat{\beta}_M) = \Phi_m C_m, \tag{2}$$

where C_m is now a vector. For sufficiently small m (relative to N), exact reconstruction of all m added individual genomes is also possible in this setting, following an algorithm we introduce.

We provide a number of simple suggestions for good practice that would help limit this attack. Firstly, precise aggregate statistics about the frequency of SNPs in the database or the frequency of co-occurrence of SNPs should never be released. Secondly, when multiple individuals are added in between two studies, then the ability to reconstruct the genomes depends on the number of SNPs being large relative to the number of individuals (Fig. 1).

In particular, if m new individuals are added, exact reconstruction is only possible if the number of SNPs $N >$ 2^m, so we suggest that similar studies should differ by more than $\log_2 N$ individuals. Another potential countermeasure could consist of randomly perturbing the GRS models before releasing them, as done in differentially private linear regression [2].

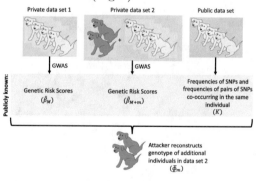

Fig. 1. The scenario under investigation.

Funding. Alan Turing Institute Research Fellowship under EPSRC Research grant (TU/A/000017); EPSRC/BBSRC Innovation Fellowship (EP/S001360/1), and EPSRC grant EP/N510129/1. It was also partly funded by a grant from CPER Nord-Pas de Calais/FEDER DATA Advanced data science and technologies 2015–20.

References

1. Hayward, J.J., et al.: Complex disease and phenotype mapping in the domestic dog. Nat. Commun. **7**, 1–11 (2016). https://doi.org/10.1038/ncomms10460
2. Wang, Y.X.: Revisiting differentially private linear regression: optimal and adaptive prediction & estimation in unbounded domain. In: Proceedings of the Conference on Uncertainty in Artificial Intelligence (UAI), pp. 93–103 (2018)

d-PBWT: Dynamic Positional Burrows-Wheeler Transform

Ahsan Sanaullah[1], Degui Zhi[2], and Shaojie Zhang[1(✉)]

[1] Department of Computer Science, University of Central Florida,
Orlando, FL 32816, USA
asanaullah2019@knights.ucf.edu, shzhang@cs.ucf.edu
[2] School of Biomedical Informatics,
University of Texas Health Science Center at Houston, Houston, TX 77030, USA
degui.zhi@uth.tmc.edu

Durbin's positional Burrows-Wheeler transform (PBWT) [1] is a scalable foundational data structure for modeling population haplotype sequences. It offers efficient algorithms for matching haplotypes that approach theoretically optimal complexity. The original PBWT paper described an array version of the PBWT, and a set of basic algorithms: Algorithms 1 and 2 for construction, Algorithms 3 and 4 for reporting all versus all long matches and set maximal matches, and Algorithm 5 for reporting set maximal matches between an out-of-panel query against a constructed PBWT panel. Recently, Naseri *et al.* [2] presented a new algorithm, L-PBWT-Query, that reports all long matches between an out-of-panel query against a constructed PBWT panel in time complexity linear to the length of the haplotypes and constant to the size of the panel. Naseri *et al.* introduced Linked Equal/Alternating Positions (LEAP) arrays, an additional data structure allowing direct jumping to boundaries of matching blocks. This algorithm offers efficient long matches, a more practical target for genealogical search. Arguably, L-PBWT-Query makes PBWT search more practical as it returns all long enough matches rather than merely the best matching ones. L-PBWT-Query represents a missing piece of the PBWT algorithms.

However, all above algorithms are based on arrays, which do not support dynamic updates. That means, if new haplotypes are to be added to, or some haplotypes are to be deleted from an existing PBWT data structure, one has to rebuild the entire PBWT, an expensive effort linear to the number of haplotypes. This will be inefficient for large databases hosting millions of haplotypes as they may face constant update requests per changing consent of data donors. Moreover, lack of dynamic updates prohibits PBWT to be applied to large-scale genotype imputation and phasing, which typically go through the panel multiple times and update individual's haplotypes in turn. It is much more efficient to allow updating the PBWT with an individual's new haplotypes while keeping others intact.

In this work we introduce d-PBWT, a dynamic version of the PBWT data structure. At each position k, instead of keeping track of sequence order using an array, we use a linked list, whose nodes encapsulate all pointers needed for traversing PBWT data structures. Our main results are: We developed efficient

© Springer Nature Switzerland AG 2020
R. Schwartz (Ed.): RECOMB 2020, LNBI 12074, pp. 269–270, 2020.
https://doi.org/10.1007/978-3-030-45257-5_33

Table 1. Summary of algorithms on PBWT and d-PBWT

	Data structure	Functions	Time complexity
Durbin [1]	PBWT	Construction	$O(MN)$
		All vs. All long matches and set maximal matches	$O(MN)$
		Set maximal match query	Avg. $O(N)$
Naseri *et al.* [2]	PBWT	Long match query (L-PBWT-Query)	Avg. $O(N)$
This work	d-PBWT	Insertion	Avg. $O(N)$
		Deletion	Avg. $O(N)$
		Set maximal match query	$O(N)$
		Long match query	$O(N)$
		Construction	$O(MN)$
		All vs. All long matches and set maximal matches	$O(MN)$
		Conversion	$O(MN)$

M is the number of sequences. N is the number of sites. c is the number of matches found. Time complexities assume $c < N$.

insertion and deletion algorithms that dynamically update all PBWT data structures. In addition, we established that d-PBWT can run Durbin's Algorithms 1–5 and L-PBWT-Query with the same time complexity as the static PBWT. While Durbin's Algorithm 5 and L-PBWT-Query are independent of the number of haplotypes in the average case, we showed that they are not in the worst case. Lastly, we developed new set maximal match and long match query algorithms with worst case linear time complexity. The long match query algorithm does not use LEAP arrays, which reduces memory consumption over L-PBWT-Query. These algorithms can also be applied to the static PBWT. Table 1 summarizes the major contributions of this work. The preprint of the full manuscript is available at https://www.biorxiv.org/content/10.1101/2020.01.14.906487v1.

References

1. Durbin, R.: Efficient haplotype matching and storage using the positional Burrows-Wheeler transform (PBWT). Bioinformatics **30**(9), 1266–1272 (2014)
2. Naseri, A., Holzhauser, E., Zhi, D., Zhang, S.: Efficient haplotype matching between a query and a panel for genealogical search. Bioinformatics **35**(14), i233–i241 (2019)

A Mixture Model for Signature Discovery from Sparse Mutation Data

Itay Sason[1], Yuexi Chen[2], Mark D. M. Leiserson[2], and Roded Sharan[1(✉)]

[1] School of Computer Science, Tel Aviv University, 69978 Tel Aviv, Israel
roded@tauex.tau.ac.il
[2] Department of Computer Science and Center for Bioinformatics
and Computational Biology, University of Maryland, College Park, MD 20740, USA

Extended Abstract

Each cancer genome is shaped by a combination of processes that introduce mutations over time. The incidence and etiology of these mutational processes may provide insight into tumorigenesis and personalized therapy. It is thus important to uncover the characteristic signatures of active mutational processes in patients from their patterns of single base substitutions [1]. Some such mutation signatures have been linked to exposure to specific carcinogens, such as tobacco smoke and ultraviolet radiation. Other mutation signatures arise from deficient DNA damage repair pathways. By serving as a proxy for the functional status of the repair pathway, mutational signatures provide an avenue around traditional driver mutation analyses. This is important for personalizing cancer therapies, many of which work by causing DNA damage or inhibiting DNA damage response or repair genes [2], because the functional effect of many variants is hard to predict. Indeed, a recent study [3] estimated a >4-fold increase in the number of breast cancer patients with homologous recombination repair deficiency – making them eligible for PARP inhibitors [4] – when using mutational signatures compared to current approaches.

Statistical models for discovering and characterizing mutational signatures are crucial for realizing their potential as biomarkers in the clinic. A broad catalogue of mutational signatures in cancer genomes was only recently revealed through computational analysis of mutations in thousands of tumors. Alexandrov et al. [1] were the first to use non-negative matrix factorization (NMF) to discover mutation signatures. Subsequent methods have used different forms of NMF [5], or focused on inferring the exposures (aka refitting) given the signatures and mutation counts [6,7]. A more recent class of approaches borrows from the world of topic modeling, aiming to provide a probabilistic model of the data so as to maximize the model's likelihood [8,9].

These previous methods are applicable for whole-genome or even whole-exome sequencing. However, they cannot handle very sparse data as obtained routinely in targeted (gene panel) sequencing assays [10]. There is only a single method that attempted to address this challenge [10] by relying on whole-genome

© Springer Nature Switzerland AG 2020
R. Schwartz (Ed.): RECOMB 2020, LNBI 12074, pp. 271–272, 2020.
https://doi.org/10.1007/978-3-030-45257-5_34

training data to interpret sparse samples and predict their homologous recombination deficiency status.

Here we report the Mix model that is the first to handle sparse targeted sequencing data *without pre-training* on rich data. Our model simultaneously clusters the samples and learns the mutational landscape of each cluster, thereby overcoming the sparsity problem. Using synthetic and real targeted sequencing data, we show that our method is superior to current non-sparse approaches in signature discovery, signature refitting and patient stratification. A full version of this paper is available at http://www.cs.tau.ac.il/~roded/mix.pdf and in bioRxiv.

References

1. Alexandrov, L.B., et al.: Signatures of mutational processes in human cancer. Nature **500**(7463), 415–421 (2013)
2. Gavande, N.S., et al.: DNA repair targeted therapy: the past or future of cancer treatment? Pharmacol. Ther. **160**, 65–83 (2016)
3. Davies, H., et al.: HRDetect is a predictor of BRCA1 and BRCA2 deficiency based on mutational signatures. Nat. Med. **23**(4), 517–525 (2017)
4. Farmer, H., et al.: Targeting the DNA repair defect in BRCA mutant cells as a therapeutic strategy. Nature **434**(7035), 917–921 (2005)
5. Fischer, A., et al.: EMu: probabilistic inference of mutational processes and their localization in the cancer genome. Genome Biol. **14**(4), 1–10 (2013)
6. Huang, X., Wojtowicz, D., Przytycka, T.M.: Detecting presence of mutational signatures in cancer with confidence. Bioinformatics **34**(2), 330–337 (2018)
7. Rosenthal, R., et al.: DeconstructSigs: delineating mutational processes in single tumors distinguishes DNA repair deficiencies and patterns of carcinoma evolution. Genome Biol. **17**(1), 31 (2016)
8. Shiraishi, Y., et al.: A simple model-based approach to inferring and visualizing cancer mutation signatures. PLOS Genet. **11**(12), e1005657 (2015)
9. Wojtowicz, D., et al.: Hidden Markov models lead to higher resolution maps of mutation signature activity in cancer. Genome Med. **11**, 49 (2019)
10. Gulhan, D.C., et al.: Detecting the mutational signature of homologous recombination deficiency in clinical samples. Nat. Genet. **51**, 912–919 (2019)

Single-Cell Tumor Phylogeny Inference with Copy-Number Constrained Mutation Losses

Gryte Satas[1,2], Simone Zaccaria[1], Geoffrey Mon[1],
and Benjamin J. Raphael[1(✉)]

[1] Department of Computer Science, Princeton University, Princeton, NJ 08544, USA
braphael@princeton.edu
[2] Department of Computer Science, Brown University, Providence, RI 02906, USA

Motivation: Single-cell DNA sequencing enables the measurement of somatic mutations in individual tumor cells, and provides data to reconstruct the evolutionary history of the tumor. Nearly all existing methods to construct phylogenetic trees from single-cell sequencing data use single-nucleotide variants (SNVs) as markers. However, most solid tumors contain copy-number aberrations (CNAs) which can overlap loci containing SNVs. Particularly problematic are CNAs that delete an SNV, thus returning the SNV locus to the unmutated state. Such mutation losses are allowed in some models of SNV evolution, but these models are generally too permissive, allowing mutation losses without evidence of a CNA overlapping the locus.

Results: We introduce a novel *loss-supported* evolutionary model, a generalization of the infinite sites and Dollo models, that constrains mutation losses to loci with evidence of a decrease in copy number. We design a new algorithm, SCARLET (Single-Cell Algorithm for Reconstructing the Loss-supported Evolution of Tumors), that infers phylogenies from single-cell tumor sequencing data using the loss-supported model and a probabilistic model of sequencing errors and allele dropout. On simulated data, we show that SCARLET outperforms current single-cell phylogeny methods, recovering more accurate trees and correcting errors in SNV data. On single-cell sequencing data from a metastatic colorectal cancer patient, SCARLET constructs a phylogeny that is both more consistent with the observed copy-number data and also reveals a simpler monoclonal seeding of the metastasis, contrasting with published reports of polyclonal seeding in this patient. SCARLET substantially improves single-cell phylogeny inference in tumors with CNAs, yielding new insights into the analysis of tumor evolution.

Availability: Software is available at github.com/raphael-group/scarlet
Preprint: A preprint of the manuscript is available at
https://www.biorxiv.org/content/10.1101/840355v1

© Springer Nature Switzerland AG 2020
R. Schwartz (Ed.): RECOMB 2020, LNBI 12074, p. 273, 2020.
https://doi.org/10.1007/978-3-030-45257-5_35

Reconstruction of Gene Regulatory Networks by Integrating Biological Model and a Recommendation System

Yijie Wang[1]([⊠]), Justin M. Fear[2], Isabelle Berger[2], Hangnoh Lee[2], Brian Oliver[2]([⊠]), and Teresa M. Przytycka[3]([⊠])

[1] Computer Science Department, Indiana University, Bloomington, IN 47408, USA
`yijwang@iu.edu`
[2] Laboratory of Cellular and Developmental Biology,
National Institute of Diabetes and Digestive and Kidney Diseases,
50 South Drive, Bethesda, MD 20892, USA
`briano@niddk.nih.gov`
[3] National Center of Biotechnology Information, National Library of Medicine, NIH,
Bethesda, MD 20894, USA
`przytyck@ncbi.nlm.nih.gov`

Extended Abstract

Gene Regulatory Networks (GRNs) control many aspects of cellular processes including cell differentiation, maintenance of cell type specific states, signal transduction, and response to stress. Since GRNs provide information that is essential for understanding cell function, the inference of these networks is one of the key challenges in systems biology. Leading algorithms to reconstruct GRN utilize, in addition to gene expression data, prior knowledge such as Transcription Factor (TF) DNA binding motifs or results of DNA binding experiments. However, such prior knowledge is typically incomplete hence resulting in missing values and current methods do not directly account for the issue of missing values [1–5]. Therefore, the integration of such incomplete prior knowledge with gene expression to elucidate the underlying GRNs remains difficult.

To address this challenge we introduce NetREX-CF – Regulatory **Net**work **R**econstruction using **EX**pression and **C**ollaborative **F**iltering – a GRN reconstruction approach that brings together a modern machine learning strategy (Collaborative Filtering model) and a biologically justified model of gene expression (sparse Network Component Analysis based model). The Collaborative Filtering (CF) model [6,7] is able to overcome the incompleteness of the prior knowledge and make edge recommends for building the GRN. Complementing CF, the sparse Network Component Analysis (NCA) model [8] can use gene expression data and biologically supported mathematical gene expression model to validate the recommended edges. Here we combine these two approaches using

---

Y. Wang and J. M. Fear—Co-first author.

R. Schwartz (Ed.): RECOMB 2020, LNBI 12074, pp. 274–275, 2020.
https://doi.org/10.1007/978-3-030-45257-5_36

a novel data integration method and show that the new approach outperforms the currently leading GRN reconstruction methods.

Furthermore, our mathematical formalization of the model has lead to a complex optimization problem of a type that has not been attempted before. Specifically, the formulation contains ℓ_0 norm that can not be separated from other variables. To fill this gap, we extend Proximal Alternating Linearized Minimization (PALM) method [9] and introduce here the Generalized PALM (GPALM) that allows us to solve a broad class of non-convex optimization problems and prove its convergence. The full version of this paper can be found on https://www.biorxiv.org/content/10.1101/2020.01.07.898031v1.full.

Acknowledgement. This research was supported by the Intramural Research Program of the NLM and the National Institute of Diabetes and Digestive and NIDDK, USA, and Precision Health Initiative of Indiana University.

References

1. Marbach, D., et al.: Predictive regulatory models in Drosophila melanogaster by integrative inference of transcriptional networks. Genome Res. **22**(7), 1334–1349 (2012)
2. Werhli, A.V., Husmeier, D.: Gene regulatory network reconstruction by Bayesian integration of prior knowledge and/or different experimental conditions. J. Bioinform. Comput. Biol. **6**(3), 543–572 (2008)
3. Mukherjee, S., Speed, T.P.: Network inference using informative priors. Proc. Natl. Acad. Sci. U.S.A. **105**(38), 14313–14318 (2008)
4. Siahpirani, A.F., Roy, S.: A prior-based integrative frame-work for functional transcriptional regulatory network inference. Nucleic Acids Res. **45**(4), e21 (2016). https://doi.org/10.1093/nar/gkw963
5. Wang, Y., et al.: Reprogramming of regulatory network using expression uncovers sex-specific gene regulation in Drosophila. Nat. Commun. **9**(1) (2018). Article number: 4061
6. Koren, Y., Bell, R., Volinsky, C.: Matrix factorization techniques for recommender systems. Computer **42**(8), 30–37 (2009)
7. Hu, Y., Koren, Y., Volinsky, C.: Collaborative filtering for implicit feed-back datasets. In: 2008 Eighth IEEE International Conference on Data Mining, ICDM 2008, pp. 263–272. IEEE (2008)
8. Liao, J.C., et al.: Network component analysis: reconstruction of regulatory signals in biological systems. Proc. Natl. Acad. Sci. U.S.A. **100**(26), 15522–15527 (2003)
9. Bolte, J., Sabach, S., Teboulle, M.: Proximal alternating linearized minimization for nonconvex and nonsmooth problems. Math. Program. **146**(1–2), 459–494 (2014). https://doi.org/10.1007/s10107-013-0701-9

Probing Multi-way Chromatin Interaction with Hypergraph Representation Learning

Ruochi Zhang and Jian Ma[✉]

Computational Biology Department, School of Computer Science,
Carnegie Mellon University, Pittsburgh, PA 15213, USA
jianma@cs.cmu.edu

Advances in high-throughput mapping of 3D genome organization have enabled genome-wide characterization of chromatin interactions. However, proximity ligation based mapping approaches for pairwise chromatin interaction such as Hi-C cannot capture multi-way interactions, which are informative to delineate higher-order genome organization and gene regulation mechanisms at single-nucleus resolution. The very recent development of ligation-free chromatin interaction mapping methods such as SPRITE and ChIA-Drop has offered new opportunities to uncover simultaneous interactions involving multiple genomic loci within the same nuclei. Unfortunately, methods for analyzing multi-way chromatin interaction data are significantly underexplored. In particular, existing methods for analyzing multi-way chromatin interaction data have limited capability of handling data noise. In addition, existing methods typically decompose each multi-way contact into pairwise ones and directly apply previous methods developed for pairwise interactions, leading to a dramatic loss of higher-order information.

Here we develop a new computational method, called MATCHA, based on hypergraph representation learning where multi-way chromatin interactions are represented as hyperedges. The overview of MATCHA is shown in Fig. 1. Specifically, MATCHA takes multi-way chromatin interaction data as input and extracts patterns from the corresponding hypergraph. The patterns are represented as embedding vectors for genomic bins that reflect the 3D chromatin organization properties. The model can further predict the probability for a group of genomic bins forming a simultaneous interaction. We demonstrate the effectiveness of MATCHA by applying it to recent SPRITE and ChIA-Drop datasets. The results suggest that MATCHA is able to achieve accurate predictions of multi-way chromatin interactions (i.e., hyperedges), which could be used for data denoising and *de novo* hyperedge prediction, reducing the potential false positives and false negatives from the original data. We also show that MATCHA is able to distinguish between multi-way interaction in a single nucleus and the combination of pairwise interactions in a cell population. In addition, the embeddings from MATCHA reflect 3D genome spatial localization and function.

MATCHA provides a promising framework to significantly improve the analysis of multi-way chromatin interaction data and has the potential to offer unique insights into higher-order chromosome organization and function.

Link to the bioRxiv preprint: https://doi.org/10.1101/2020.01.22.916171

© Springer Nature Switzerland AG 2020
R. Schwartz (Ed.): RECOMB 2020, LNBI 12074, pp. 276–277, 2020.
https://doi.org/10.1007/978-3-030-45257-5_37

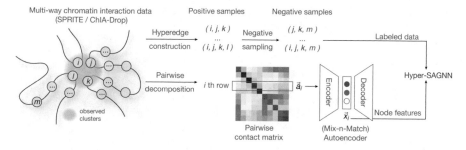

Fig. 1. Overview of MATCHA. There are four main components of MATCHA (1) Constructing hypergraphs based on multi-way chromatin interaction data where non-overlapping genomic bins are defined as nodes, and bins in the same multi-way interactions are connected by hyperedges. (2) Generating node features for the hypergraph based on decomposed pairwise contact matrix from multi-way interaction data. The decomposed pairwise contact matrix further passes through the Mix-n-Match autoencoder, which makes the non-linear transformation of the contact matrix. (3) Generating labeled data for the training of the hypergraph representation learning model, where positive samples are defined as existing hyperedges and negative samples are unobserved ones. The negative samples are generated through an efficient and biologically meaningful negative sampling strategy. (4) Training the hypergraph representation learning model Hyper-SAGNN that takes both labeled data and node features as input.

Author Index

Printed in the United States
By Bookmasters